# The Physics and Chemistry of Liquid Crystal Devices

## THE IBM RESEARCH SYMPOSIA SERIES

Computational Methods in Band Theory
Editors: P.M. Marcus, J.F. Janak, and A.R. Williams

Computational Solid State Physics
Editors: F. Herman, N. W. Dalton, and T. R. Koehler

Sparse Matrices and Their Applications
Editors: D. J. Rose and R. A. Willoughby

Complexity of Computer Computations
Editors: R. E. Miller and J. W. Thatcher
Associate Editor: J. D. Bohlinger

Computational Methods for Large Molecules
and Localized States in Solids
Editors: F. Herman, A. D. McLean, R. K. Nesbet

Ion Implantation in Semiconductors
and Other Materials
Editor: Billy L. Crowder

Stiff Differential Systems
Editor: Ralph A. Willoughby

Optimal Estimation in Approximation Theory
Editors: Charles A. Micchelli and Theodore J. Rivlin

Topics in Surface Chemistry
Editors: Eric Kay and Paul S. Bagus

Advances in Digital Image Processing:
Theory, Application, Implementation
Editor: Peter Stucki

The Physics and Chemistry of Liquid Crystal Devices
Editor: Gerald J. Sprokel

# The Physics and Chemistry of Liquid Crystal Devices

Edited by
Gerald J. Sprokel
*IBM Research Laboratory*
*San Jose, California*

Springer Science+Business Media, LLC

Library of Congress Cataloging in Publication Data

Symposium on the Physics and Chemistry of Liquid Crystal Devices, San Jose, Calif., 1979.
The physics and chemistry of liquid crystal devices.

(IBM research symposia series)
"'Proceedings of the Symposium on the Physics and Chemistry of Liquid crystal Devices, held at the IBM Research Laboratory, San Jose, California, February 7–8, 1979."
Includes index.
1. Liquid crystal devices—Congresses. I. Sprokel, Gerald J. II. Title. III. Series: International Business Machines Corporation. IBM research symposia series.
TS518.S95 1979 681'.754 80-12097

ISBN 978-1-4899-2185-7 ISBN 978-1-4899-2183-3 (eBook)
DOI 10.1007/978-1-4899-2183-3

Proceedings of the Symposium on the Physics and Chemistry of Liquid Crystal Devices, held at the IBM Research Laboratory, San Jose, California, February 7–8, 1979.


Originally published by Plenum Press, New York in 1980
Softcover reprint of the hardcover 1st edition 1980

# Foreword

Over 100 scientists met at the IBM Research Laboratory in San Jose, California for a symposium on the Physics and Chemistry of Liquid Crystal Devices. The two-day meeting was intellectually stimulating with excellent oral presentations and with person-to-person discussions.

The applications of liquid crystals have developed dramatically in the past ten years. In these few years, they have moved from being a laboratory curiosity to products in the market place.

The first commercial application (1940's) of liquid crystals was the preparation of a light polarizer. The second commercial application was their use as temperature sensors. The third major application of liquid crystals dealt with commercial displays. Other current applications include polymeric and graphitic fibers and light attenuators.

The future of liquid crystals looks very promising indeed. One can expect to see new fibers of qualities which will be superior to those presently known. Graphitic fibers or other physical forms of graphitic materials will be used as catalytic surfaces for chemical synthesis. In the display area, one can expect to see television screens using liquid crystals. Larger displays than are now used in wrist watches and pocket calculators will become available.

Liquid crystals using color displays will become commercially practical. Watches, calculators and television screens will have color.

Even though this conference did not address the field of lyotropic liquid crystals, one can suggest that their properties and structures are just beginning to be studied carefully. Two uses of lyotropic liquid crystals will be in oil recovery and in our understanding of life processes.

Few happenings in science and engineering have been as dramatic as the use of liquid crystals in commercial products. Many applications have taken only a year or so to move from the laboratory curiosity to the commercial product. We have only begun to take advantage of the liquid crystalline state of matter as we develop new products for the consumer market.

Glenn H. Brown

Director of Liquid Crystal Institute
Regents Professor of Chemistry
Kent State University

# Preface

This volume of the IBM Research Symposia Series contains most of the papers presented at the Symposium on the Physics and Chemistry of Liquid Crystal Devices held at the IBM Research Laboratory, San Jose, California, on February 7 and 8, 1979.

While the symposium is only one of the continuing series of conferences organized by IBM Research, San Jose, it was the first industry sponsored conference on Liquid Crystal Display Devices in the U.S.A. The only previous L.C.D. device conference was held at the Brown Boveri Research Centre, Baden, Switzerland in September 1975.

Invitation to the conference was by direct mail; there were no announcements in the Technical press. The conference audience was about 125 while approximately 140 people attended the conference dinner at the San Jose Hyatt House. The dinner speaker was Professor Glenn H. Brown of the Liquid Crystal Institute at Kent State University. Attendees represented the U.S.A., Japan and most West European Countries (West Germany, Switzerland, England and the Netherlands), which gave the conference an international flavor.

The conference was organized in four sessions:

| | |
|---|---|
| Physics I | A. H. Sporer, chairman |
| Physics II | R. C. Durbeck, chairman |
| L.C. Devices | B. G. Huth, chairman |
| Materials | G. J. Sprokel, chairman |

Each session had 1 or 2 invited papers which were followed by contributed papers. In total, 28 papers were presented, 24 of which are collected here. The remaining papers were not available for publication.

The amazing rapid development of liquid crystal display devices and their general acceptance by the public are now well known. Papers on twisted nematic devices and their remaining problems dominated the conference. However, guest-host displays, laser addressed smectic displays and photo sensor-liquid crystal displays were discussed. Physics I was entirely devoted to the liquid crystal-glass interface. Physics II dealt mainly with matrix addressing. Two papers on surface alignment studies by optical techniques emphasize the interest in this area. The device session contained papers about non-random matrix addressing as now exemplified by the Texas Instruments analog watch and the RSRE analog meter display. The material session had interesting papers on the relation between device performance and material parameters. As was shown by several papers, mixtures are needed to optimize device performance. The session ended with a review of machine analysis of liquid crystal mixtures.

To end with a note of acknowledgment, I was encouraged by the IBM Research management to organize this meeting and I want to express my appreciation for having been selected for this assignment. However, without the continuous strong support of Dr. R. C. Durbeck, manager of I/O Technology, the conference might well have floundered. I want to express my appreciation for the efforts of my colleagues in this laboratory and elsewhere while preparing for this conference. And lastly, a special note of appreciation is due to Linda Ferguson, manager of the Manuscript Processing Center, and to the Graphics Group of the late Bradley Hall, recently and unexpectantly departed, for all the typing and artwork involved in producing this book.

I acknowledge my appreciation to the authors and invited speakers and to all who contributed to this successful conference.

# Contents

PHYSICS

## LIQUID CRYSTAL DEVICES

## LIQUID CRYSTAL MATERIALS

# SURFACE ORIENTATION AND COMPLIANCE EFFECTS ON TWIST-CELL PERFORMANCE

Dwight W. Berreman

Bell Laboratories
Murray Hill, New Jersey 07904

## INTRODUCTION

In a recent article Nehring, Kmetz and Scheffer described effects of weak anchoring on equilibrium configurations of twist cells.[1] In order to obtain tractable equations for analytic solution, the three bulk elastic constants were made equal, the field-and-strain-free orientation at the surface was assumed to be parallel to the surface, and only nematic liquids were discussed. That treatment gives useful insights into the nature of the problem but leaves a number of interesting questions unanswered.

In this paper some results of numerical computations of dynamic configuration changes in twist cells will be described. We have used the full Leslie-Ericksen hydrodynamic equations[2,3] for cholesterics in laminar flow between parallel surfaces, except for the justifiable omission of inertial terms. Use of cholesterics removes the difficulty mentioned in Ref. 1 of obtaining a 90° twist with very weak anchoring. The effects of using models for surface energy other than the symmetric $\sin^2(\underline{n},\underline{n}_o)$ form used in previous work[1,4] are investigated. Perhaps the most interesting result is the discovery that a separate viscosity must accompany surface elasticity.

## OUTLINE OF THE EQUILIBRIUM PROBLEM

Geometric parameters to describe a twist cell are illustrated in Fig. 1. In the absence of an applied voltage, the elastic strain energy of the bulk and at the surface is minimized at

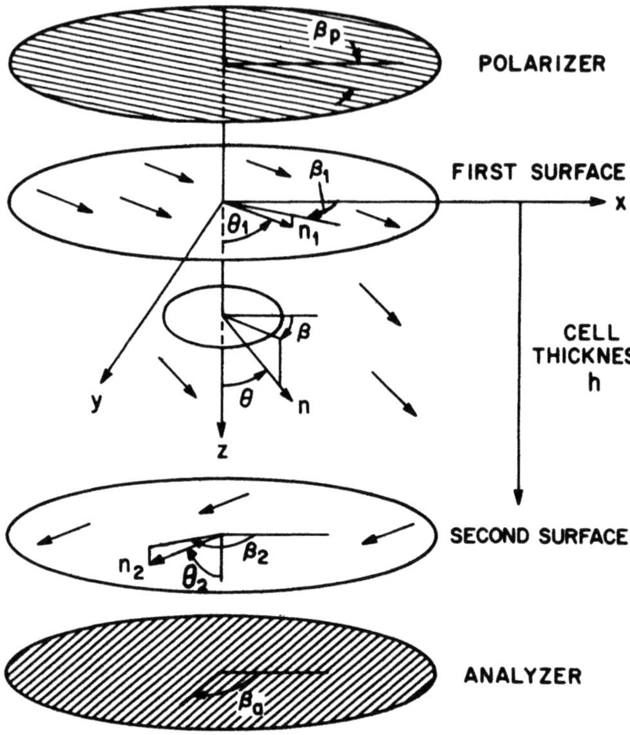


Fig. 1. Schematic diagram of a twist cell showing tilt angle Θ, azimuth β, cell thickness h and relative positions of polarizer and analyzer.

equilibrium. The bulk energy density is given by the Oseen-Frank expression[5,6,7]

$$U_B = \frac{k_{11}}{2}(\nabla\cdot\underline{n})^2 + \frac{k_{22}}{2}(\underline{n}\cdot\nabla\times\underline{n}+q_o)^2 + \frac{k_{33}}{2}(\underline{n}\times\nabla\times\underline{n})^2 .$$

The three elastic constants $k_{ii}$ define resistance to the three deformations shown in Fig. 2. Integration of this function from zero to h yields the total bulk energy. Integration in closed form is greatly simplified if the three elastic constants are equal. $q_o$ is the unstrained cholesteric helicity.

If the surface anchoring is not perfectly rigid, additional energy is stored at the surfaces in straining the directors away from the direction of lowest energy. The form of the surface energy function previously studied is[1,4]

$$U_{sa} = \frac{k_s}{2} \sin^2(\underline{n},\underline{n}_o) \; .$$

See Fig. 3a.

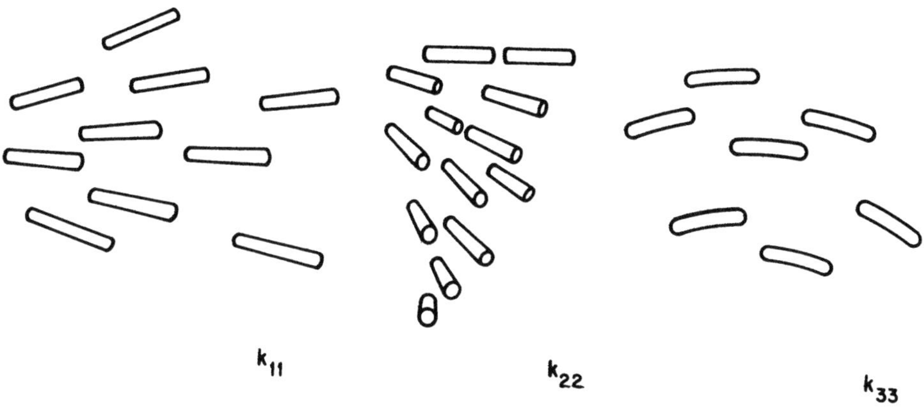


Fig. 2. Depiction of the three independent types of bulk strain and their associated elastic constants, $k_{11}$ for splay, $k_{22}$ for twist and $k_{33}$ for bend.

This function has the symmetry of nematic liquid crystals in bulk, but there is reason to doubt that the surface layer has that symmetry. We show other possible energy functions of great simplicity but without the same symmetry in Figs. 3b and 3c.

The analytic form of function (b) is

$$U_{sb} = \frac{k_s}{2} (\underline{n},\underline{n}_o)^2 \; .$$

That of function (c), is

$$U_{sc} = \frac{k_s}{2} (\underline{n}-\underline{n}_o)^2 \; .$$

If an electric displacement field is applied, a negative electrostatic energy term is introduced:[8]

$$U_E = \frac{-D_z^2}{8\pi[\varepsilon_{\parallel}\cos^2(\underline{n},\underline{z})+\varepsilon_{\perp}\sin^2(\underline{n},\underline{z})]} \; ,$$

where $D_z$ is the normal component of the displacement field. This energy leads to deformation in equilibrium configurations, at least if the displacement field is sufficiently large. The first form of surface energy, ($U_{sa}$) leads to two critical displacement fields or voltages in cells having parallel equilibrium orientation at the surfaces.[1] At the first voltage, the molecules start to tilt away from parallel. At and above the second, the configuration becomes parallel to the z axis everywhere.

If the surface energy has either of the other forms (b or c) or any form that is not symmetric about homeotropic alignment, then there is no true critical field above which absolutely homeotropic alignment occurs. When the field is removed, alignment returns to the twisted configuration without the ambiguity present when alignment is homeotropic. Even with surface energy of form (a), neither of the critical fields exists if the equilibrium alignment at the surface is tilted, but the amount of deformation change above the upper or below the lower critical field is small if the tilt is small.

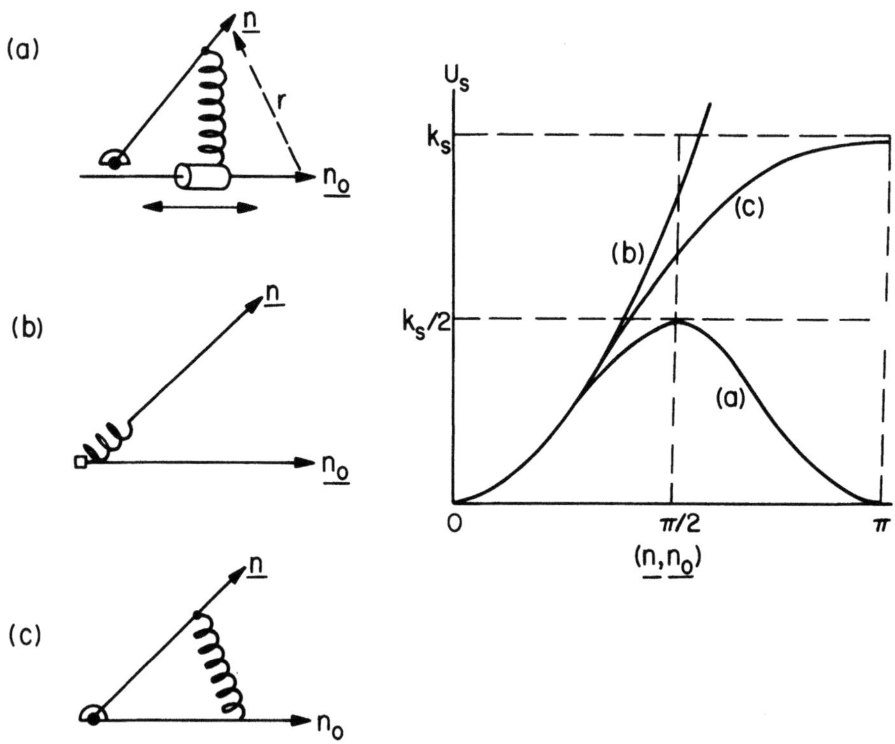


Fig. 3. Three models of surface energy showing mechanical analogs and functional form: (a) $\sin^2$ function $U_{sa}$, (b) parabolic function $U_{sb}$, (c) intermediate function $U_{sc}$.

OUTLINE OF THE DYNAMIC PROBLEM

Five independent viscosity parameters are required to describe the flow of nematics and cholesterics.[2,3] The set of five may be chosen in various ways.[9] One particularly simple set determines resistance to the motions shown in Fig. 4. The first three, $\eta_1$, $\eta_2$ and $\eta_3$, are like the viscosity of ordinary isotropic liquids but depend on the relative direction of shear and director orientation. These three may be regarded as principal axis of a viscosity ellipsoid. The fourth, $\eta_{12}$, is a measure of deviation from true ellipticity, giving a difference in viscosity as directors lean toward or opposite the shear flow. The final viscosity, $\gamma_1$, determines resistance to pure rotation without flow. In some cases a fair approximation to twist cell behavior may be obtained ignoring flow and considering only $\gamma_1$.[10] When numerical computations are undertaken, however, one might as well do the problem correctly and include the flow.[11,12]

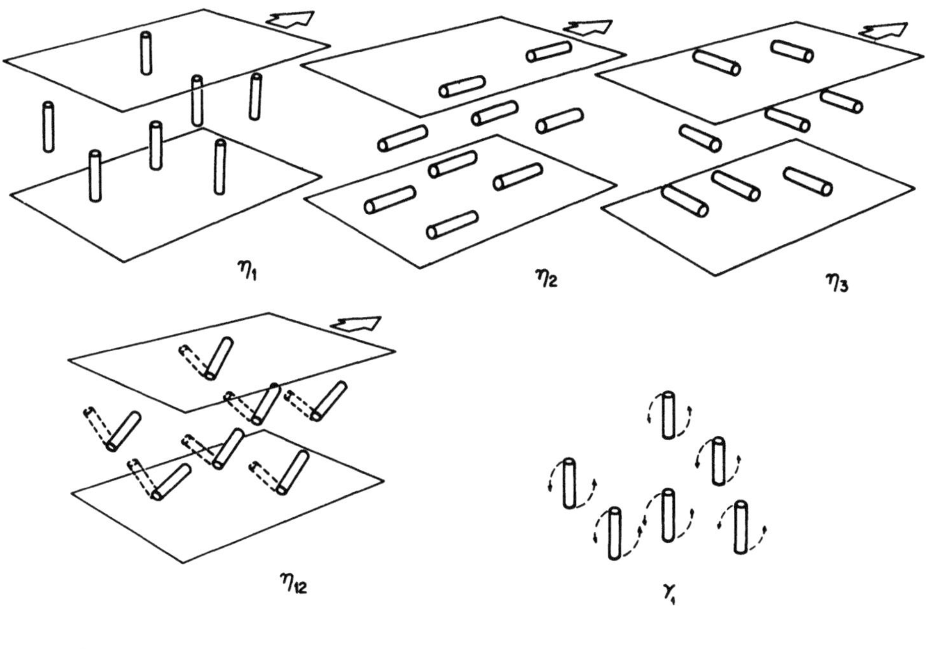


Fig. 4. Depiction of the five independent viscosity parameters. Principal viscosities $\eta_1$, $\eta_2$, and $\eta_3$ resemble ordinary isotropic shear viscosity but depend on orientation $\eta_{12}$ is a sort of correction yielding symmetric deviation of viscosity from ellipticity for forward and backward tilt. $\gamma_1$ gives resistance to director rotation without flow.

If the surface anchoring is assumed to be elastic, one immediately finds that surface reorientation is instantaneous and violently oscillatory when fields are suddenly applied unless a viscosity independent of the five bulk parameters is introduced. Since the surface is assumed to rotate but not to flow, a simple viscosity that resists surface rotation, $\gamma_s$, may be introduced. It differs in dimensionality from the bulk viscosity, $\gamma_1$, however, just as the surface elasticity differs from the bulk elasticities. The surface relaxation equation is of the following form for small perturbations:

$$\gamma_{\underline{w}} \dot{\underline{n}} = -k_s(\underline{n}-\underline{n}_o) .$$

This expression yields a surface relaxation time $(\gamma_s/k_s)$. In the bulk, the second derivative of orientation with respect to z replaces the simple direction orientation difference in the surface term. Ignoring flow, the bulk relaxation has a characteristic time $\gamma_1 h^2/k$, where h is the liquid crystal film thickness and k is a sort of average of the three bulk elastic constants. If flow is included, the same thickness dependence occurs, but all the viscosities enter and motion is very nonlinear.

## ILLUSTRATIVE COMPUTATIONS

To illustrate the effects of viscous weak anchoring, we have computed the dynamic and optical behavior of cells with rigid anchoring and with weak anchoring. The material chosen for an example is E-7, a widely used mixture of high dielectric anisotropy made by BDH.[13] Anchoring based on surface energies of types a and c were used. Surface time constants comparable to bulk time constants were used. Longer surface time constants give behavior that depends on history in a rather obvious way, but computation time for such behavior is prohibitively long. Shorter time constants give oscillation at the surfaces when fields are changed. These have not been observed in real cells, so we suppose surface time constants are generally longer. Bulk liquid crystal physical constants that are relevant to dynamic and optical computations are listed in Table 1.[13,14] Since we could find no useful viscosity measurements on E-7, we used the values for MBBA[9] for purposes of illustration. More complete bulk viscosity measurements on a variety of liquid crystals would be very helpful. There are no data at all on surface viscosity. We and others have observed slow surface reorientation over minutes or hours in some cases. More than one relaxation mechanism may explain real surface viscosities.

We have chosen cell thickness such that the optical path difference for an ordinary and an extraordinary ray in a parallel, untwisted cell would be two wavelengths for normally incident light. Thus,

$$h = 2\lambda / |\underline{n}_e - \underline{n}_o| \ .$$

This criterion gives a wide viewing angle. The wavelength chosen is that at which optical anisotropy has been measured. Polarizer and analyzer were assumed to transmit light polarized parallel to the direction of directors at the respective surfaces when there is no surface torque. Normal transmission of light in a single pass (not reflected) is given. Reflection loss at outside glass surfaces is ignored. Transmission is relative to incident plane-polarized light intensity.

The liquid crystal was assumed to be doped with an optically active material that gives it a quarter turn of cholesteric twist in the cell thickness. This prevents untwisting of the liquid crystal even if anchoring is weak. It also prevents walls due to twist reversal.

Figure 5 shows the transmission (T) and the tilt angle from normal, $\Theta_m$, at the midplane of a cell with directors rigidly attached at the boundaries with tilt $\Theta_b$=86°. Other parameters are given in Table 1. The cell starts at time zero at equilibrium with no field. Three volts are applied for 0.05 sec and then switched off. Note that equilibrium is almost reached after about 0.04 sec with the field on. The wiggles in transmission are interference phenomena. When the field is turned off, the angle $\Theta_m$ dips briefly below zero because of backflow[11,12,15] and returns almost to its original configuration in about 0.04 sec. It is interesting to notice how much more sudden the optical transition is than the tilt response. Transmission gives a very indirect measure of director behavior.

Figures 6 and 7 show response of cells having weak anchoring with constants given in Table 1, using the symmetric energy function (a) and the asymmetric form (c), respectively. The surface tilt as well as midplane tilt are variable.

The computation for the symmetric function $U_{sa}$ must be terminated before equilibrium is reached with the field on, after about 0.03 sec, as shown in Fig. 6. This is because the configuration comes so close to uniformly homeotropic alignment that minor rounding errors by the computer will otherwise lead to random directions of return after the field is turned off. Thermal fluctuations probably would have the same effect on real cells. As with the rigid boundary cell, we are using three volts, but the weak anchoring reduces the critical voltage considerably.[1] Notice that turn-on time is faster but turn-off is much slower than with the rigid-boundary cell because of the

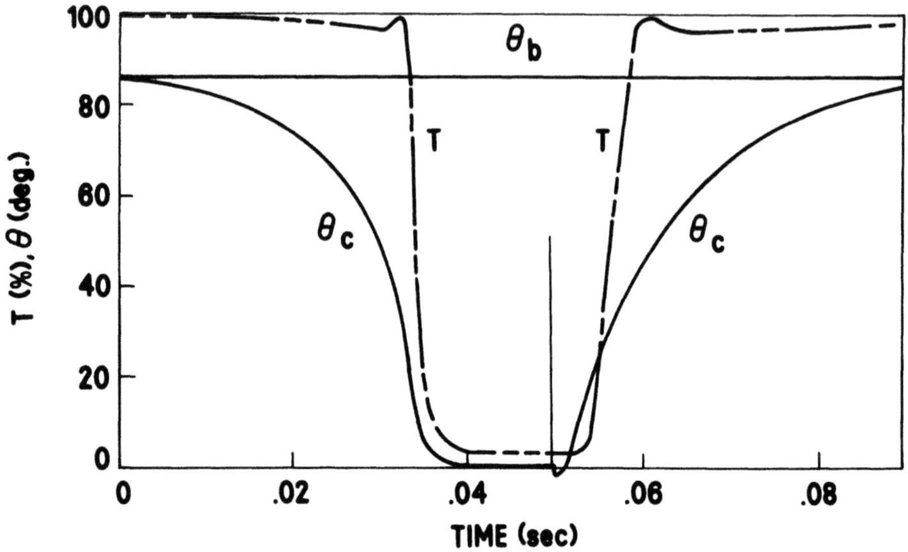


Fig. 5. Computed director tilt angle at midplane ($\Theta_c$) and transmittance with crossed polarizer and analyzer (T) as a function of time in a 90° twisted cholesteric cell with boundaries rigidly fixed at $\Theta_b$=86°. ($k_s$ and $\eta_s$ much greater than values listed).

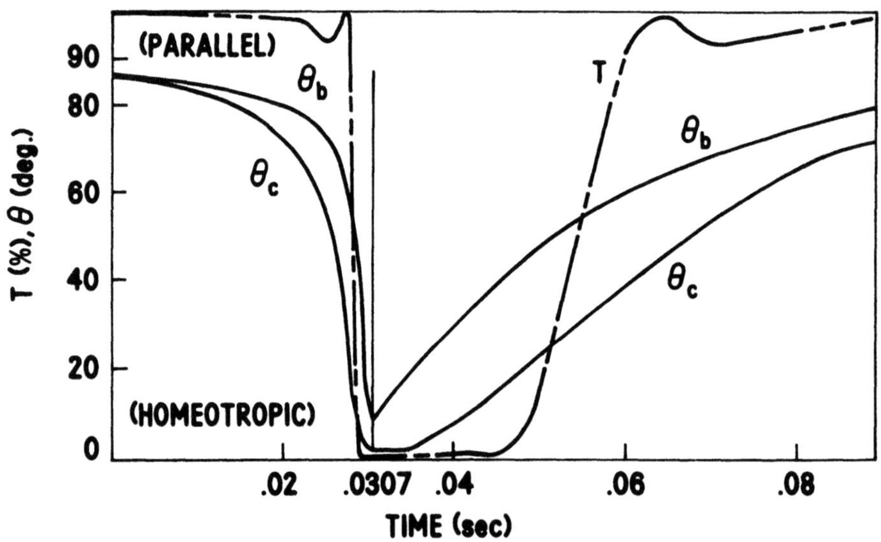


Fig. 6. Computed director tilt angle at midplane ($\Theta_c$) and at surface ($\Theta_b$), and transmittance for 90° twisted cholesteric cell with surface energy function $U_{sa}$. Surface tilt at zero field is 86°. (Listed $k_s$ and $\eta_s$.)

near-unstable-equilibrium state reached when the field is on. If equilibrium surface tilt farther from 90° is used, this ambiguity disappears and the cell behavior is much like that shown in Fig. 7.

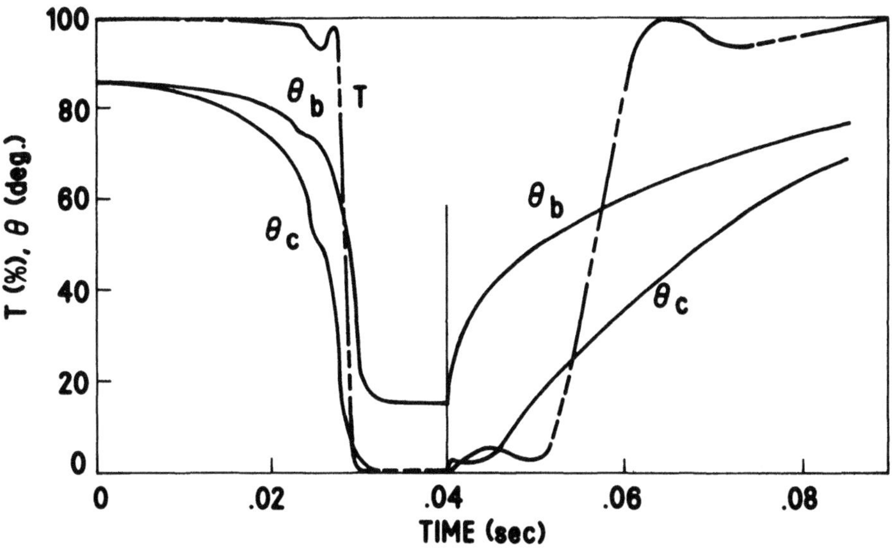


Fig. 7. Same as Fig. 6 but intermediate surface energy function $U_{sc}$ is used.

Figure 7 represents evolution of configuration in a cell with all the same parameters as that given in Fig. 6, except that the surface energy has the functional form $U_{sc}$. In this case there is a restoring torque on the boundaries when the field is on so that three volts only brings them to an angle of 15.5°, and they rapidly and unambiguously return toward 86° after the field is removed. The critical voltage for small perturbation of cells with parallel (90°) boundaries would be the same for either of the last two cases, but the cell with energy function $U_{sc}$ behaves predictably after large fields are applied while the other does not. The behavior of a cell with the parabolic surface energy function $U_{sb}$ would be practically identical to that with $U_{sc}$, except that for a given value of field and $k_s$, the surface tilt $\Theta_b$ would be somewhat larger. If some of the molecules at a cell surface were bonded as suggested in Fig. 3b and others as in Fig. 3a, the effective energy function on average might be very similar to that in Fig. 3c in the range of interest, even though the simple mechanical model shown seems less realistic than the other two.

If larger values of $\gamma_s$ are used without altering other parameters, changes of $\Theta_b$ are slower. In the case of very slow surface response, the cell may respond to rapid fluctuations in applied field as if the surface attachment were rigid, but the surface tilt will have a value determined by averaging the effects of fields applied over approximately the preceeding surface time constant.

Table 1. Assumed Parameters of Liquid Crystal in CGS Gaussian Units

| | | |
|---|---|---|
| Elastic Cts.[14] | $K_{11}$ | $1.85 \times 10^{-7}$ |
| | $K_{22}$ | $1.30 \times 10^{-7}$ |
| | $K_{33}$ | $2.02 \times 10^{-7}$ |
| Viscosities[9] | $\gamma_1$ | 0.95 |
| | $\eta_1$ | 1.21 |
| | $\eta_2$ | 0.238 |
| | $\eta_3$ | 0.416 |
| | $\eta_{12}$ | 0.065 |
| Surface Elastic Ct. | $k_s$ | 0.024 |
| Surface Viscosity | $\eta_s$ | $3.1 \times 10^{-5}$ |
| Dielectric Cts.[14] | $\parallel$ | 18.8 |
| | $\perp$ | 5.3 |
| Refractive Index[13] | $n_e$ | 1.724 |
| | $n_o$ | 1.513 |
| Wavelength | $\lambda$ | $6328.0 \times 10^{-8}$ |
| Cell Thickness | h | $6.0 \times 10^{-4}$ |

REFERENCES

1. J. Nehring, A. R. Kmetz and T. J. Scheffer, J. Appl. Phys. 47:850 (1976).
2. F. M. Leslie, Arch. Rat. Mech. Anal. 28:235 (1968).
3. J. L. Ericksen, Trans. Soc. Rheol. 5:23 (1961).
4. A. Rapini and M. Papoular, J. Phys. (Paris) 30:C4-54 (1969).
5. C. W. Oseen, Trans. Faraday Soc. 29:883 (1933).
6. F. C. Frank, Discuss. Faraday Soc. 29:945 (1958).
7. P. G. deGennes, "The Physics of Liquid Crystals," (Clarendon Press, Oxford, second printing, 1975) p. 143.
8. W. Helfrich, Mol. Cryst. Liq. Cryst. 21:187 (1973).
9. W. H. de Jeu, Phys. Lett. A (in press).
10. G. Baur, F. Windscheid and D. W. Berreman, Appl. Phys. 8:101 (1975).
11. C. Z. van Doorn, J. Appl. Phys. 46:3738 (1975).
12. D. W. Berreman, J. Appl. Phys. 46:3746 (1975).
13. G. W. Gray, "Advances in Liq. Cryst. Materials for Chem. Applications," (Publ. by BDH Chemicals, Ltd., Dorset, England, 1978).
14. K. Suzuki, K. Toriyama, T. Ishibashi and K. Odawara, 7th Int. Con. Liq. Cryst., Bordeaux, 1978, Paper DP-41 (to be published).
15. P. Pieranski, F. Brochard and E. Guyon, J. Phys. (Paris) 34:35 (1973).

# MOLECULAR ORIENTATION OF MONOMOLECULAR LIQUID CRYSTAL LAYERS ON VARIOUS SUBSTRATES

S. Meiboom and M. Sammon

Bell Laboratories
Murray Hill, New Jersey 07974

## INTRODUCTION

It is well known that liquid crystals will assume a specific orientation at boundary surfaces. Depending on the nature of the liquid crystal and of the surface, the director can be oriented either perpendicular to the surface ("homeotropic" alignment), parallel to it ("homogeneous" alignment), or at an intermediate angle ("inclined" alignment). The actual orientation is no doubt the result of a complex interplay of interface forces - electrostatic, van der Waals, steric effects, etc. The work reported here is an attempt to learn more about the respective roles of these forces.

The approach taken is to deposit very thin layers of liquid crystal material on a substrate and to measure the orientation of these mono- or few-molecular layers. We investigate how the orientation depends on the thickness of the liquid crystal layer, and on the nature of the substrate.

## EXPERIMENTAL

Basically, a thin layer of liquid crystalline material is deposited by evaporation in a high vacuum on a fused silica window. Its molecular orientation is determined by attenuated total reflection (ATR) spectroscopy. Use is made of the fact that most liquid crystal molecules are highly absorbing in the near ultra-violet, and strongly pleochroic, i.e., the transition moment for the first strong absorption band in the UV is pretty well oriented along the long axis of the molecule. Thus, a comparison of the reflection coefficients for the two polarizations of the incident light gives an indication as to the orientation of the liquid crystal molecules.

A sketch of the apparatus is given in Fig. 1. The liquid crystal material is deposited onto the inner surface of the fused quartz window by evaporation in a high vacuum (~$10^{-9}$ Torr). A quartz crystal oscillator is used to measure the amount deposited. This method is essentially sensitive to the weight per unit area. However, in the results given below, this is expressed in Å units, assuming the bulk density of the liquid crystal. The photomultiplier tube in channel A measures the light intensity reflected from the window. Incoming light intensity is monitored by channel B, via a beam splitter in the incident beam. A PDP8/E minicomputer does all the necessary processing, and controls the wavelength scan of the monochromator. Plots of the absorption spectra for the two polarizations can be obtained on an X-Y recorder. The digital data can also be transmitted to a central computer for additional processing and plotting.

The relations between incident light polarization and liquid crystal orientation are illustrated in Fig. 2. At the top of the figure are sketches of the window surface, with the deposited

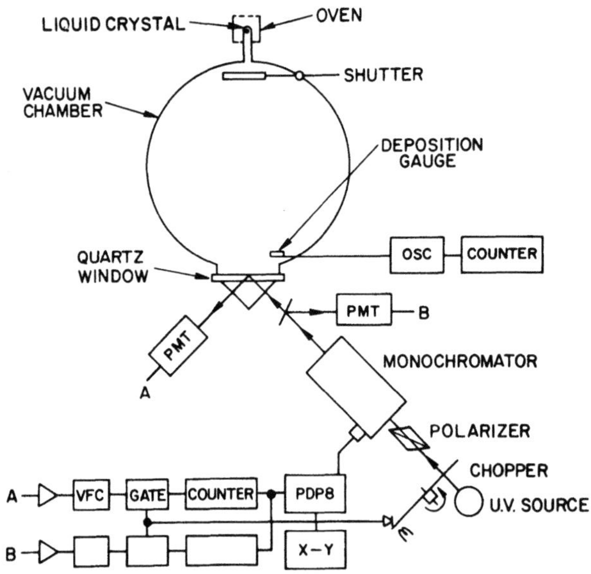


Fig. 1. Sketch of vacuum chamber and ATR spectrometer. The diameter of the chamber is about 30 cm. The deposition gauge is a 5 MHz quartz oscillator crystal, the frequency of which is multiplied by 10 and counted. The PDP8 minicomputer collects the difference of the digitized photomultiplier (PMT) outputs for chopper open and closed, for each of the PMT's.

liquid crystal molecules assumed perpendicular to the interface in the left part of the figure, and parallel to it in the right part. If the incident light is polarized with the electric vector in the plane of incidence (P-polarization (some authors define the "plane of Polarization" to contain the magnetic field, while others define it to contain the electric field. We adopt here the latter convention.)), the evanescent wave has its electric vector perpendicular to the interface, and the liquid crystal will absorb strongly in the first case (left side figure), but weakly in the second case (right side figure). The reverse will be true for S-polarization (i.e., electric vector perpendicular to plane of incidence). In the lower half of Fig. 2 are calculated curves of the reflection coefficients for the various cases. They were computed for a single Lorentzian absorption band, with a peak molar extinction coefficient of 17.000 liter $mole^{-1}$ $cm^{-1}$, and for a liquid crystal thickness of 1 nm; an order parameter of 0.7 has been assumed for the liquid crystal. The curves were calculated using a computer program written by D. W. Berreman.[1] It is obvious that for the liquid crystal orientation perpendicular to the interface, the P-polarization shows the larger absorption dip, while for parallel orientation the S-polarization is the stronger one. In the following discussion we shall use this in a qualitative way to decide on the molecular orientation.

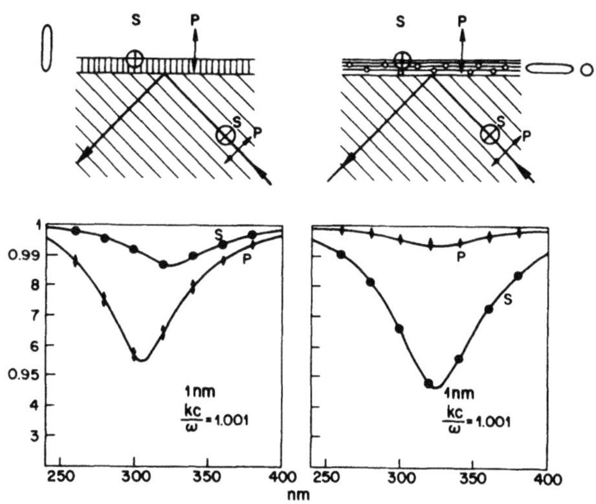


Fig. 2. Polarization dependence of attenuated total reflection. In the left half of the figure, the liquid crystal molecules are oriented perpendicular to the interface, in the right half parallel to it. The bottom graphs give calculated absorption bands for P-polarization (electric vector in plane of incidence) and S-polarization (electric vector perpendicular to plane of incidence). For details, see experimental section of the text.

In all the experiments reported here the substrate window consisted of fused silica (suprasil II). In some runs the inside surface was covered with a thin layer of poly-tetrafluoroethylene (PTFE). This was deposited by plasma polymerization, as described by Sprokel.[2] During deposition the thickness of the layer was monitored with a 5 MHz quartz oscillator. The thickness so obtained was checked in a number of samples by ellipsometry.

Before each run the vacuum system was baked at 200°C for about two hours.

All measurements reported here were done with the substrate window at room temperature. Oven temperature during evaporation was about 150°C for CBOOA liquid crystal (cyanobenzylidene-octyloxyaniline), with which the measurements reported here were made. The material was obtained from Eastman Kodak Co., and was thoroughly degassed by repeated melt-freeze in vacuum. Otherwise it was used as received. It should be pointed out that this material is crystalline at room temperature. At temperatures above the melting point the vapor pressure is high enough for appreciable reevaporation to occur, which makes measurements difficult. For this reason all runs reported here were made with the substrate at room temperature, where the vapor pressure is negligible. It is believed that the structure of the very thin layers used is not crystalline but resembles that of the liquid. In fact, we observe that crystallization sets in only when a thickness of about 150Å is reached. This is evident from the sudden onset of a strong, nearly wavelength independent scattering. All measurements were made before this occurred.

## RESULTS

A set of observed spectra are shown in Fig. 3. They are for cyanobenzylidene-octyloxyaniline (CBOOA) deposited on a fused silica substrate. The spectra shown are computer plots of the raw data, without any corrections or enhancements. The spectra in the left half of the figure apply to P-polarization, those at the right to S-polarization. The spectra at the bottom are for deposition on bare fused silica substrate; those at the top for fused silica covered with 20Å of PTFE. The second and third from the top are for 10Å and 5Å of PTFE on fused silica, respectively. The individual traces in each of the figures are for increasing thickness of deposited liquid crystal. The applicable numbers are given in the caption of Fig. 3.

For the present purpose we consider only the broad absorption band centered at 380 nm. Note that in the top figures (20Å PTFE) it is much stronger for the P-polarization than for the S-polarization. We conclude that the long axes of the liquid

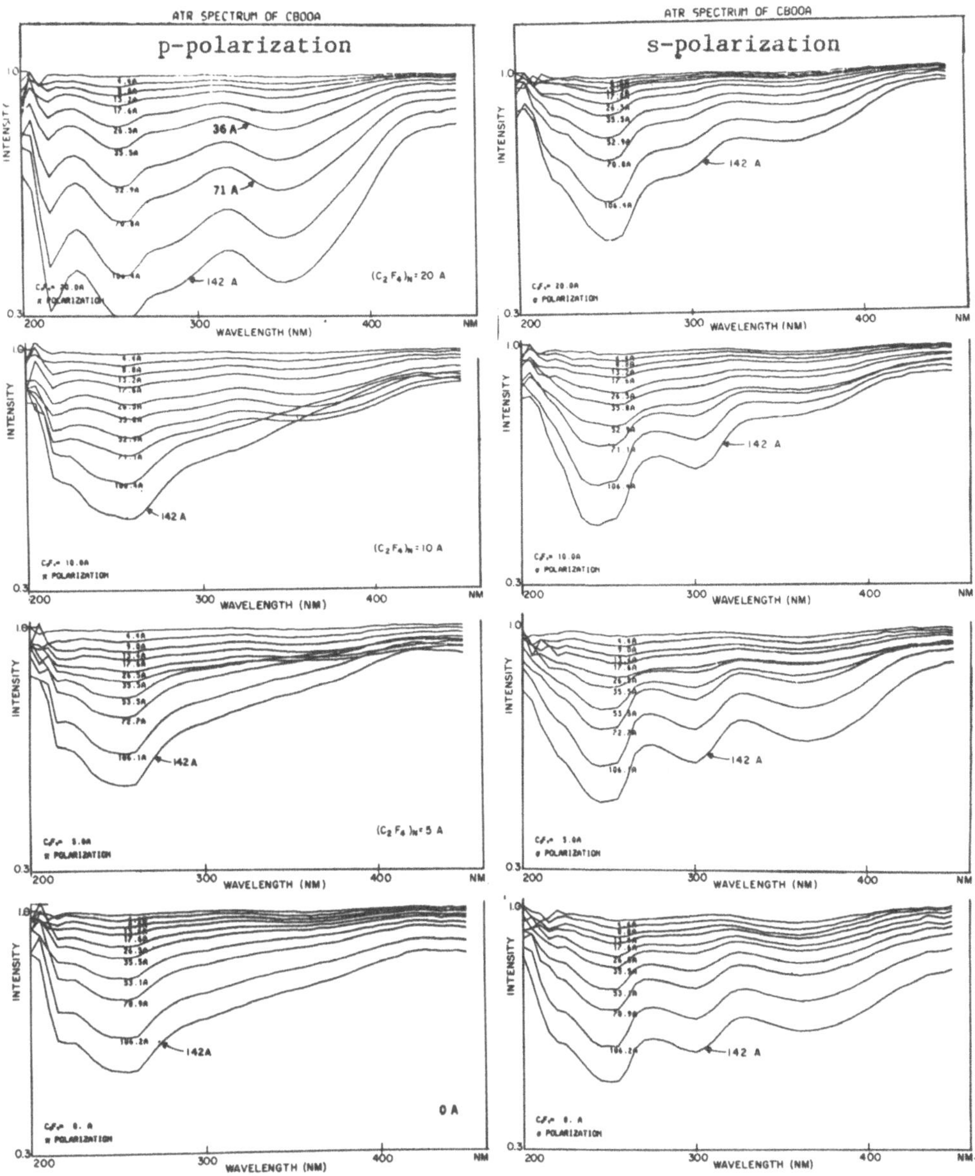


Fig. 3. Attenuated total reflection spectra of thin layers of CBOOA on PTFE-covered fused silica. Wavelength is plotted on the x-axis, and reflection coefficient on the y-axis. The thickness of the PTFE layer is, from top to bottom: 20Å, 10Å, 5Å and 0Å. The individual traces in each of the figures are for liquid crystal thicknesses of 4.4; 9; 13; 18; 27; 36; 53; 71; 106 and 141Å, respectively.

crystal molecules are perpendicular to the surface of the substrate. The reverse is true for the bottom figures (bare fused silica), and we conclude that the long axes of the molecules are here parallel to the surface. In the intermediate figures the behavior is more complex, and to show it in more detail we have plotted in Fig. 4 the absorption intensity at the maximum of the band (380 nm) as function of liquid crystal thickness. The four plots in the left half of Fig. 4 give these curves for PTFE thicknesses of 20, 10, 5 and 0Å, from top to bottom. Curves for P-polarization are indicated by squares; for S-polarization by circles. The four graphs in the right half of Fig. 4 show the same data on an expanded y-scale, to show detail at small liquid crystal thickness.

In interpreting the plots in Fig. 4, we recall that if the curve for P-polarization is below that for S-polarization, the molecular orientation is perpendicular to the surface. The molecular orientation is parallel to the surface if the curves are reversed. With this in mind the following facts stand out:

With increasing liquid crystal deposition thickness, the curves always start off with the one for S-polarization below the one for P-polarization. (See expanded curves at the right side of Fig. 4.) This means that the liquid crystal molecules always lie down parallel to the surface at the start of the deposition. On the bare fused silica substrate (bottom curves), the molecules remain in that orientation as the liquid crystal deposition is continued. For the PTFE covered surface (top figures), however, the curves cross at a liquid crystal thickness of about 10Å, and at thicknesses above 20Å all the molecules appear to have adopted a perpendicular orientation.

These observations allow a straightforward explanation: the length of a CBOOA molecule is about 25Å, while its transverse dimensions average 5Å. Thus, for a deposition thickness less than 5Å, one has essentially isolated liquid crystal molecules on the substrate surface and would expect that these will maximize their binding energy with the substrate by lying down. However, if the number of liquid crystal molecules is increased beyond that required for a dense monomolecular film - i.e., beyond 5Å - then, on PTFE, it apparently becomes energetically more profitable for the molecules to stand up. Presumably, the strongly polar -CN group of the CBOOA molecule contributes the major interaction with the substrate, and it is energetically advantageous to maximize their number near the surface. This interpretation, though somewhat tentative at this time, is supported by the fact that similar measurements with hexyloxyazoxybenzene as the liquid crystal show that the molecules of this material always lie down, independent of PTFE and liquid crystal layer thickness. This

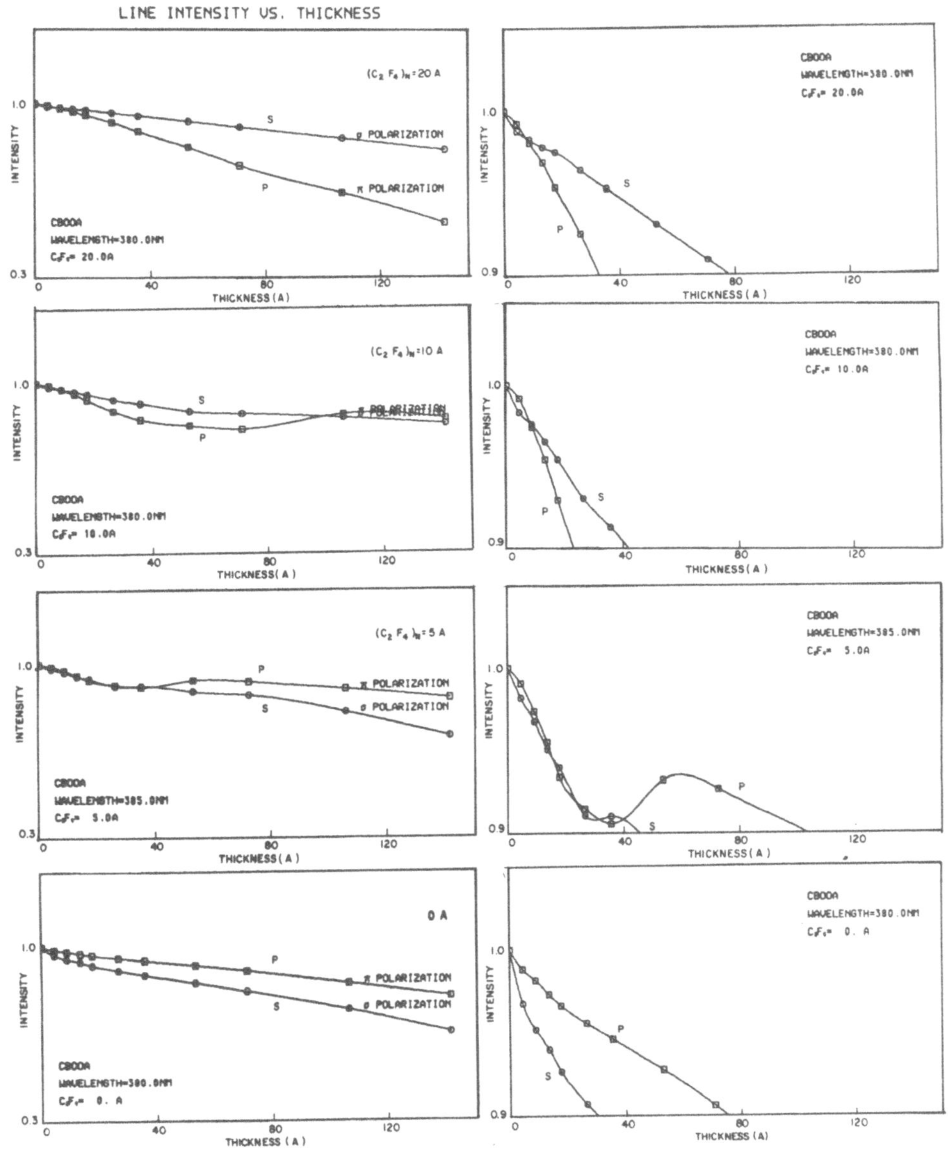


Fig. 4. Absorption band intensity as function of liquid crystal thickness. The points were taken from Fig. 3 at a wavelength of 380 nm. Circles indicate s-polarization, squares p-polarization. The plots in the right half of the figure give the same data as those in the left half, but the Y-scale has been expanded by a factor of 7, to show detail at low liquid crystal thickness.

compound has polar groups nearer the center, not at the end of the molecule.

Another interesting feature of Fig. 4 are the results for the substrate with 10Å thick PTFE (the plots in the second row of the figure). In this case the behavior for thin liquid crystal layers - less than about 50Å - is similar to those on the thicker (20Å) PTFE. However, at about 100Å liquid crystal thickness an inversion occurs, and from there on at least part of the liquid crystal molecules turn parallel to the surface (we cannot decide from the present measurements whether in fact some molecules are perpendicular and some parallel to the surface, or whether all molecules are at some intermediate inclination). A similar effect occurs for the 5Å thick PTFE (the plots in the third row), though here the transition is at a smaller liquid crystal thickness. Our tentative interpretation is that the thin PTFE layers are patchy. Then, for liquid crystal layers thinner than the crossover thickness, the molecules on top of the patches will be perpendicular to the surface, while those between patches will be parallel. (We disregard here the case of very thin layers discussed above, when all molecules are parallel.) With increasing liquid crystal thickness, however, the elastic bend energy in the transition between the two orientations becomes gradually more important, and at some transition thickness a single orientation, parallel to the surface, may become energetically favored. Conceivably, a compromise inclined orientation could also be the results.

## CONCLUSION

In discussing the orientational forces between liquid crystal and substrate, it is helpful to differentiate them as short-range and long-range ones. We define as short range those that decrease with distance in an essentially exponential manner. Somewhat paradoxically, the electrostatic forces must be considered as short range. This is the result of the overall charge neutrality of the medium: the rapidly increasing screening with distance results in an exponential fall-off of the net field. Other short-range forces that may play a role are steric effects, chemical bonding and hydrogen bonding. The long-range forces are the result of van der Waals (dispersion) interaction. For macroscopic bodies these, in general, fall off with between the third and fifth power of the distance. In such a model, the overall liquid crystal orientation is the result of the interplay between short-range (surface) forces, long-range van der Waals forces, and liquid crystal orientational elasticity. The consequences of such an approach have been investigated by Dubois-Violette and de Gennes.[3] For such models it can be concluded that under the conditions of the measurements reported here, the short-range forces are dominant. This is so because the thickness of the liquid crystal layer is too small for the long-range forces to play a decisive role.

There is some experimental evidence to support this view: we have made observations on a system in which a silane surfactant (octadecyltriethoxysilane) was applied to the fused silica, rather than the PTFE. Here vacuum-deposited CBOOA showed invariably parallel alignment. In contrast, use of the same surfactant in a bulk sample showed perpendicular (homeotropic) alignment under the microscope. As the van der Walls interaction is expected to produce perpendicular alignment (Richmond and White),[4] its dominance in the bulk only could explain the observations.

## ACKNOWLEDGMENTS

We are indebted to D. W. Berreman for use of his computer program, help and suggestions, and also to D. E. Aspnes for introducing us to the theory and practice of ellipsometry, H. Schonhorn for instructing us in surface chemistry and G. Smolensky for guidance in plasma deposition. Our use of PTFE was the result of a discussion with G. J. Sprokel, of IBM Laboratory, San Jose, California.

## REFERENCES

1. D. W. Berreman, J. Opt. Soc. Am. 62:502-510 (1972).
2. G. J. Sprokel, Mol. Cryst. Liq. Cryst. 42:233-248 (1977).
3. E. Dubois-Violette and P. G. de Gennes, J. de Physique 36:L255-258 (1975); J. of Colloid Interface Sci. 57:403-410 (1976).
4. P. Richmond and L. R. White, Mol. Cryst. Liq. Cryst. 27:217-230 (1973).

# AN R.F. PLASMA TECHNIQUE FOR PRODUCING TWISTED NEMATIC LIQUID CRYSTAL CELLS

G. J. Sprokel

IBM Research Laboratory
San Jose, California 95193

## INTRODUCTION

In a previous communication[1] it was shown that thin $SiO_2$ films deposited by oxidizing a silane derivative in an $Ar/O_2$ plasma will cause parallel alignment for alkylcyanobiphenyl nematic or smectic liquid crystals. Subsequently it was found that alkylcyanophenyl-cyclohexanes and commercial mixtures of phenylesters also align parallel with the substrate surface. From $\varepsilon(H)$ measurements the tilt angle was found to be zero to within 2°, i.e., the standard deviation of the measurement. The Fredericksz threshold is usually in the 5-6K Gauss region.

The experimental conditions in commercial "tunnel" or "diode" r.f. plasma systems lead to isotropically deposited films. Therefore, even though these films cause strong parallel alignment, the alignment direction in the plane cannot be expected to be uniform over large areas. Indeed most cells show a mosaic structure under the polarizing microscope. Uniform alignment of liquid crystal material throughout the cell is obtained by deliberately introducing some anisotropy in the aligning film. Janning's[2] method, evaporation of SiO from a Knudsen cell onto a substrate at a small angle of incidence, is now widely used in the industry. The anisotropy of SiO films evaporated at oblique angles was studied by Priest et al.[3] about 10 years before its application to liquid crystal alignment.

While such films indeed produce uniformly aligned cells, the tilt angle is typically in the 15-35° range.[4,5] There is no longer a well-defined threshold and multiplexing such cells becomes difficult. A technique using two consecutive evaporations at different incidence angles and in orthogonal directions along the substrate surface was introduced by Meyerhof.[6] Observed tilt angles are in the 1-10° range but the process becomes cumbersome.

A method was found to introduce anisotropy in an r.f. plasma deposited film and such films do establish a preferred orientation for liquid crystals while keeping the tilt angle at approximately 0°. This paper is a first account of these results.

Theoretical work on the structure of obliquely evaporated films was started by Henderson.[7,8] Using a Monte Carlo technique, it was shown that the anisotropy is largely topological and is caused by a shadowing effect of already deposited material. In this model the mean-free path of the material being evaporated must be at least comparable to the dimensions of the evaporator. Thus the bell jar pressure has to be in the microtorr region. However, typical r.f. plasma deposition systems operate in the region of 0.1 to 0.5 Torr. The system described below was designed to operate in the low millitorr range; nevertheless, the mean-free path is still only of the order of centimeters. The shadow effect could still be operative although there is some doubt whether it is the reason for the anisotropy.

## A BEAM R.F. PLASMA SYSTEM

A plasma beam is formed by passing a suitable gas mixture through a pair of perforated electrodes which confine the r.f. discharge. Figure 1 shows the physical layout schematically.

The two electrodes form a parallel plate capacitor. The top electrode is at ground, the lower electrode is connected through a tunable matching network to a 1 kW r.f. generator (Plasmatherm HFS1000E). Aluminum rather than stainless steel was used to fabricate these parts since aluminum has a lower sputtering coefficient and any metal sputtered off tends to deposit as $Al_2O_3$.

Gases are mixed in a manifold and enter the chamber through a diffuser. A deflector directs the gas stream into the discharge region. A turbomolecular pump mounted directly on the chamber provides the required pumping speed.

If only Argon is admitted to the chamber and the discharge is started, a beam of ionized Argon emerges from the ground electrode and extends into the exit orifice for appropriate values of pressure, pumping speed and r.f. power dissipation. There is only a weak Argon glow outside the beam.

If oxygen is added, most of the Ar* excitation energy is used for dissociation of $O_2$ and the intensity of the Argon glow

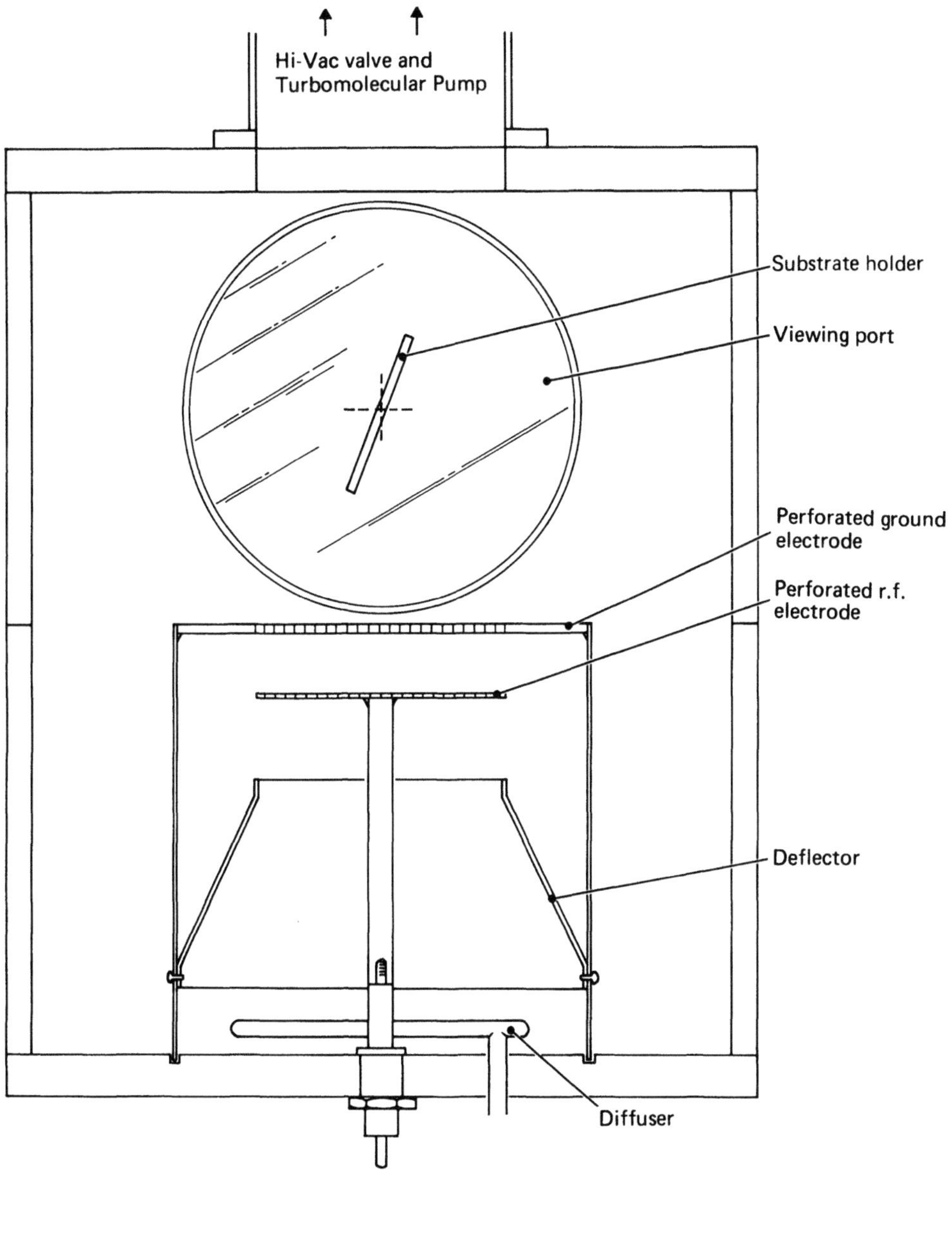


Fig. 1. Plasma Chamber Schematic.

diminishes. Initially a Tektronix "Fast-Scan" Spectrometer was used to monitor the changes in the Argon spectrum as other gases are added. Subsequently it was found that the system can be characterized adequately by the intensity of just one of the Ar* lines. A filter at 8150Å is now used to monitor the Ar* concentration at the desired location in the beam.

Substrates are mounted in a frame placed in the beam some distance from the ground electrode. The frame can be rotated around an axis perpendicular to the beam direction such that the angle between the substrate surface and the beam direction can be adjusted from 0° to 90°. The frame can hold two substrates in orthogonal positions as shown in Fig. 2. Other frames defining other angles are used for special purposes. As will be shown, the angle of incidence determines to a large extent the anisotropy of the deposit. The distance between the ground electrode and the substrates is a compromise. Etching rates and deposition rates increase as the distance is shortened, but the effect of the beam incidence angle decreases.

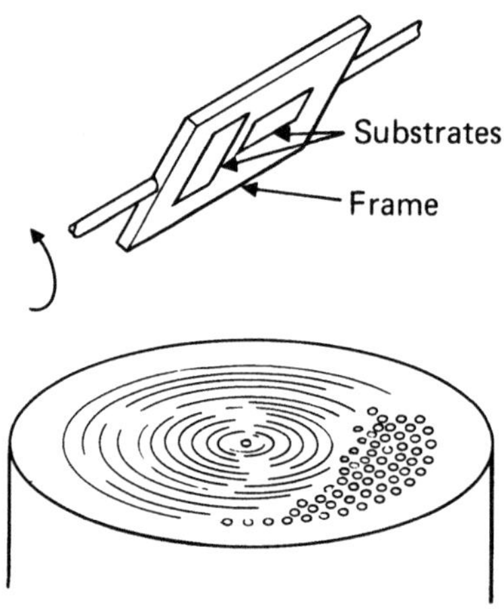


Fig. 2. Detail of sample holder. The configuration shown is used for TN cells.

The aligning layer is an amorphous $SiO_2$ film obtained by oxidizing a silane derivative. The preferred source for Si is bis-(dimethylamino)- dimethyl silane. Its vapor pressure at room temperature makes it convenient to control the composition of the gas mixture in the manifold using only a Granville Phillips valve. There are, however, other silanes and siloxanes that can be used.

The deposition step is usually preceded by an in situ cleaning step to remove residual organics. Cleaning is carried out in an $Ar/O_2$ plasma. Typical conditions are: 60% Ar, 40% $O_2$; 5-15 millitorr; 300-500W r.f. input. The removal of the organic residue occurs mostly by oxidation. Sputter cleaning in pure Argon can also be used.

The change over from oxygen etching to $SiO_2$ deposition is made without breaking vacuum or shutting down the r.f. generator. The sample is placed at the desired incidence angle and the vapor of the silane compound is admitted until the chamber pressure is 20-30 millitorr.

Excess argon is essential during the deposition step and it is commonly used in the cleaning step as well. Under appropriate conditions the deposited $SiO_2$ film is clear and glassy. If the argon concentration is insufficient, a brown powder accumulates on the substrate surface. This has no aligning properties. The need for excess Ar* suggests that the film is formed at, or very near to, the surface since it is largely the Ar* excitation energy which drives the reaction.

Typical cleaning and deposition take about 20 min. The deposited film is of the order of 100Å.

## BEAM INCIDENCE ANGLE AND CELL CHARACTERISTICS

### Tilt Angle

The beam incidence angle has no effect on the director tilt angle; all cells are essentially parallel aligned and have 0° tilt. The tilt angle is calculated from the permittivity of the sample held in a strong magnetic field directed either parallel with or perpendicular to the substrate surface. The cell capacitance is determined from phase-gain data over an appropriate frequency interval as described in some detail in an earlier publication.[9] Table 1 shows typical results. The two cells represent two consecutive runs made under identical conditions. Cells were filled with n-pentylcyanobiphenyl (BDHK15) and the permittivity was measured in the range 200-5000 Hz for each value of H. Data in parentheses are the standard deviations of the experimental data. This standard deviation is typically 0.5-0.7%, which leads to an uncertainty of 2° in the experimental value of the tilt angle. Within this limit changing the beam incidence angle from normal to 5° has no effect on the tilt. Plotting the data in Table 1 results in a conventional Fredericksz transition curve. From Fig. 3 it is seen that the threshold is about 5k Gauss at room temperature (24°C). Sample-to-sample variations as indicated by Fig. 3 are typical.

Table 1. Permittivity versus Magnetic Field

| H Field | kGauss | Sample 8-15 | Sample 8-16 |
|---|---|---|---|
| no H | | 6.20(.05) | 6.18(.03) |
| H⊥E | 12.0 | 6.25(.05) | 6.19(.007) |
| H∥E | 5.1 | 6.40(.03) | 6.18(.006) |
| | 6.1 | 7.28(.05) | 6.91(.08) |
| | 7.2 | 8.46(.06) | 8.26(.07) |
| | 7.9 | 9.43(.05) | 9.41(.04) |
| | 8.8 | 10.21(.07) | 10.31(.02) |
| | 9.2 | 10.83(.08) | 10.97(.002) |
| | 9.8 | 11.33(.09) | 11.46(.02) |
| | 10.4 | 11.74(.06) | 12.05(.03) |
| | 10.6 | 11.88(.04) | 12.32(.02) |
| | 11.0 | 12.17(.05) | 12.61(.03) |
| | 11.5 | 12.35(.07) | 12.76(.04) |
| | 11.6 | 12.48(.05) | 12.88(.04) |
| | 12.0 | 12.70(.04) | 13.12(.05) |

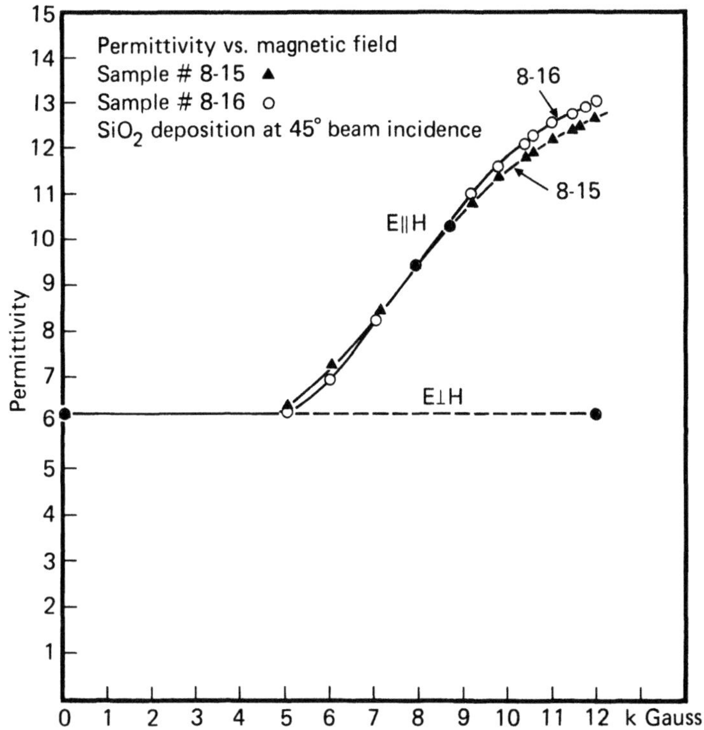


Fig. 3. ε(H) for two 45° incidence cells.

Preferred Orientation

Although the tilt angle does not change as the beam incidence angle is decreased from normal incidence to oblique, the molecular alignment in the plane changes remarkably. Gradually a preferred orientation is established which is best exemplified by transmission photographs using a polarizing microscope. The two substrates used to make the cells shown in Figs. 4, 5 and 6 were processed simultaneously in mutually orthogonal directions as indicated in Fig. 2. Thus if a preferred alignment direction is established, the completed cell should show twisted nematic behavior.

At normal beam incidence there is little evidence of a preferred orientation, but even at small deviations from normal a preferred orientation begins to show. Figures 4, 5 and 6 show the textures of cells made at 70°, 45° and 30° beam incidence respectively. The orientation of the cell on the microscope stage is shown schematically.

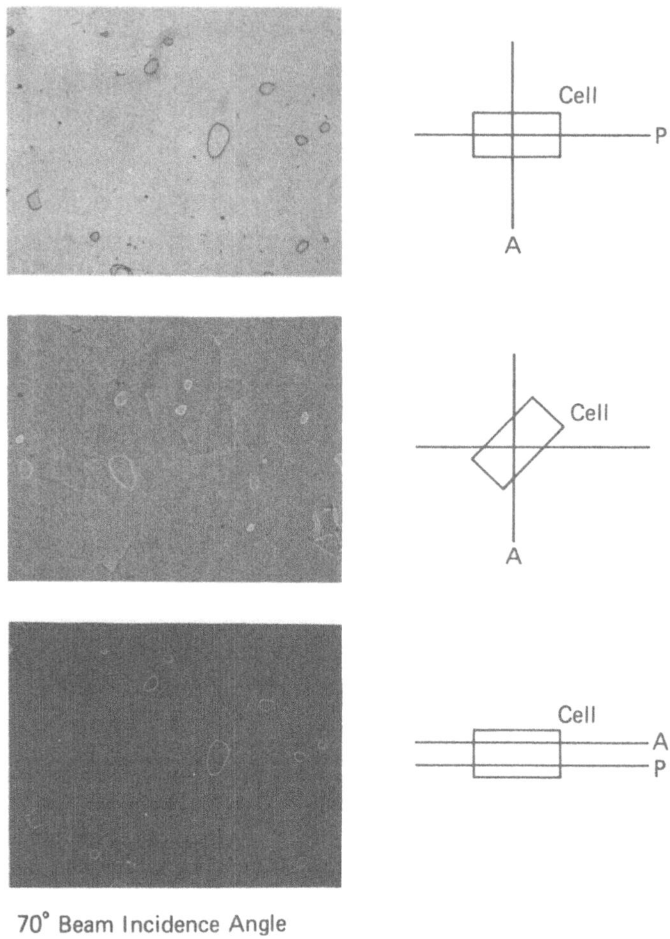


Fig. 4. Texture of 10 micron cell filled with K15. Magnification 160×. Aligning layer deposited at 70° incidence. Top: crossed polarizers, cell parallel to A and P. Middle: crossed polarizers cell at 45°. Bottom: parallel polarizers, cell parallel to A and P.

The cell prepared at 70° beam incidence (Fig. 4) shows a large number of nematic disclination loops. If the cell is turned 45° between crossed polarizers, the intensity changes but little. If the microscope polarizers are aligned parallel, the transmission decreases by about 3 f-stops (about 8×). Clearly this cell shows twisted nematic alignment, but there are many areas of different orientation delineated by disclination loops. At the disclination the main orientation of the director is perpendicular to the surface; the lines are black at crossed polarizers but white for parallel polarizers.

As the incidence angle is made smaller, the irregular shape of the disclination lines changes. At 45° beam incidence (Fig. 5) all disclinations run in the beam direction. Again, turning the cell between crossed polarizers changes the transmission only slightly. It appears that two preferred orientations are established. The director is either in the beam direction or perpendicular to it. The parallel disclinations separate areas of reverse twist.

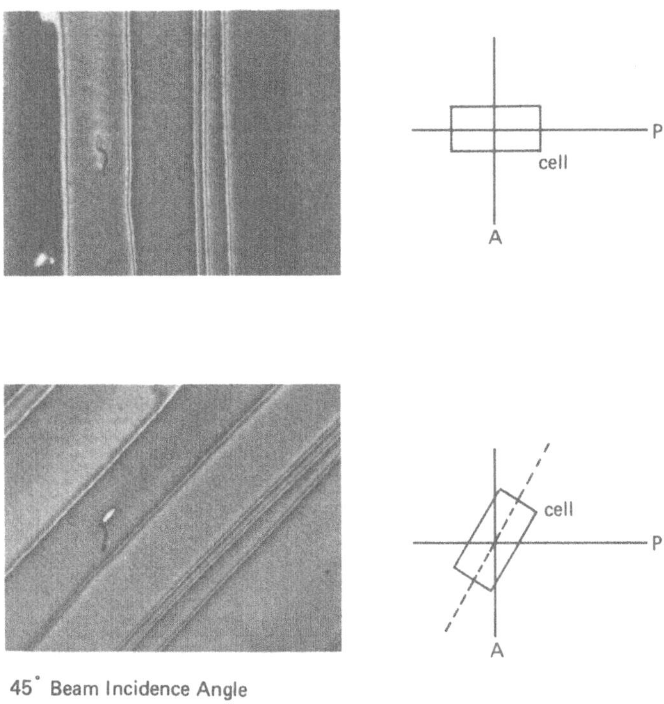


Fig. 5. Texture of 10 micron cell filled with K15. Magnification 160×. Aligning layer deposited at 45°. Disclination lines parallel to beam direction.

Finally at sufficiently small incidence angle the cell becomes uniformly aligned as illustrated in Fig. 6. About 30° is the largest beam incidence angle which can be used. At smaller angles the alignment remains uniform, but the deposition time increases markedly.

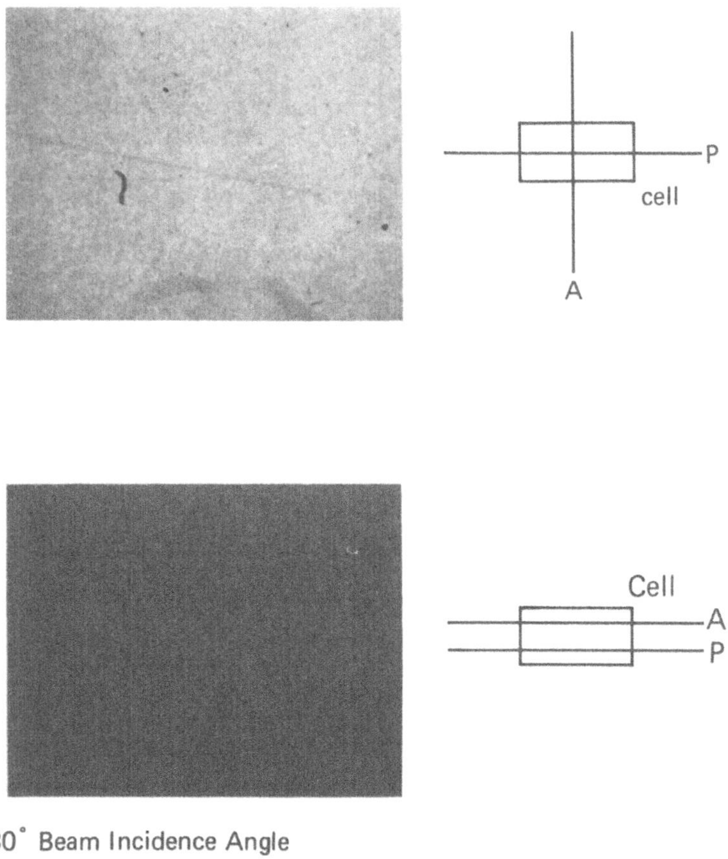


Fig. 6. Texture of 10 micron cell filled with K15. Magnification 160×. Aligning layer deposited at 30°.

Optical Response

The effect of many reverse twist regions is clearly apparent in the optical response. Figure 7 shows the pulse response of the cell used for Fig. 4 having many regions of different orientations. Figure 8 shows the response of the cell used for Fig. 6 which is essentially uniformly parallel oriented. In both figures the voltage across the cell is a 0.1 kHz square wave applied for 100 ms. The upper trace shows the amplitude of the driving signal. The lower trace is the photo diode output. To calibrate this, the top left trace of each figure shows the transmission of the cell for crossed polarizers (clear) and parallel polarizers (black), with the applied voltage well below threshold. The distance between the parallel traces, of course, represents the intensity of the microscope illuminator and the transmission of the clear cell.

All other oscilloscope traces were made with analyzer and polarizer crossed and the deposition direction parallel to the polarizer. The three traces forming the right column in each figure show the optical response from threshold to saturation. The uniformly aligned cell, Fig. 8, has a sharp threshold at 4.8 Vptp and saturates at about 8 Vptp. Increasing the amplitude beyond this only sharpens the rise time but there is no further increase in light attenuation. The mosaic cell in Fig. 7 shows a threshold at about 5 Vptp but there are sharp transients at turn on. The light transmission in fact increases for a short time as the voltage is applied (Fig. 7 right bottom trace). The transient is noticeable to the unaided eye and the cell appears to flicker. The uniformly aligned cell turns dark without any flicker. The turn on at 20 Vptp is illustrated for both cells in the left column, middle trace. For the uniformly aligned cell the rise time is about 1 ms and the transient is smooth. For the mosaic cell one cannot assign a unique rise time. Finally, in both figures the lower trace of the left column shows the pulse repetition rate used for all photographs. The repetition rate is 1 pps to allow complete decay between pulses.

The decay part of the curves in Figs. 7 and 8 are characteristic for the T.N. cells as discussed by van Doorn[10] and Berreman.[11] It appears that the "back-flow dip" is much less pronounced (see Fig. 2 in Ref. 10). A quantitative analysis has not yet been made.

The threshold voltage for the uniform cell is shown in Fig. 9. A continuous sine wave at 1 kHz is applied and light transmission between crossed polarizers is plotted versus the rms value of the applied voltage. Threshold voltage is 1.1V and saturation is reached at 2V rms.

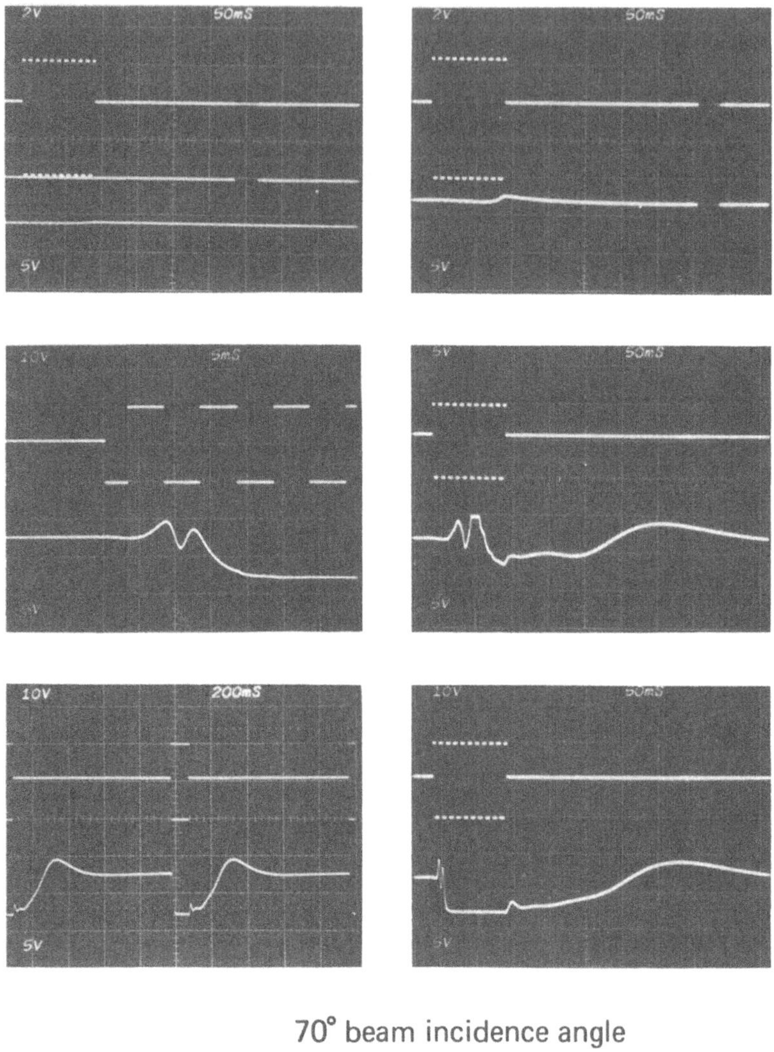


Fig. 7. Electro-optical response of cell used for Fig. 4. Top left trace calibrates photodiode output. Mosaic cells show a sharp transient at turn on.

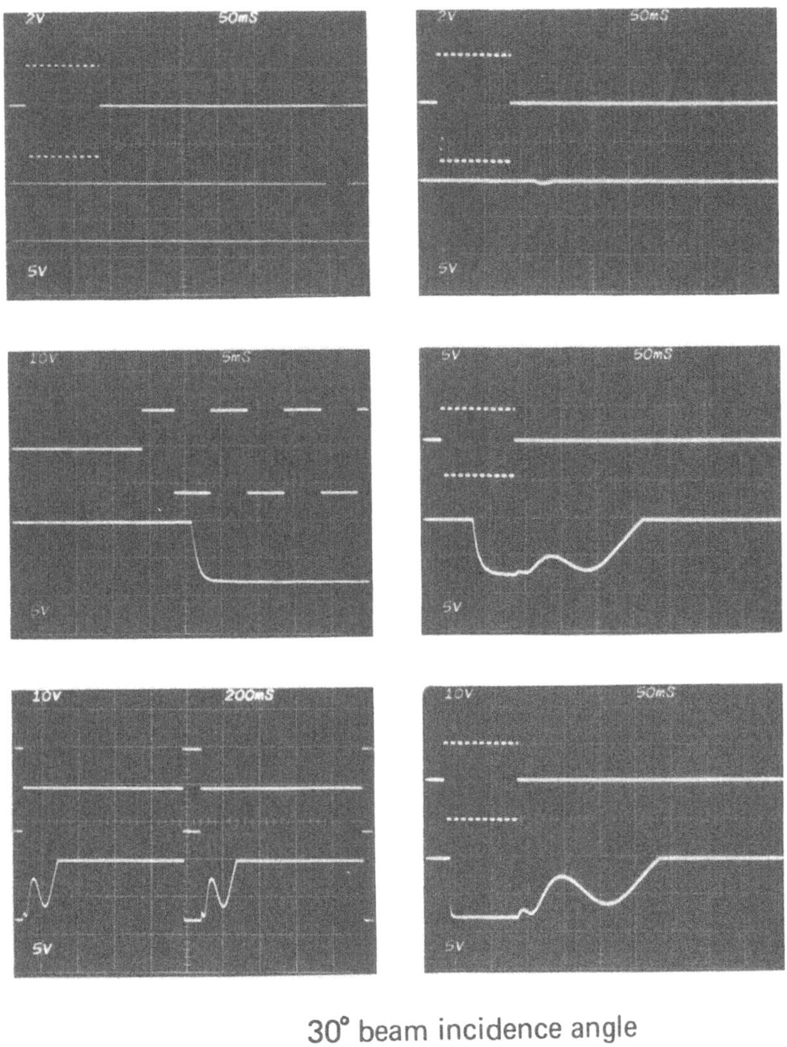


Fig. 8. Electro-optical response of cell used for Fig. 6. Uniformly aligned cells turn on without transient effects.

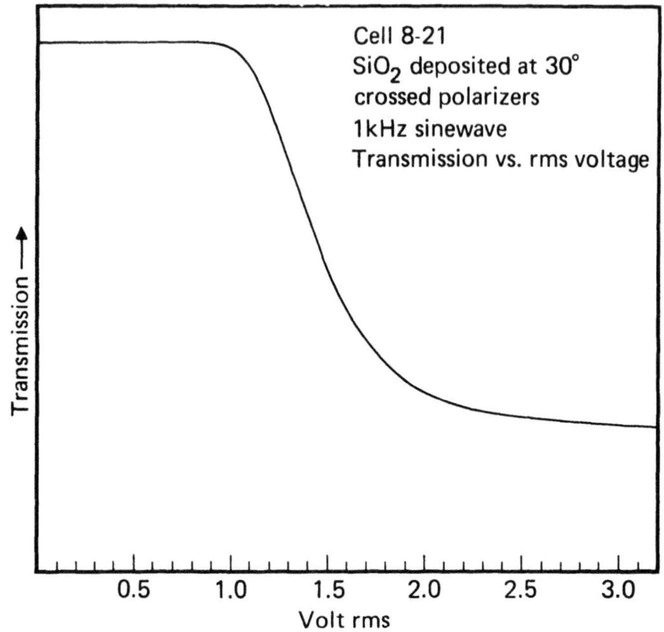


Fig. 9. Threshold voltage for cell used for Fig. 6. Light transmission between crossed polarizers is plotted against rms value of applied 1 kHz sine wave.

DISCUSSION AND CONCLUSIONS

The experimental evidence presented here indicates that anisotropic $SiO_2$ films can be deposited in a beam r.f. plasma system. The films cause parallel alignment for commonly used liquid crystal material and for appropriate beam incidence angles uniform parallel alignment is obtained in a single deposition. The plasma system built for this purpose is only of moderate size, the beam is 5" in diameter. Larger systems could readily be made for commercial application.

T.N. cells made from substrates on which an anisotropic film has been deposited show a pronounced threshold which should facilitate multiplex addressing of arrays.

There are two important differences between r.f. plasma deposited $SiO_2$ films and obliquely evaporated SiO films.

For r.f. plasma deposited films the tilt angle is essentially zero for all materials tested so far, i.e., biphenyls, phenylcyclohexanes and diesters, and for all beam incidence angles. It is well-known from the work of Scheffer et al.,[4]

Crossland et al.[5] and others not referenced here that the tilt angle for anisotropic SiO films depends both on the incidence angle and on the liquid crystal material.

Much remains to be learned about the interaction of surfaces and liquid crystal material and more specific about the aligning properties of treated surfaces. Crossland[12] has shown that thick SiO films are somewhat birefringent if deposited at oblique angles. The stress anisotropy of such films was documented by Priest et al.[3] The aligning films in T.N. cells are far too thin for direct optical measurements. Plasma deposited films as used in this work have been examined using phase contrast microscopy. Weak modulations, indicative of ridges running across the beam direction can be observed, but the evidence is not regarded as sufficiently conclusive.

The mechanism of film growth is probably quite different. In SiO evaporation it is likely that molecular aggregates of the composition SiO travel in straight lines from the heater to the substrate and there is likely very little reorganization along the substrate surface. In the r.f. plasma beam system it is more likely that fragments of the organo silicon compound already at the surface are oxidized in situ after further fragmentation by Ar* and O*. Topological anisotropy of the deposit could be induced by the flow of the plasma along the substrate surface thus bypassing the requirement for long mean-free path. Further experiments are needed to justify this hypothesis.

## REFERENCES

1. G. J. Sprokel, Mol. Cryst. Liq. Cryst. 42:233 (1977).
2. J. L. Janning, APL 21(4):15 (1972), U.S. Patent 3,834,792 (1974).
3. J. Priest, H. L. Caswell and Y. Budo, JAP 34(2):347 (1963).
4. T. J. Scheffer and J. Nehring, JAP 48:1783 (1977).
5. W. A. Crossland, J. H. Morissy and B. Needham, Appl. Phys. 9:2001 (1976).
6. D. Meyerhof, APL 29(11):691 (1976).
7. D. J. Henderson, M. H. Brodsky and P. Chaudhari, APL 25(11):641 (1974).
8. S. Kim, D. J. Henderson and P. Chaudhari, Thin Solid Films 47:155 (1977).
9. G. J. Sprokel and R. M. Gibson, JECS 124(9):557 (1977).
10. C. Z. van Doorn, JAP 46(9):3738 (1975).
11. D. W. Berreman, JAP 46(9):3746 (1975).
12. W. A. Crossland, APL 26(11):598 (1975).

# LIQUID CRYSTAL - SURFACE INTERACTIONS BY SURFACE PLASMONS

J. D. Swalen and G. J. Sprokel

IBM Research Laboratory
San Jose, California 95193

The molecules forming liquid crystals interact among themselves to form the liquid crystal but also are oriented by wall interaction and an applied electric or magnetic field. A surface treatment can produce either parallel or perpendicular alignment of the molecules.[1] Optical measurements with polarized light ascertain these specific alignments. Measurements of the apparent permittivity as a function of magnetic field through the Frederiks transition can be used to scale an angular distribution from the Frank-Oseen equation of continuum theory. For strong anchoring of molecules in a parallel orientation to a surface and for high magnetic or electric fields applied perpendicularly to a cell, the molecules will align with the field in most of the slab and two thin regions will be located at each wall. Above a critical field for the Frederiks transition there will be a transition region for the molecules to rotate from, for example, parallel on the wall to a perpendicular alignment in the cell.[2] Computer calculations[1] show that high fields are necessary for this perpendicular alignment and this alignment is not accomplished until well into the cell, i.e., several micrometers.

The reorientation of the molecules has only been inferred by detailed investigation of the integral of the dielectric function. Recent studies of surface layers with surface plasmons have shown that now a very narrow region next to a surface can be measured specifically, the optical constants and thicknesses of thin overlayers on a metal.[3-5] We have applied this technique of surface plasmons to study the electrode surfaces of liquid crystal cells to observe the molecular properties as implied from the changes in the refractive indices. First the method will be described and then our results on liquid crystal cells whose, electrodes were overcoated for parallel and perpendicular alignment of the molecules, will be given. The application of an electric field to a parallel oriented cell causes a reorientation. Shifts in the surface plasmon can be used to infer details of this reorientation.

The surface plasmon is an electromagnetic wave which propagates on the surface of a metal, i.e., at a metal-dielectric interface. They involve the quasi-free electrons in the metal. The optical fields exponentially decay perpendicularly from the interface. Figure la shows schematically this decaying field. Note that the field decays much more rapidly in the metal than in the dielectric. To give one a feeling for the extent, the length for the field to become 1/e of its value at the surface in the metal is about 10 nm, while in the dielectric, this length is about 300 nm. The sharp cusp in the field at the interface is also seen in Fig. la. When a surface layer is added, the field

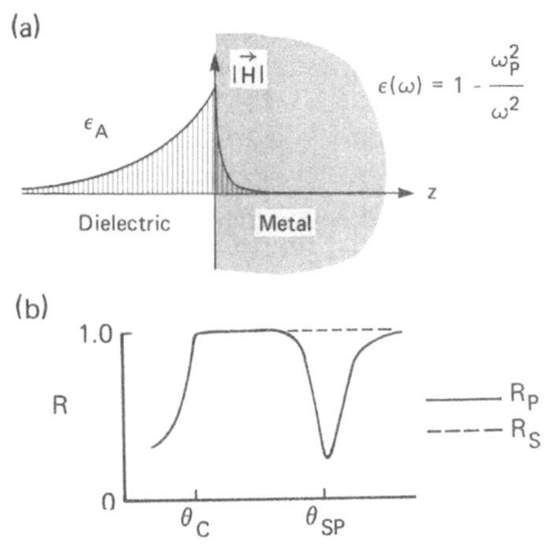


Fig. la. Schematic showing the optical field (magnetic component) at an interface which supports surface plasmons. The dielectric function in the dielectric medium is $\varepsilon_a$ and the diectric function in the metal can be approximated by the Drude-Lorentz expression given in the upper right hand corner. Notice that the field extends much farther into the dielectric than the metal.

b. The reflectivity in an ATR configuration. The $\theta_c$ is the critical angle and $\theta_{sp}$ is the angle at which the surface plasmon is excited. Reflectivity extends from zero to one. Notice that the reflectivity from s waves, i.e., those waves with their electric vector perpendicular to the plane of incidence do not excite a surface mode. Only the p waves with their electric vector in the plane of incidence do.

readjusts to the new boundary conditions and the surface mode is displaced.

A convenient way to observe the surface plasmon is by reflectivity. Because the propagation vector, k, for the surface plasmon is greater than that for free propagation, a high index prism can be used to couple to this surface wave. As another way of saying it, the surface mode is in the nonradiative region and cannot be excited directly by optical irradiation. In Fig. 1b is shown a schematic for the attenuated total reflectivity beyond the critical angle, $\theta_c$. Notice that the s wave with its electric field polarized perpendicular to the plane of incidence does not exhibit a reflectivity minimum, showing that no excition of a surface wave occurs.

The construction of the liquid crystal cells is shown schematically in Fig. 2. On a 60° prism of Schott glass LaSF5 (n=1.87625 at 632.8 nm wavelength of an He-Le laser) gold films. approximately 50 nm thick, were evaporated in a vacuum system. Simultaneously a glass slide was evaporated for the counter electrode.

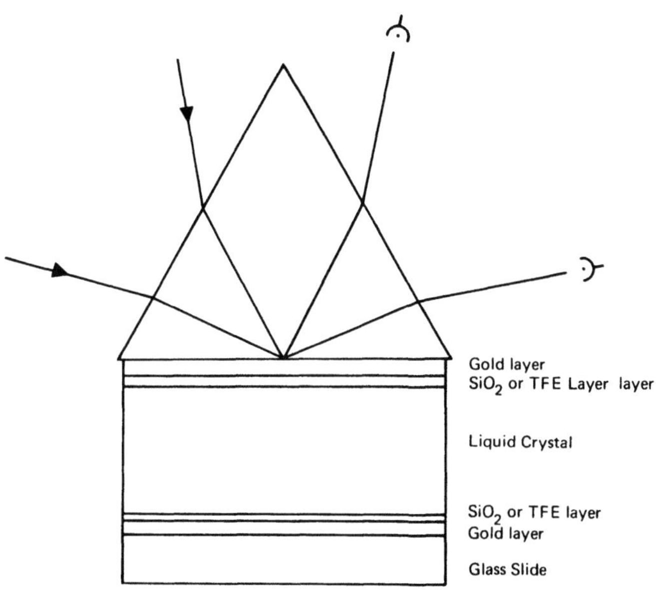


Fig. 2. Schematic showing the prism and cell for surface plasmon experiments. Light is shown entering from the left at two different angles and two corresponding detectors are shown on the right of the diagram. The thicknesses of all the layers are greatly exaggerated compared to the dimensions of the prism so as to depict everything of importance.

The prism in air was then rotated through a series of angles while it was irradiated with an He-Ne laser polarized in the plane of incidence, i.e., the plane of Fig. 2. At the angle at which the surface plasmon was excited, energy was extracted from the light beam and a reflectivity minimum occurred. This is shown in Figs. 3 and 4 at an external angle of approximately 86°. From an analysis with the complex Fresnel equations,[4,5] the optical constants and thicknesses of the gold films were determined.

According to the method described in Ref. 1, polyfluorocarbon films were deposited on the gold using an argon-tetrafluoroethylene plasma. The $SiO_2$ overlayers were deposited from a plasma discharge of oxygen and argon with a slow addition of bis(dimethylamino) dimethyl silane. The ATR minima from the surface plasmons were now shifted (see Figs. 3 and 4) from the overlayers placed on the gold films. Unfortunately one cannot both determine the refractive index and thickness of an overlayer without additional information.[4] Hence we assumed the bulk refractive indices (1.35 for TFE[6] and 1.457 for $SiO_2$[7]). This fortunately does not cause a large error because any error in refractive index is compensated for by a thickness change. Consequently the thickness is only known approximately, based on our assumption of the values for n. Currently we are conducting guided wave experiments and interferometric measurements on the films to determine n and t to greater accuracy. From the reflectivity shown in Fig. 3, the thickness of the TFE layer was found to be 23.6 nm, and from that in Fig. 4, the thickness of the $SiO_2$ layer was found to be 18.5 nm. A slightly better fit to the data was obtained in the $SiO_2$ case if the gold were reduced in thickness by some 10 to 20%. Apparently some etching of the gold takes place in the Si-O-Ar plasma environment. Although this reduction in gold thickness was applied in subsequent calculations for the liquid crystal cells, the main conclusions are relatively insensitive to this "correction".

The liquid crystal cells were then assembled with epoxy spacers and filled with the liquid crystal 4-cyano-4' pentylbiphenyl, K15, obtained from BDH chemicals.

In the TFE coated cell, in which the molecules were known to be oriented perpendicularly, the ATR minimum shifted to 5°. The calculated reflectivity curve shown in Fig. 5 agrees well with the experimental curve. Since it was felt that the refractive index of K15 was known[8,9] better in the ordinary direction, i.e., the direction perpendicular to the long axis of the molecule, this index was assumed - n=1.535 at 632.8 nm - and the other index the extraordinary one along the molecular axis, was calculated to be 1.741, this direction being perpendicular to the electrode surfaces.

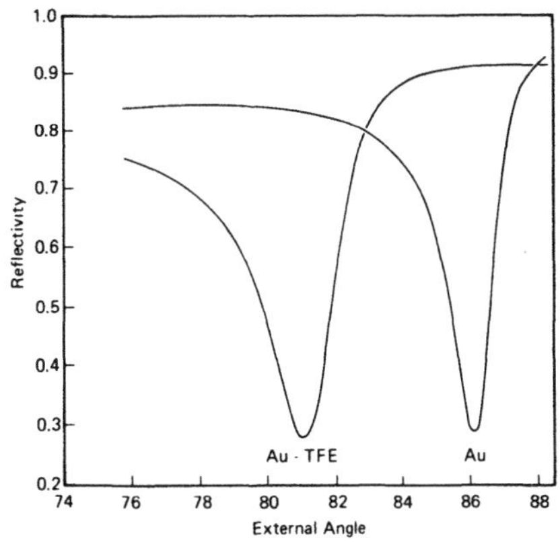


Fig. 3. The experimental ATR measurements for a gold film and for the same gold film overcoated with a 23.6 nm film of polymerized tetrafluoroethylene (TFE).
ε = -12.566 - 1.472i

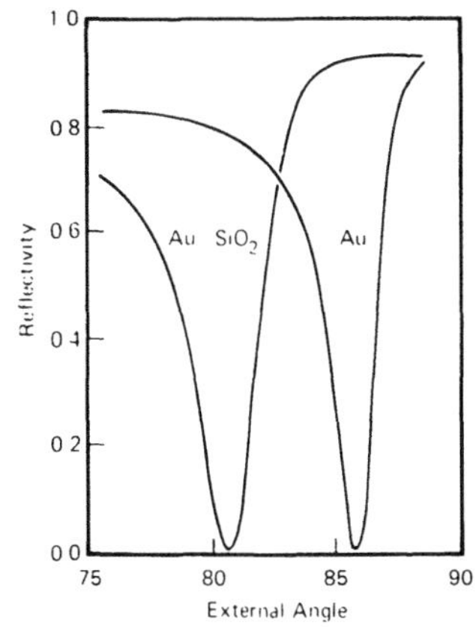


Fig. 4. The experimental ATR measurements for a gold film and for the same gold film overcoated with an 18.5 nm film of silicon oxide.
ε = -11.527 - 1.096i

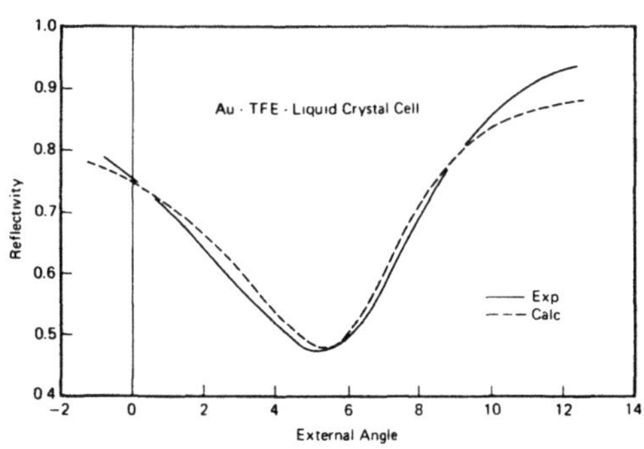


Fig. 5. The ATR reflectivity for the same film as shown in Fig. 3 but now with the liquid crystal cell assembled and filled with the liquid crystal 4-cyano 4'pentylbiphenyl (K15).

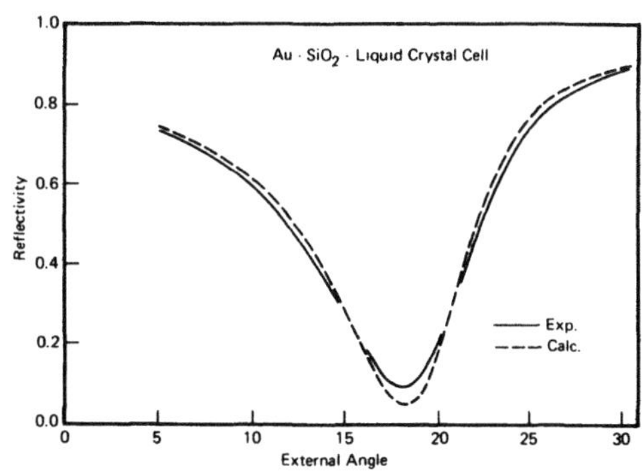


Fig. 6. The ATR reflectivity of the same film as shown in Fig. 4 but now with liquid crystal in the cell.

In the $SiO_2$ coated cell the molecules were oriented with their long axis along the surface of the coated electrodes in the plane of incidence. As shown in Fig. 6, the reflectivity minimum is now at 18°. Assuming the ordinary refractive index oriented perpendicularly to the surface, we calculated the extraordinary index to be 1.742, very close to TFE case, but now oriented 90° from it.

The close agreement for the two evaluations of the extraordinary refractive index indicates that the two orientation are orthogonal with no evidence of any tilt. Our value is somewhat larger than that reported (1.72 for our wavelength and temperature).[8,9] However, the reported values were obtained from the ordinary value and the isotropic value,[8] or from measurements on an Abbe refractometer where the surfaces of the prisms were treated by rubbing them with quarternary ammonium salts to align the molecules.[9] Although the latter value should be fairly accurate, the alignment is probably not as complete as we obtained with our surface treatments.

An aligning electric field should have no influence on the perpendicularly oriented cell because the molecules are already aligned in the field direction. In the parallel aligned cell, on the other hand, a reorientation of the molecules, from parallel at the surface to approaching a perpendicular orientation in the middle of the cell, should occur. When we applied electric fields to the $SiO_2$ overcoated cells, approximate shifts in the ATR minima of about 3° at 15V ac and about 7° at 24V dc took place (see Fig. 7). One can clearly see that the cell is being aligned perpendicularly as we approach the ATR minimum for the TFE cell which is completely aligned perpendicularly.

In order to analyze this shift, we modeled the system as a four layered system with thickness ratios of 1:4:8:∞ at director angles of 0, 30, 60 and 90°. Admittedly this layered model is crude but soluble, while the continuum model, to the best of our knowledge, has not been solved. A least squares fit was made to the ATR data with only one parameter - the thickness of the first layer. All other layers were scaled accordingly. Our results were that the layers were approximately 25 nm, 100 nm and 200 nm thick, respectively, at 24V and about twice as thick at 15V. The model for 24V is shown in Fig. 8. The dielectric functions were anisotropic in each layer, but the components were calculated from the limiting values and the angles.

$$\varepsilon = \varepsilon_1 \cos^2\theta + \varepsilon_2 \sin^2\theta \ ,$$

where $\theta$ is the director angle from the electrode surface. The parallel values are listed in Fig. 8 and the perpendicular values

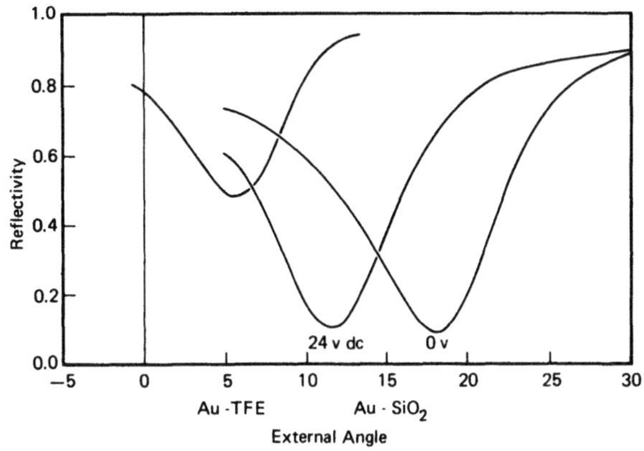


Fig. 7. The measured ATR for the two cells shown in Figs. 5 and 6 and the cell with an $SiO_2$ overcoat and a field of 24v.

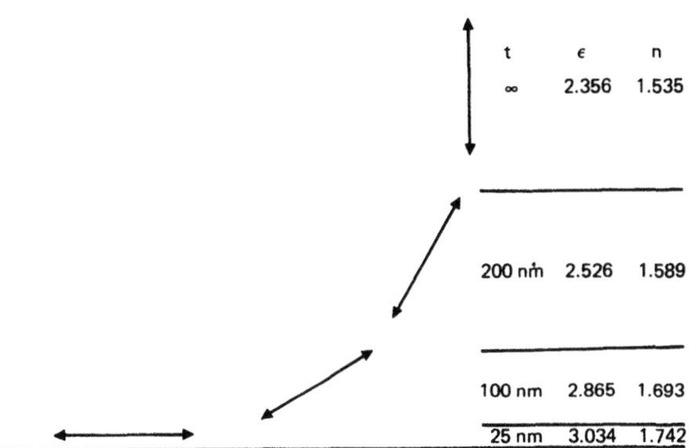


Fig. 8. Schematic showing the orientation of molecules from their uniaxial direction parallel to the surface to perpendicular within the cell. The two intermediate angles are 30 and 60 degrees, respectively. On the right side are shown the thicknesses of each layer, its dielectric function, and its refractive index. Only the parallel components is given for e and n. The other component is given at the layer with the complementary angle.

given at the complimentary orientation. Plotting 1/t vs. voltage gave a threshold of approximately 4V.

As a way of concluding, we believe the use of ATR techniques to excite a surface plasmon mode on an electrode surface is an accurate and convenient method to probe the surface region in the 1 to 30 nm range. These results presented here give us new information about the reorientation of liquid crystals away from an aligning surface. Why it exhibits so much sharper turning radii than calculated from the integral of the dielectric function and the Frank-Oseen equation is not clear at this time. Hopefully this work will stimulate new theoretical thoughts to give us a better understanding of liquid crystals.

## REFERENCES

1. G. J. Sprokel, Mol. Cryst. Liq. Cryst. 42:233 (1977).
2. P. G. de Gennes, "The Physics of Liquid Crystals," (Caredon Press, Oxford, 1974).
3. J. G. Gordon II and J. D. Swalen, Opt. Commun. 22:374 (1977).
4. I. Pockrand, J. D. Swalen, J. G. Gordon II and M. R. Philpott, Surface Sci. 74:237 (1978).
5. J. D. Swalen, J. Phys. Chem. 83:1438 (1979).
6. P. G. Cummins, D. A. Dunmur and D. A. Laidler, Mol. Cryst. Liq. Cryst. 30:109 (1975).
7. H. A. Tarry, Gallard-Schlesinger Chemical Mfg. Corp., 584 Mineola Ave., Carle Place L.I., N.Y. 11514 Technical Data Sheets on "The Refractive Indices of Cyanobiphenyl Liquid Crystals."
8. Handbook of Chemistry and Physics, R. C. Weast, editor, (CRC Press, Cleveland, 54th Edition 1973-1974) p. C-769.
9. Ibid., E-222.

# ANCHORING PROPERTIES OF LIQUID CRYSTALS AT THE SUBSTRATES: TEMPERATURE DEPENDENCE OF PRETILT

Sadashi Shimoda, Hitoshi Mada and Shunsuke Kobayashi

Department of Electronic Engineering
Faculty of Technology
Tokyo University of Agriculture and Technology
Koganei, Tokyo 184, Japan

## I. INTRODUCTION

The preparation of well-aligned nematic liquid crystal (LC) films is of considerable technical interest. For obtaining a uniform alignment, various techniques such as rubbing,[1,2] oblique evaporations,[3-6] and coating by an organic surfactant[7] were developed. The resulting initial molecular tilt at the surface, the pretilt angle, is the significant parameter exerting influence upon the optical and the physical performances.[8]

The origin of the pretilt angle has been discussed by several authors. Guyon et al.[9] and Berreman[10] have used an elastic model to connect the surface structure, and Goodman et al.[11] stated that columnar-growth structures arrayed in rowlike order are the origin of the pretilt. These statements, which seem like reasonable explanations for the origin of the pretilt angle, are considered to be insufficient to explain the temperature dependence of pretilt angles.

Several works on the temperature dependence of pretilt angles have been reported. For example, Bouchiat and Langevin-Cruchon[12] have measured pretilt angles at the free surface of MBBA as a function of temperature. The theoretical study of this dependence at the free surface has been given by Parsons.[13] The temperature dependence of the pretilt angle at the interface between glass plates and LC films was obtained first by Kahn[14] with twisted nematic samples and then by Toda et al.[15] with non-twisted ones. A theoretical explanation for the non-twisted nematic samples has recently been proposed by one of the present authors, H.M.[16]

This theory employed a purely macroscopic point of view and used a continuum theory which had been modified to include a surface order parameter $\bar{Q}_{ij}$ and an easy axis.

In this paper, we describe the measurement of the temperature dependence of the pretilts for twisted as well as non-twisted nematic samples, using some conventional nematic LCs. In order to have a unified understanding of the temperature dependence of the pretilts of the different samples, an adequate normalization of both the pretilts and the nematic temperature range has been made. Then the differences in the changes of the temperature-dependent pretilt angles between non-twisted and twisted samples were compared.

Furthermore, the theory proposed by Mada was applied for the twisted nematic samples, and a comparison was made between the results obtained using this theory and those from experiment.

## II. EXPERIMENTAL RESULTS

Our experiment consisted of two parts. The first was the experiment on the non-twisted nematic samples, in which MBBA, 7CB, E-8, E-18, ZLI-1132 and ZLI-1083 were used, as shown in Table 1. The other was that on the twisted nematic samples, in which 7CB was used. Pretilts below 10° have been realized by the double evaporation of $SiO_x$ having an appropriate incidence angle,[6] those with medium value, say, 20-40° by a single 85° oblique evaporation; and those with a high tilt, about 50°, by an application of the surfactant followed by a single 85° oblique evaporation of $SiO_x$. The thickness of all samples was about 8 μm.

Table 1. Used Liquid Crystals in this Experiment

| Liquid Crystal | Nematic Range (deg.) | Types |
|---|---|---|
| MBBA | 22.8~46.8 | Schiff |
| 7CB | 28.5~41.9 | Biphenyl |
| E-8 | -12.0~71.0 | Biphenyl |
| E-18 | -9.0~59.0 | Biphenyl |
| ZLI-1132 | -6.0~70.0 | PCH |
| ZLI-1083 | -3.0~51.0 | PCH |

The pretilt angle was measured by the magneto-capacitive null (MCN) method for the non-twisted samples and by the capacitance-voltage (C-V) method for the twisted ones. The MCN method was not applied to the twisted samples because the accuracy is still not satisfactory. Thus, in order to obtain pretilt angles, the theoretical C-V curves of the twisted cells, which were calculated with the assumption of the rigid boundary, were best-fitted to the observed curves in the region of low voltage by adjusting the pretilt angle. Since it may be doubted whether this assumption still holds good even in the region of high voltage, the fitting has been made only in the region of low voltage. The two methods of measuring the pretilt angle, i.e., the C-V method and the MCN method, gave results in good agreement with each other.

## 1. Non-Twisted Case

The experimental results of the temperature dependence of pretilt angles are shown in Figs. 1 and 2. All results given in Fig. 1 are for 7CB and MBBA, which were aligned on the $SiO_x$ layers deposited by single or double oblique evaporation. Data shown in Fig. 2 are for E-8, E-18, ZLI-1132 and ZLI-1083, which were aligned on the films of $SiO_x$ prepared by a single oblique evaporation and by the evaporation after coating with a surfactant. There is a wide nematic temperature range of E-8 and ZLI-1132, but we could not measure the data in the high temperature range because of the limitation of the apparatus. All the measured pretilt angles decrease with increasing the temperature. In order to know whether this decreasing behavior is dependent on surface treatments, all measured pretilt angles were normalized by those at 13°C below the clearing point $T_c$. The value 13°C below the clearing point has been taken as a common temperature for the nematic range of all LCs. This result is shown in Fig. 3. In spite of the differences in surface treatments, the temperature dependences of the normalized values could almost be fitted to a single curve for each LC. Similar results were obtained for E-8, ZLI-1132 and ZLI-1083. Furthermore, the nematic temperature range should also be normalized in order to determine its effect on LC materials. The lines shown in Fig. 4 were drawn in this way. These results show that while the magnitude of the pretilt angle at a low temperature was determined properly by the combination of the surface treatment and LC materials, the decreasing tendency of the pretilt angle with temperature has no dependence on the surface treatment of $SiO_x$, but is different for each nematic. Among the LC materials examined ZLI-1132 and E-8 have a weaker decreasing tendency of the pretilt angle.

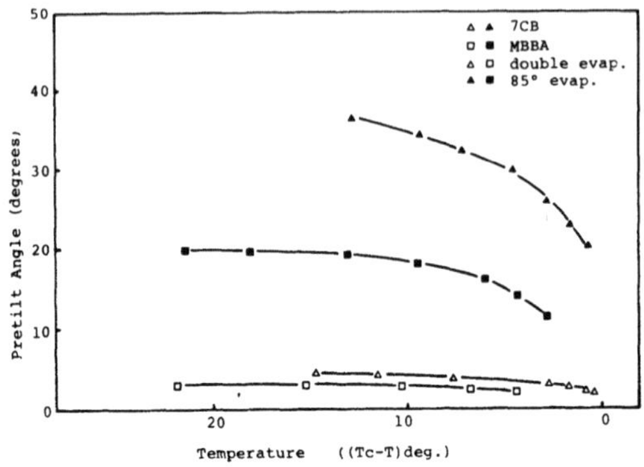


Fig. 1. The temperature dependence of pretilt angles for non-twisted samples using 7CB and MBBA. $T_c$ is the clearing point.

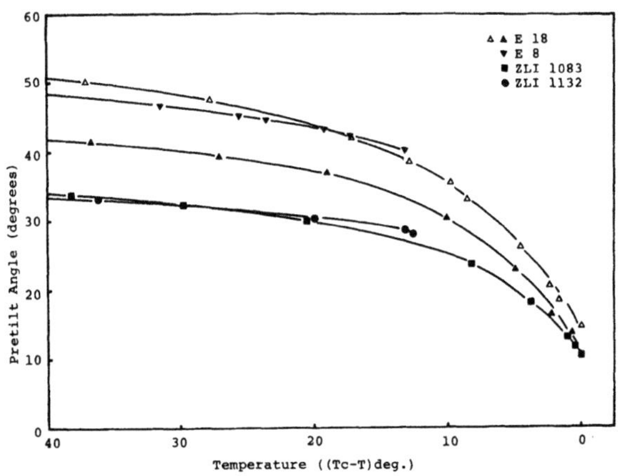


Fig. 2. The temperature dependence of pretilt angles for non-twisted samples using E-8, E-18, ZLI-1083 and ZLI-1132.

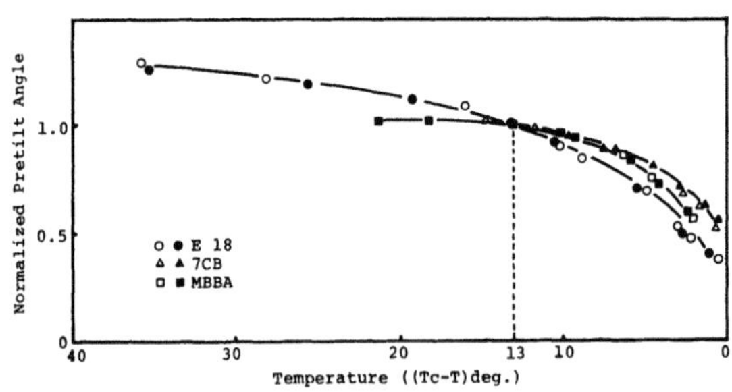


Fig. 3. The temperature dependence of pretilt angles normalized at the pretilt angle of each cell at 13°C below the clearing point. Solid and open points show the preparation of 85° single evaporation and double evaporation or the coating surfactant, respectively.

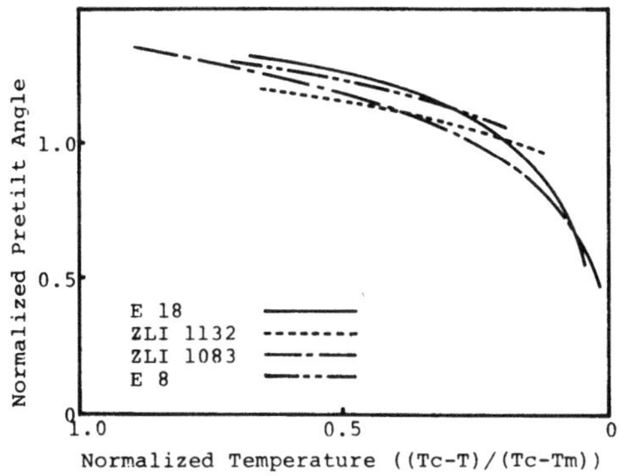


Fig. 4. The temperature dependence of pretilt angles normalized not only pretilt but also temperature. $T_m$ shows the melting point. Normalized temperature 0 and 1 show the clearing and melting point, respectively.

## 2. Twisted Case

In order to compare the temperature dependence of the pretilt angle for the twisted and the non-twisted samples, twisted cells were made using the same treated glass plates which had been used for the measurements by the MCN method. Thus, as soon as the temperature dependence of the pretilt angle for the non-twisted cell was measured, the cell was taken apart, rinsed with an organic solvent to remove LC, and dried, then the twisted cell was reassembled.

The durability of the evaporated substrates was proven using a non-twisted cell, as shown in Fig. 5 where it is shown the reproducibility of the pretilt angles and their temperature dependence of a non-twisted cell assembled after disassembling the cell subjected to the pretilt measurement: this cycle was repeated three times. We prepared four samples for measurement using the single 85° evaporation; their values of pretilt angles at 30°C were between 36° and 41°. As the tendency of the temperature dependence of the pretilt angles is the same in all samples, only one representative of these results is shown in Fig. 6. The pretilt angle decreases with rising temperature for both non-twisted and twisted samples. Furthermore, by twisting by 90°, a slight increase of the pretilt angle was observed in all temperature ranges.

In order to compare the non-twisted and twisted sample, pretilt angles in all samples are normalized by the angle at the temperature of 30°C as shown in Fig. 7. The temperature dependence of all measured values could almost be fit to a single curve. It is clear that the twisted cell has a weak decreasing tendency

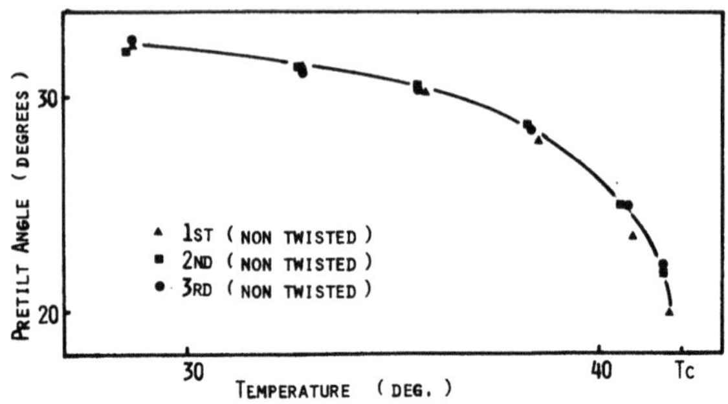


Fig. 5. The temperature dependence of pretilt angles, showing the lack of influence of the organic solvent. This measurement was performed three times for the non-twisted samples.

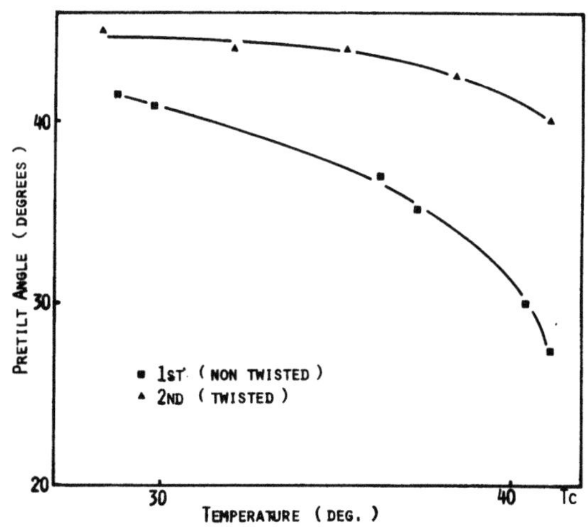


Fig. 6. The effect of the 90° twist on the temperature dependence of pretilt. Solid squares are data of the first measurement for non-twisted cell and triangles for twisted cell using same glass substrates.

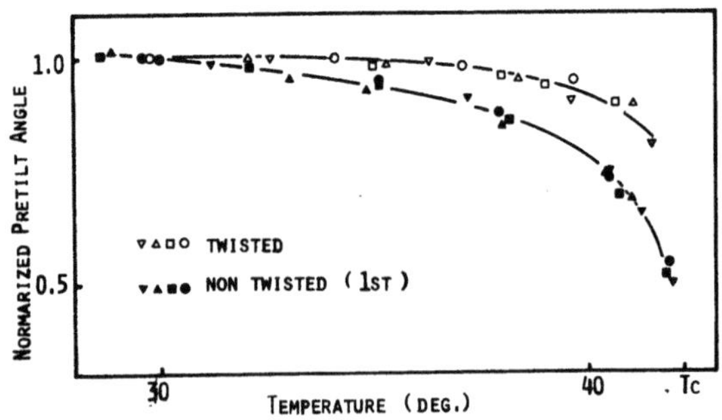


Fig. 7. The temperature dependence of pretilt angles for non-twisted and twisted samples. These are normalized at the pretilt angle of each cell at 30°C.

of the pretilt angle compared to that of the non-twisted cell. This behavior will be discussed theoretically in §III and IV.

## III. ANALYTICAL CONSIDERATION

In order to express the surface energy, it is assumed that there is an easy axis on the surface or interface. The existence of the easy axis is the only respect in which the surface differs from the bulk. The interaction between the LC and the interface in contact with it may be mainly due to the van der Waals-Lifshitz forces,[16,17] which are related to the imaginary part of the dielectric permittivity at frequencies above the infrared, i.e., light absorption. In this paper, the easy axis is defined by a unit vector whose direction coincides with the optical axis of the anisotropic surface. For a single oblique evaporation of 85°, it was found that the optical axis tilted from the surface by about 5°.[18]

The bulk and the surface energy can be expressed by the power series of an order parameter in this analysis. Therefore the surface tensor order parameter $\bar{Q}_{ij}$ was introduced to obtain the surface energy; $\bar{Q}_{ij}$ is the tensor order parameter with the easy axis as its principal axis at the surface.

As well known, the tensor order parameter in the bulk introduced by de Gennes[19] is

$$Q_{ij} = S\left(n_i n_j - \frac{1}{3}\,\delta_{ij}\right), \tag{1}$$

where S is the scalar order parameter which is a function only of the temperature, and $n_i$ and $\delta_{ij}$ are the director and Kronecker's $\delta$, respectively. Similarly, $\bar{Q}_{ij}$ is defined as follows:

$$\bar{Q}_{ij} = S^d\left(d_i d_j - \frac{1}{3}\,\delta_{ij}\right), \tag{2}$$

where $d_i$ is the easy axis and $S^d$ is the order parameter directed along the easy axis at the surface. Using the boundary condition, i.e., $d_iQ_{ij}d_j - d_i\bar{Q}_{ij}d_j$, $S^d$ can be written as follows:

$$S^d = \frac{3}{2}\,S^s\left\{(\underset{\sim}{n}\cdot\underset{\sim}{d})^2 - \frac{1}{3}\right\}, \tag{3}$$

where $S^s$ is the scalar order parameter at the surface. It should be noted that $S^d$ is not only a function of the temperature but also of the director and of the easy axis. Here, for simplicity, we assume that the scalar order parameter is constant throughout the bulk, and therefore $S^s$ can be replaced by S in Eq. (3).

The surface deformation energy can be expanded in powers of the gradient of surface tensor order parameter $\bar{Q}_{ij,k}$. Finally, the surface deformation energy per unit area can be given by

$$\bar{\Lambda}_d = \frac{1}{2}(\underset{\sim}{n}\cdot\underset{\sim}{n})^2\left\{\bar{k}_1\left(d_i d_j n_{j,}\right)^2 + \bar{k}_2\, d_j n_{j,}\right\}^2 , \tag{4}$$

where $k_i$s are surface elastic constants and vary as $S^2$ with temperature. It should be noted that the deformation in the bulk $n_{i,j}$ exerts an influence on the surface deformation energy $\bar{\Lambda}_d$.

On the other hand, at the surface, the director tends to orient its long axis along the easy axis. The energy of orientation, which we call here the easy axis field energy, can be expressed as follows, by analogy with an electric or magnetic field in the bulk:

$$\bar{\Lambda}_e = -\frac{1}{2}\pi_{ij}d_i d_j = -\frac{1}{2}\Delta\beta\bar{Q}_{ij}d_i d_j , \tag{5}$$

where $\Delta\beta$ is the anisotropy of the easy axis field energy per molecule. Using Eqs. (2) and (3), Eq. (5) becomes

$$\bar{\Lambda}_e = -\frac{1}{2}\Delta\pi\left\{(\underset{\sim}{n}\cdot\underset{\sim}{d})^2 - \frac{1}{3}\right\}, \tag{6}$$

with $\Delta\pi = S\Delta\beta$.

It is noted that $\bar{\Lambda}_d$ becomes minimum when the director is perpendicular to the easy axis; on the other hand, $\bar{\Lambda}_e$ becomes maximum when the two are parallel. Thus the pretilt is considered to originate from the competition and balance between $\bar{\Lambda}_d$ and $\bar{\Lambda}_e$.

The total free energy per unit area at the surface $\bar{\Lambda}$ is

$$\bar{\Lambda} = \bar{\Lambda}_d + \bar{\Lambda}_e . \tag{7}$$

Furthermore, the total energy F throughout the system can be given by

$$F = \int_V \Lambda\left(n_i, n_{i,j}\right)dv + \int_S \bar{\Lambda}\left(n_i, n_{i,j}\right)ds , \tag{8}$$

where $\Lambda$ is free energy per unit volume in the bulk. Minimizing F with respect to variations of $n_i$, the equilibrium equations are given by

$$h_i = \left(\frac{\partial\Lambda}{\partial n_{i,j}}\right)_{,j} - \frac{\partial\Lambda}{\partial n_i} , \tag{9}$$

$$\bar{h}_i = \left(\frac{\partial\bar{\Lambda}}{\partial n_{i,j}}\right)_{,j} - \frac{\partial\bar{\Lambda}}{\partial n_i} - \frac{\partial\Lambda}{\partial n_{i,j}}\nu_j , \tag{10}$$

where $h_i$ is the molecular field in the bulk, $\bar{h}_i$ is the molecular field at the surface, and $\nu_i$ is a unit vector normal to the surface. It should be noted that $\bar{h}_i$ is influenced by the free energy in the bulk $\Lambda$. The torque exerted by the molecular field without taking into account the backflow is written as follows:

$$\underset{\sim}{\Gamma} = \underset{\sim}{n} \times \underset{\sim}{h} = 0 , \tag{11}$$

$$\underset{\sim}{\bar{\Gamma}} = \underset{\sim}{n} \times \underset{\sim}{\bar{h}} = 0 . \tag{12}$$

Consequently the stable conformation is obtained by solving Eqs. (11) and (12) with arbitrary boundaries.

In the non-twisted case, the solution for the stable conformation has been given already. When the easy axis is perpendicular or parallel to the surface, the temperature dependences of the pretilt angle $\theta_0$ are as follows:

$$\frac{\alpha}{S} = - \frac{Kx^2-2x+1}{(1-Kx)^2} \text{ , perpendicular} \tag{13}$$

$$\frac{\alpha}{S} = \frac{Kx^2-2x+1}{(1-Kx)^2} \text{ , parallel} \tag{14}$$

where $K=(k_{11}-k_{33})/k_{11}$, $x=\sin^2\theta_0$, and $\alpha=S\Delta\pi/\bar{k}_2C^2$ is constant with temperature and is determined by the combination of the nature of the surface and that of the LC materials.

Now, we elaborate on the solution for the twisted case; the temperature dependence of the pretilt angle can also be obtained by solving Eqs. (11) and (12). Here, for simplicity, the easy axis $\underset{\sim}{d}$ was selected to be parallel to the local surface. Therefore, the director $\underset{\sim}{n}$ and the easy axis $\underset{\sim}{d}$ can be written as follows, with the coordinate system shown in Fig. 8:

$$\underset{\sim}{n} = \begin{pmatrix} \cos\theta\cos(45°-\phi) \\ \cos\theta\sin(45°-\phi) \\ \sin\theta \end{pmatrix} \qquad \underset{\sim}{d} = \begin{pmatrix} \cos(45°-\phi_1) \\ \sin(45°-\phi_1) \\ 0 \end{pmatrix} \tag{15}$$

Furthermore, since the twisted angle was 90°, the azimuth angle at the boundary $\phi_1=45°$. Consequently, the easy axis $\underset{\sim}{d}$ is as follows:

$$d = \begin{pmatrix} 1 \\ 0 \\ 0 \end{pmatrix} \tag{16}$$

From Eqs. (4), (5), (10), (12), (15) and (16), the surface torque equations are then given by

$$\left\{k_{11}+k_{33}\left(1+\sin^2\theta_0\right)\right\}\cos^2\theta_0 \left.\frac{d\theta}{dz}\right|_{z=0} + \bar{k}_2\sin\theta_0\cos\theta_0$$

$$\left\{\left(\sin^2\theta_0-\cos^2\theta_0\right)\left(\frac{d\theta}{dz}\right)^2_{z=0} - \sin\theta_0\cos\theta_0\left(\frac{d^2\theta}{dz^2}\right)_{z=0}\right.$$

$$\left. = \cos^2\theta_0\left(\frac{d\phi}{dz}\right)_{z=0}\right\} + \Delta\pi\sin\theta_0\cos\theta_0 = 0 \ . \tag{17}$$

The torque equations in the bulk are given from Eqs. (9), (11) and (15) as

$$g(\theta)\,\frac{d^2\phi}{dz^2} + \frac{d}{d\theta}\,g(\theta)\,\frac{d\theta}{dz}\,\frac{d\phi}{dz} = 0 \ , \tag{18}$$

$$g(\theta)\,\frac{d\phi}{dz} = A_1 \ , \tag{19}$$

$$f(\theta)\,\frac{d^2\theta}{dz^2} + \frac{1d}{2d\theta}\,f(\theta)\,\left(\frac{d\theta}{dz}\right)^2 - \frac{1d}{2d\theta}\,g(\theta)\,\left(\frac{d\phi}{dz}\right)^2$$

$$+ \varepsilon_0\Delta\varepsilon E_z^2\sin\theta\cos\theta = 0 \ , \tag{20}$$

$$f(\theta)\,\left(\frac{d\theta}{dz}\right)^2 + g(\theta)\,\left(\frac{d\phi}{dz}\right)^2 + \varepsilon_0\left(\varepsilon_\perp\cos^2\theta+\varepsilon\,\sin^2\theta\right)E_z^2 = A_2^2 \tag{21}$$

in which

$$f(\theta) = k_{11}\cos^2\theta + k_{33}\sin^2\theta \ ,$$

$$g(\theta) = \cos^2\theta\left(k_{22}\cos^2\theta+k_{33}\sin^2\theta\right) \ ,$$

and $A_1$ and $A_2^2$ are the constants of integration.

Here, for simplicity, the following assumptions are made: 1) the surface energy is larger than the bulk energy, so the first and second terms in Eq. (17) may be neglected, 2) the best fitting of the C-V method was performed in the low voltage region; therefore, in Eqs. (20) and (21), the term involving $E_z$ can be neglected, 3) $\phi_1$ is independent of the temperature. Under these assumptions, Eqs. (17), (20) and (21) become as follows:

$$\bar{k}_2\sin\theta_0\cos\theta_0\left\{\left(\sin^2\theta_0-\cos^2\theta_0\right)\left(\frac{d\theta}{dz}\right)^2_{z=0} - \sin\theta_0\cos\theta_0\right.$$

$$\left.\left(\frac{d^2\theta}{dz^2}\right)_{z=0} - \cos^2\theta_0\left(\frac{d\phi}{dz}\right)^2_{z=0}\right\} + \Delta\pi\sin\theta_0\cos\theta_0 = 0 \qquad (22)$$

$$f(\theta)\frac{d^2\theta}{dz^2} + \frac{1d}{2d\theta}f(\theta)\left(\frac{d\theta}{dz}\right)^2 - \frac{1d}{2d\theta}g(\theta)\left(\frac{d\phi}{dz}\right)^2 = 0 \qquad (23)$$

$$f(\theta)\left(\frac{d\theta}{dz}\right)^2 + g(\theta)\left(\frac{d\theta}{dz}\right)^2 = A_2^2 \qquad (24)$$

Finally, the stable condition can be obtained by solving simultaneous equations, i.e., Eqs. (18), (19), (22) and (24).

The particular solutions are obtained from

$$\sin\theta_0\cos\theta_0 = 0\ , \qquad (25)$$

Giving $\theta_0=0°$. In these cases, we have no temperature dependence under any conditions.

Putting $\cos^2\theta_0=y$, the general solution of the simultaneous Eqs. (18), (19), (22), (23) and (24) is

$$\frac{\Delta\pi}{k_2} = A_1^2\left\{\frac{K'K''x^3+(K'^2-2K'K'')x^2+(K'-K''-1)y+1}{y(1+K'y)^2(1+K''y)^2}\right\}$$

$$+ A_2^2\left\{\frac{K'y^2+2y-1}{(1+K'y)^2}\right\}, \qquad (26)$$

where K' and K" are $(k_{11}-k_{33})/k_{33}$ and $(k_{22}-k_{33})/k_{33}$, respectively, and are independent of temperature. The left-hand side of Eq. (26) varies as 1/S with temperature. Therefore Eq. (26) can be written as follows:

$$\frac{\alpha}{S} = \frac{K'K''y^3+(K'^2-2K'K'')y^2+(K'-K''-1)y+1+Ay(K'y^2+2y-1)(1+K''y)^2}{y(1+K'y)^2(1+K''y)^2}\ , (27)$$

Where $A=A_2^2/A_1^2$ and $\alpha=S\Delta\pi/A_1^2\bar{k}_2$. The final equation will give the temperature dependence of the pretilt angle for twisted cases.

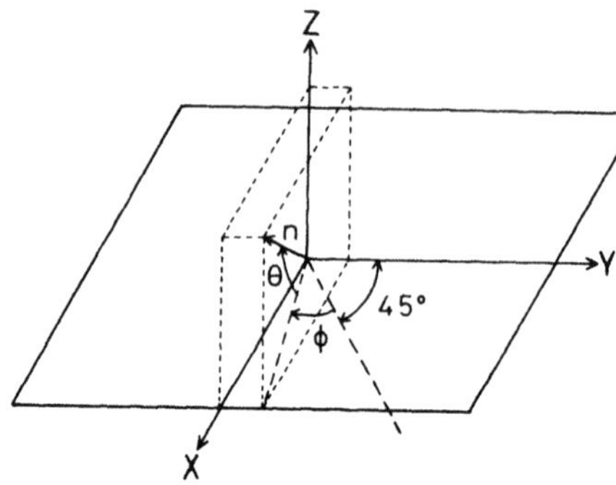


Fig. 8. Coordinate system.

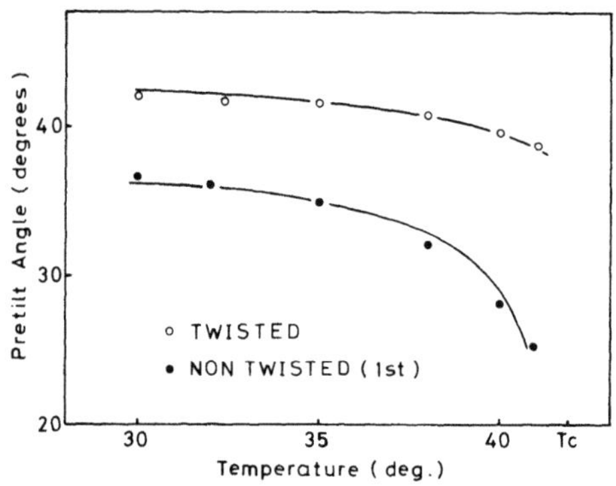


Fig. 9. The comparison of the temperature dependence of pretilt angles. Plotted points are experimental results and solid lines are theoretical ones.

## IV. DISCUSSION

The theoretical and experimental results obtained above will be compared and discussed in this section. First of all, using Eq. (27), the theoretical curves were best fitted to the observed curves by adjusting A and $\alpha$. The order parameters and elastic constants $k_{ii}$ used in this computation were measured by optical and capacitive methods, respectively. Parameters A=10 and $\alpha$=1.50 were selected by a least squares method. Here, the quantity A means the ratio of $(d\theta/dz)^2$ and $(d\phi/dz)^2$ at the surface. Thus it shows the magnitude of the deformation near the surface. Since A=10, it is considered that a large deformation away from the direction of the z-axis will occur near the surface. For the non-twisted case, using Eq. (14), the theoretical curves were also best fitted to the observed curves to determine the value of $\alpha$ by a least squares method. In this way, $\alpha$=0.07 was obtained.

These results of fitting are shown in Fig. 9. As shown in this figure, a better agreement was obtained for the twisted case than for the non-twisted one, because of the availability of two parameters, A and $\alpha$.

In making the comparison between Eqs. (14) and (27), the second term of the right-hand side in Eq. (26) corresponds to the term of the right-hand side in Eq. (14). By twisting the samples, the first term in Eq. (26) appears in the equation for the temperature dependence of pretilt angles. This is again understood more easily from the fact that the first term involves $k_{22}$ and second does not. Here, a more interesting problem appears in two terms: i.e., the first term has the characteristic of increasing the pretilt angle with temperature, while the second term decreases with temperature. From these facts, it is obvious that the twisting makes the pretilt angle increase with temperature. So the decreasing tendency for twisted samples is weaker than that for non-twisted one.

As shown in §III, in our analysis, the case of $\theta_0=0°$, does not have the temperature dependence of pretilt angles for twisted samples. However, an experiment which would prove this was impossible to perform because of the inaccuracy of the C-V method for cells having $\theta_0 \simeq 0°$.

## V. CONCLUSION

In this work, it is shown that the temperature dependence of the pretilt angle for twisted samples could be explained by a theory formulated by introducing the concept of the easy axis field and by considering the surface energy. Good agreement between experimental and theoretical results was obtained, so that the theory seems effective for phenomenological analysis of surface properties.

Using this theory, the following fact was explained qualitatively: the decreasing tendency of the pretilt angle for the twisted case is different from that for the non-twisted case. In the case of 7CB the former has a weaker decreasing tendency than the latter. Furthermore, it is shown experimentally that data for the same material fall on a single line, even though there were differences in the pretilt at low temperature.

In order to make a more qualitative discussion, it will be necessary to find a method to measure the coefficients $\bar{k}_1$ and $\Delta\pi$, which determine the value of $\alpha$, and to measure the surface order parameter $S^s$. Furthermore, the direction of the easy axis must be determined for the oblique evaporated surface exactly and this result must be introduced in the analytical procedure. Moreover, it has been assumed that the twisted angle at the surface $\phi_1$ is a constant with temperature; it should be examined whether this is correct or not.

## REFERENCES

1. P. Chatelain, Bull. Soc. Fr. Miner, 66:105 (1943).
2. D. W. Berreman, Phys. Rev. Lett. 28:1683 (1972).
3. J. L. Janning, Appl. Phys. Lett. 21:173 (1972).
4. L. A. Goodman, D. Meyerhofer and S. Digiovanni, IEEE Trans. Elect. Dev. ED-23:1176 (1976).
5. K. Fahrenschon and M. F. Schiekel, J. Electro. Soc. 124:953 (1977).
6. D. Meyerhofer, J. Appl. Phys. 48:1179 (1977).
7. F. J. Kahn, Appl. Phys. Lett. 22:386 (1973).
8. A. Miyaji, M. Yamaguchi, A. Toda, H. Mada and S. Kobayashi, IEEE Trans. Elect. Dev. ED-24:811 (1977).
9. E. Guyon, P. Pieranski and M. Boix, Appl. Eng. Sci. 1:1 (1973).
10. D. W. Berreman, Mol. Cryst. Liq. Cryst. 23:215 (1973).
11. L. A. Goodman, J. T. McGinn, C. H. Anderson and F. Degeronimo, IEEE Trans. Elect. Dev. ED-24:795 (1977).
12. M. A. Bouchiat and D. Langevin-Cruchon, Phys. Lett. 34A:331 (1971).
13. J. D. Parsons, Mol. Cryst. Liq. Cryst. 31:79 (1975).
14. F. J. Kahn, Mol. Cryst. Liq. Cryst. 38:109 (1977).
15. A. Toda, H. Mada and S. Kobayashi, Japan. J. Appl. Phys. 17:261 (1978).
16. H. Mada, Mol. Cryst. Liq. Cryst. 51:43 (1979).
17. K. Okano and J. Murakami, Proc. 7th Intern. Liquid Crystal Conf. (to be published).
18. W. A. Crossland, Appl. Phys. Lett. 26:598 (1975).
19. P. G. deGennes, "The Physics of Liquid Crystals," (Claredon, Oxford 1974), p. 59.

# OPTICAL CHARACTERISTICS OF LIQUID CRYSTAL DISPLAYS

G. Baur

Fraunhofer-Institut für
Angewandte Festkörperphysik (IAF)
D-7800 Freiburg, W. Germany

## INTRODUCTION

The information rate to be displayed by liquid crystals is limited by the multiplexing capability of the liquid crystal display. Therefore electrooptic effects with a strong nonlinearity of the electrooptic characteristics and short duty cycle response for a wide range of viewing angles are required.

Most of the electrooptic effects are based on light scattering or on light absorption by polarizers or by dissolved dyes. Electrooptic effects which belong to the second group are tunable birefringence, the twisted nematic effect and the guest-host effect.

At present the most useful candidate for applications in wristwatch-, calculator- and clock-displays is the Twisted Nematic (TN) effect. Displays based on this effect are produced in quantities of many millions per month as reflective, transflective or transmissive displays. They are achieving great popularity as low voltage, low power devices, which are visible over a great range of ambient light levels. Perhaps the most serious disadvantage of these displays is the difficulty of achieving satisfactory multiplex addressing over a reasonable temperature range. A considerable improvement should be attainable if material and cell parameters are optimized.

THEORY

If an electric field is applied to a TN cell with a twist of 90° and pretilt angle of 0° the deformation starts at the well-known threshold voltage[1]

$$V_o = \pi \left( \frac{K_{11} + (K_{33} - 2K_{22})/4}{\varepsilon_o \Delta\varepsilon} \right)^{1/2} .$$

Somewhat above this voltage the cell becomes transparent in the case of parallel polarizers and nontransparent in the case of crossed polarizers. The optical response is a complicated function of material and cell parameters.

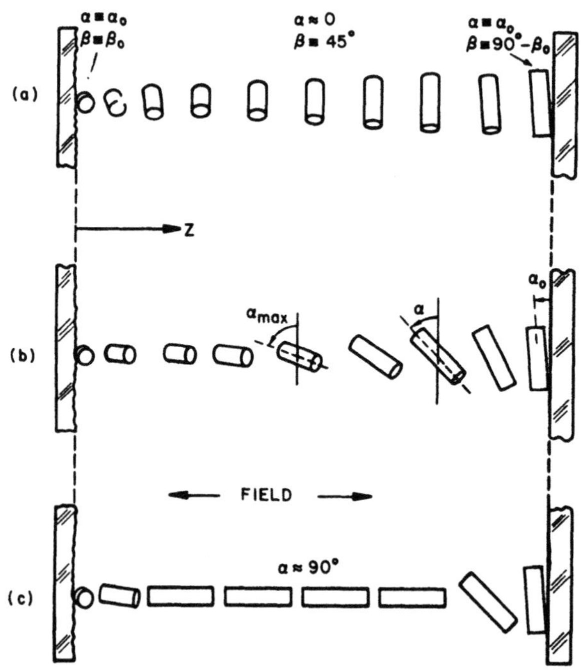


Fig. 1. Diagram of the orientation of the liquid crystal axis in a cell (a) with no applied field, (b) with about twice the critical field, and (c) with several times the critical field. Note slight permanent tilt ($\alpha_o$) and turn ($\beta_o$) at the surfaces.

From a theoretical point of view it seems reasonable first to calculate tilt ($\alpha$) and twist ($\beta$) angles in a sequence of sublayers as a function of the applied voltage and second the optical transmission for different angles of observation.[2,3,4] A definition of these angles is given in Fig. 1. The deformation profile is dependent on the dielectric constants $\varepsilon_{\parallel}$ and $\varepsilon_{\perp}$, the elastic constants for splay, twist and bend $K_{11}$, $K_{22}$, $K_{33}$, the total twist $(90-2\beta_o)$, the tilt angle at the surface of the substrates $\alpha_o$, and the applied voltage. The optical response depends in addition on the refractive indices $n_e$ and $n_o$ and the ratio of wavelength to cell thickness. For display applications a finite tilt at the surfaces is required to avoid areas of opposite tilt. Therefore the deformation profiles are calculated for various combinations of $K_{33}/K_{11}$, $\Delta\varepsilon/\varepsilon_{\perp}$ and $\alpha_o$, using Berreman's program. All calculations are performed for 10 μm cells and a pretilt $\alpha_o=1°$.

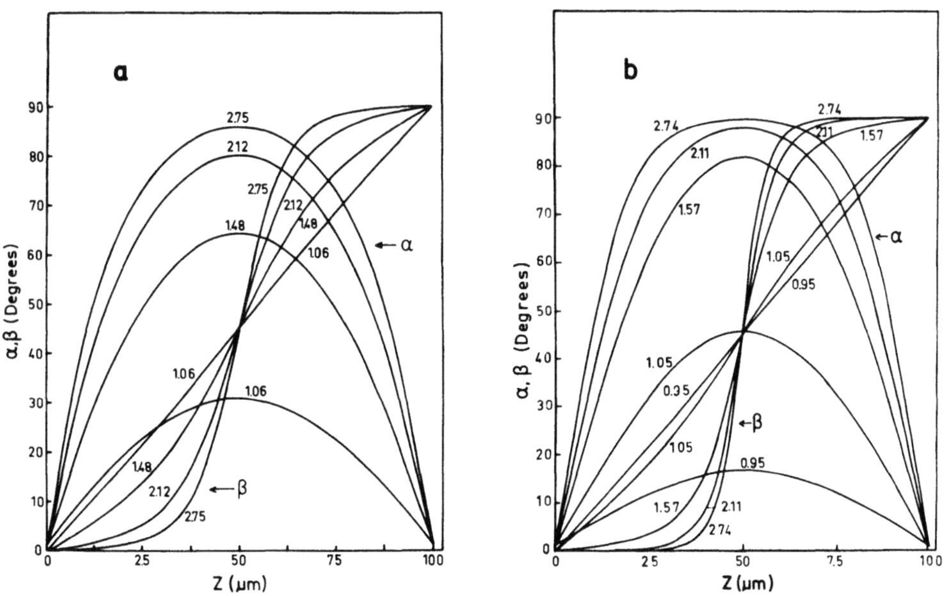


Fig. 2. Tilt ($\alpha$) and twist ($\beta$) angle as function of the position z within the layer for a 10 μm cell. The curves are calculated with $\varepsilon_{\parallel}=10.5$, $\varepsilon_{\perp}=10.0$, $K_{11}=10.87\cdot10^{-7}$ and with $K_{22}=9.5\cdot10^{-7}$, $K_{33}=15.37\cdot10^{-7}$ in Fig. 2a resp. $K_{22}=4.0\cdot10^{-7}$, $K_{33}=5.0\cdot10^{-7}$ in Fig. 2b (cgs units). Parameter is the the reduced voltage $V/V_o$.

The results are shown in Figs. 2 and 3. The tilt and twist angles α and β are plotted as functions of position within the cell thickness. In Fig. 2, the curve parameter is the reduced voltage while $\Delta\varepsilon/\varepsilon_\perp$ is kept constant. Note that for $V/V_0$ near and greater than unity the tilt is increasing with decreasing $K_{33}/K_{11}$. A similar plot for a fixed $K_{33}/K_{11}$ is presented in Fig. 3. The curve parameter is $\Delta\varepsilon/\varepsilon_\perp$. In the case of larger $\Delta\varepsilon/\varepsilon_\perp$, a larger $V/V_0$ is necessary in order to achieve a comparable tilt α in the center of the cell. Note, that the half width of the α-curves is larger for large $\Delta\varepsilon/\varepsilon_\perp$, while the shape of the β-curves is not too much affected.

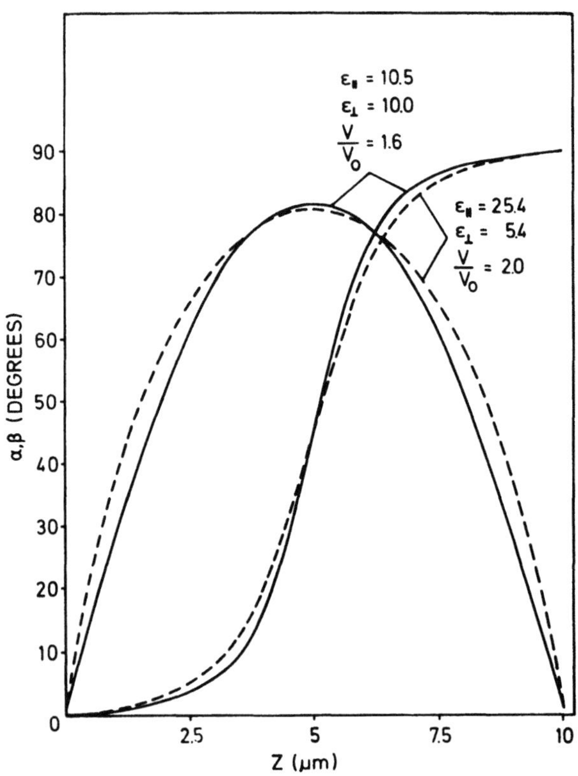


Fig. 3. Tilt (α) and twist (β) angle as function of the position z within the layer for a 10 μm cell. Solid line: $\Delta\varepsilon/\varepsilon_\perp$=0.05; $V/V_0$=1.6; dashed line $\Delta\varepsilon/\varepsilon_\perp$=3.7; $V/V_0$=2.0. The ratio of $K_{33}/K_{11}$ is 0.5.

In Figs. 4 and 5 an angle in the center of the cell and another one near the surface, are plotted as a function of the applied voltage. In Fig. 4, the parameter is $K_{33}/K_{11}$. It is obvious that the deformation is larger for smaller $K_{33}/K_{11}$ at the same reduced voltage.

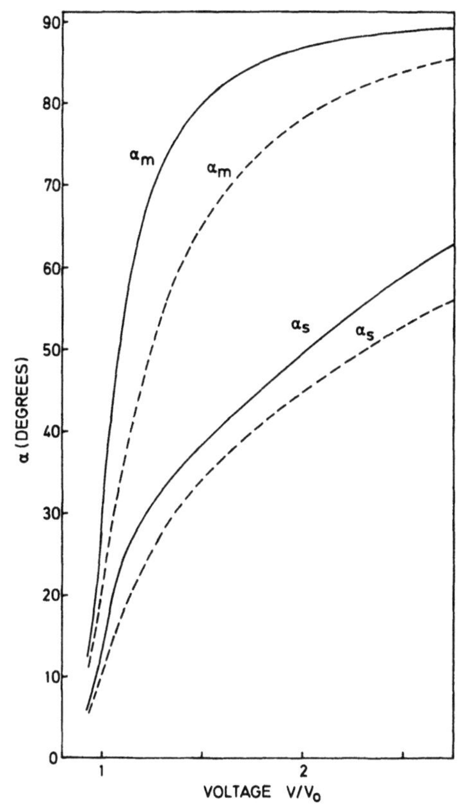


Fig. 4. Tilt angle $\alpha_m$ in the center and $\alpha_s$ about a quarter of the way from the surface as function of the reduced voltage $V/V_o$. The curves are calculated with $\varepsilon_{\parallel}$=10.5, $\varepsilon_{\perp}$=10.0. The parameter $K_{33}/K_{11}$ is 0.5 for the solid line and 1.5 for the dashed line.

If $\Delta\varepsilon/\varepsilon_{\perp}$ is the curve parameter and $K_{33}/K_{11}$ is kept constant, the deformation in the center of the cell is larger for smaller $\Delta\varepsilon/\varepsilon_{\perp}$, at the same $V/V_o$. This is shown in Fig. 5.

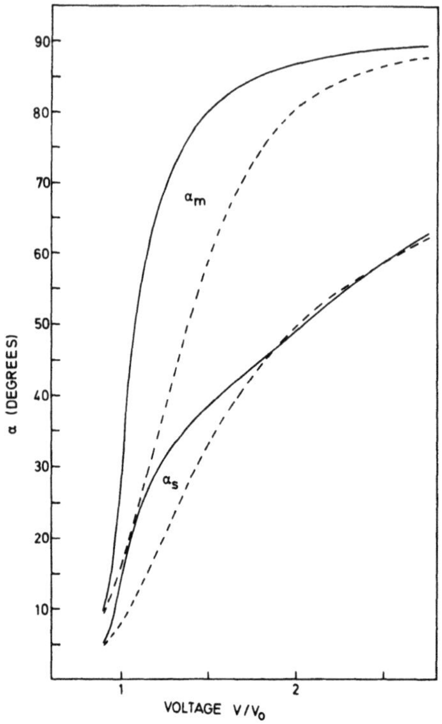


Fig. 5. Tilt angle $\alpha_m$ in the center and $\alpha_s$ about a quarter of the way from the surface as a function of the reduced voltage $V/V_o$. The curves are calculated with $K_{33}/K_{11}=0.5$. The parameters for the solid lines are $\varepsilon_{\parallel}=10.5$, $\varepsilon_{\perp}=10.0$, and $\varepsilon_{\parallel}=25.4$, $\varepsilon_{\perp}=5.4$ for the dashed lines.

With these sets of angles and with fixed $n_o$ and $\lambda$ the transmission curves for two combinations of $\Delta n$ and d are calculated. The first combination is d=10 µm and $\Delta n=0.2$. The transmission curves for normal incidence are presented in Fig. 6. The curve parameter is $K_{33}/K_{11}$. The transmission characteristics is steepest for smallest values of $K_{33}/K_{11}$.

To characterize the performance of a display it is necessary to know the transmission for a large range of viewing angles.[5,6] Therefore the transmission curves for 6 different viewing angles in 4 different planes of observation are evaluated. The viewing angles are characterized by the off-axis angle $\theta$ and an angle defined by the plane of polarization of the incident light and the plane of observation, as shown in Fig. 7.

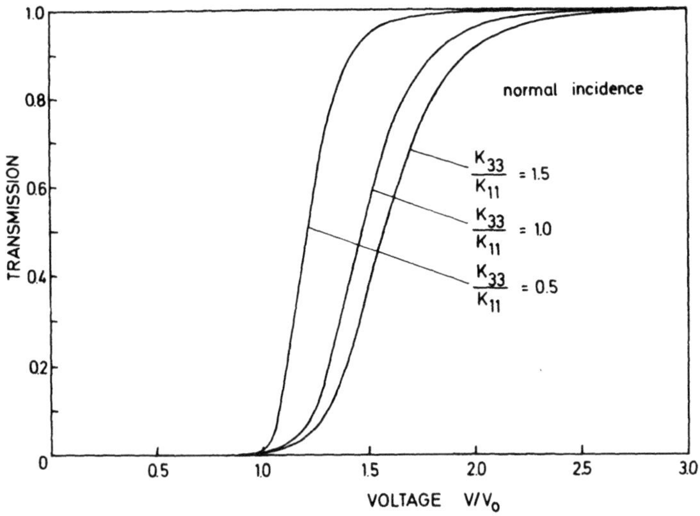


Fig. 6. Calculated optical transmission of a TN-cell between parallel polarizers for normal incidence. Parameter is $K_{33}/K_{11}$. Curves are calculated with $\varepsilon_{\parallel}=10.5$, $\varepsilon_{\perp}=10.0$, $n_e=1.728$, $n_o=1.523$.

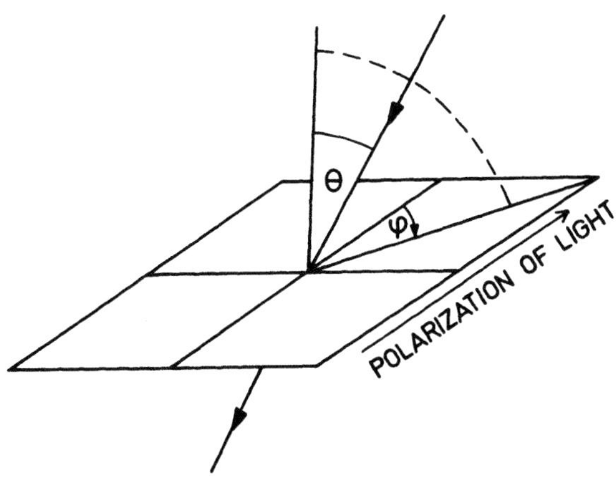


Fig. 7. Schematic diagram to explain the viewing angles, chosen for the calculation of the transmission characteristics. $\Theta$ is the off-axis angle, $\varphi$ the angle between the plane of polarization of incident light and the plane of observation.

Transmission curves for $\theta=\pm45°$ combined with $\varphi=0°$, $\pm45°$ and $90°$ are given in Fig. 8. $K_{33}/K_{11}$ is 1.5 in the upper set of graphs and 0.5 in the lower set. It is obvious that the slopes become steeper for smaller values of $K_{33}/K_{11}$. In Fig. 9, results are plotted for $K_{33}/K_{11}=0.5$ while $\Delta\varepsilon=0.5$ in the upper part and $\Delta\varepsilon=20$ in the lower part. For normal incidence there is no significant change in the steepness of the transmission curves due to a variation of $\Delta\varepsilon$ ranging from 0.5 to 20. Nevertheless, for large $\Delta\varepsilon/\varepsilon_\perp$, these curves are shifted towards higher values of $V/V_o$ and the distance between curves for normal incidence ($\theta=0$; $\varphi=0$) and off-axis incidence ($\theta=-45°$; $\varphi=-45°$) becomes larger.

Results for a variation of $\Delta n$ from 0.2-0.1 are given in Fig. 10. In this case the characteristic curves are steeper for smaller $\Delta n$ but there is a residual transparency of about 10% in the off state. The curves are calculated for 633 nm laser-light. The optimized refractive index anisotropy for a 10 μm TN cell should therefore have a value between $\Delta n=0.2$ and $\Delta n=0.1$.

Our calculations are summarized in Table 1. From a theoretical point of view the optical characteristics of a TN display can be optimized by using a material with a small $K_{33}/K_{11}$ in a cell with low tilt angle at the surfaces. In addition the cell thickness and the refractive index anisotropy must be adjusted.

Table 1. Calculated Optical Performance $\dfrac{V_{90\%\ \text{Transmission}}\ (\theta,\varphi)}{V_{10\%\ \text{Transmission}}\ (\theta,\varphi)}$ for different combinations of $K_{33}/K_{11}$, $\Delta\varepsilon/\varepsilon_\perp$, $\Delta n$. The cell thickness d=10 μm, the pretilt $\alpha_o=1°$.

| $\frac{K_{33}}{K_{11}}$ | $\frac{\Delta\varepsilon}{\varepsilon_\perp}$ | $\Delta n = 0.2$: $\frac{V_{90(0.0)}}{V_{10(0.0)}}$ | $\Delta n = 0.2$: $\frac{V_{90(0.0)}}{V_{10(45.45)}}$ | $\Delta n = 0.1$: $\frac{V_{90(0.0)}}{V_{10(0.0)}}$ | $\Delta n = 0.1$: $\frac{V_{90(0.0)}}{V_{10(45.45)}}$ |
|---|---|---|---|---|---|
| 1.5 | 3.7 | 1.38 | 1.84 | 1.43 | 2.06 |
| 1.5 | 0.05 | 1.50 | 1.81 | 1.43 | 1.81 |
| 1.0 | 3.7 | | | | |
| 1.0 | 0.05 | 1.48 | | | |
| 0.5 | 3.7 | 1.29 | 1.57 | 1.28 | 1.67 |
| 0.5 | 0.05 | 1.32 | 1.47 | 1.26 | 1.35 |

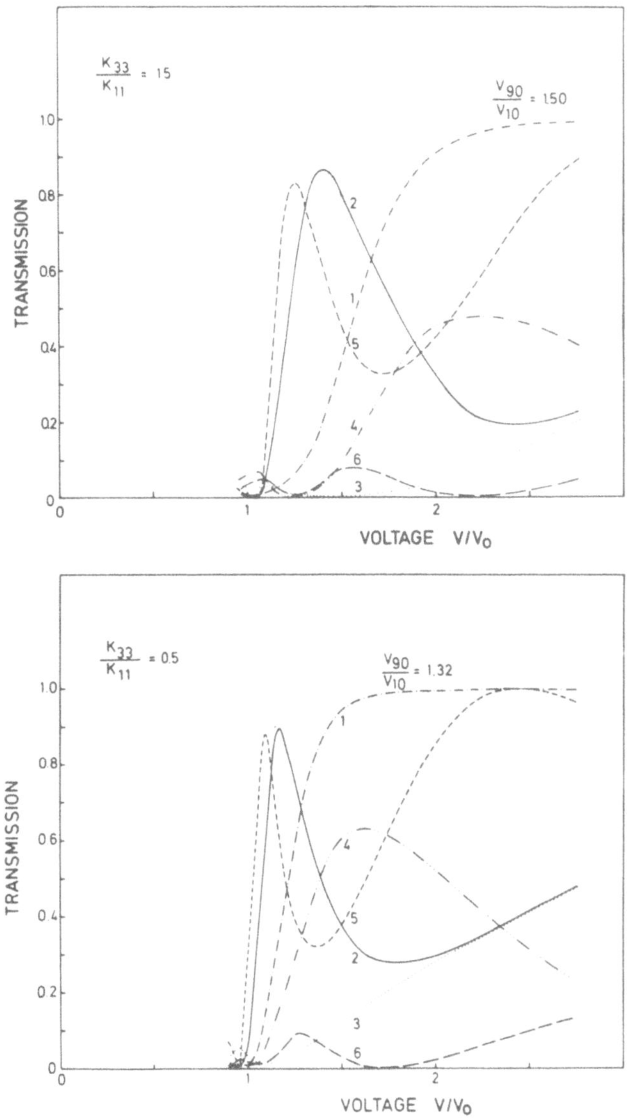


Fig. 8. Transmission characteristics of a TN-cell, calculated with $\varepsilon_{\parallel}$=10.5, $\varepsilon_{\perp}$=10.0, $n_e$= 1.728, $n_o$=1.523, d=10 μm. Upper part: $K_{33}/K_{11}$=1.5, lower part: $K_{33}/K_{11}$=0.5. The viewing angles are: θ=0° for curves 1, θ=-45° and φ=0° for curves 2, θ=+45° and φ=0° for curves 3, θ=±45° and φ+45° for curves 4, θ=-45° and φ=-45° for curves 5, θ=+45° and φ=-45° for curves 6.

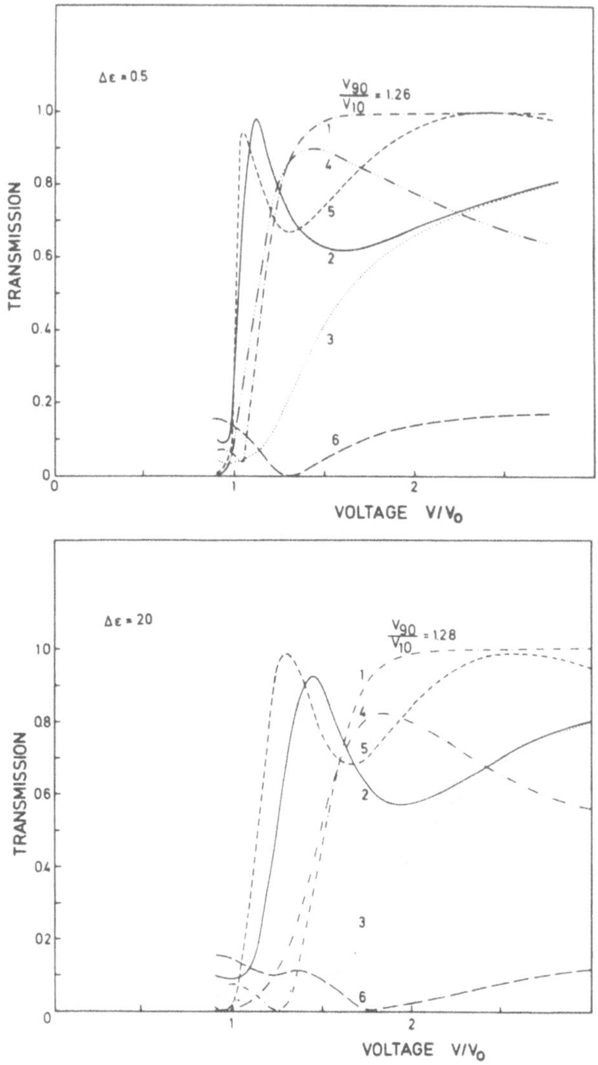


Fig. 9. Transmission characteristics of a TN-cell, calculated with $K_{33}/K_{11}=0.5$, $n_e=1.62$, $n_o=1.52$, d=10 μm. Upper part: $\varepsilon_{\parallel}=10.5$, $\varepsilon_{\perp}=10.0$; lower part: $\varepsilon_{\parallel}=25.4$, $\varepsilon_{\perp}=5.4$. The viewing angles are $\theta=0°$ for curves 1, $\theta=-45°$ and $\varphi=0°$ for curves 2, $\theta=+45°$ and $\varphi=0°$ for curves 3, $\theta=\pm45°$ and $\varphi=+45°$ for curves 4, $\theta=-45°$ and $\varphi=-45°$ for curves 5, $\theta=+45°$ and $\varphi=-45°$ for curves 6.

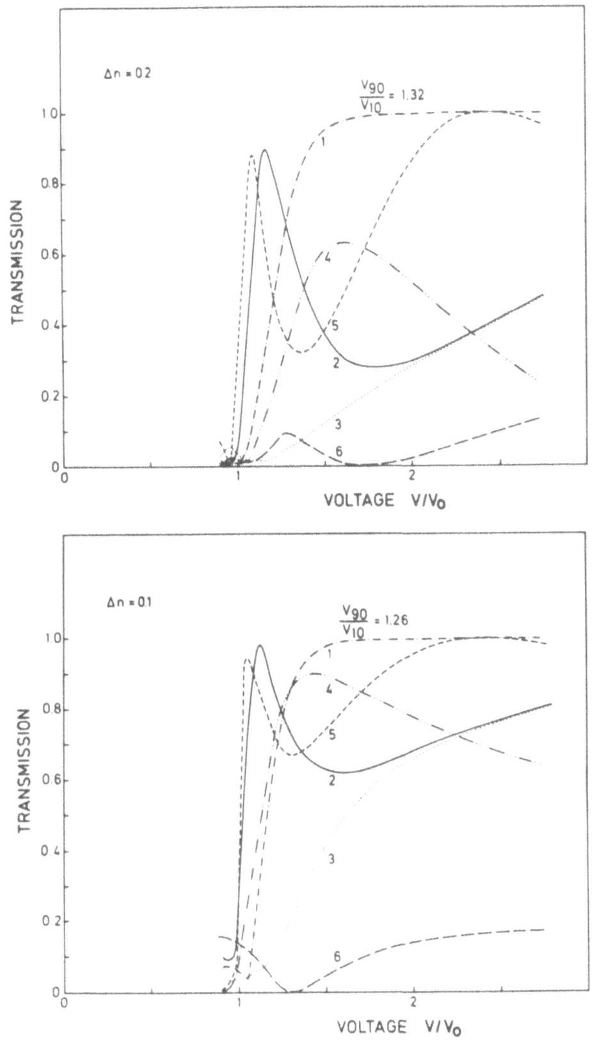


Fig. 10. Transmission characteristics of a TN-cell, calculated with $K_{33}/K_{11}=0.5$, $\varepsilon_{\parallel}=10.5$, $\varepsilon_{\perp}=10.0$, d=10 μm. Upper part: $n_e=1.728$, $n_o=1.523$, lower part: $n_e=1.62$, $n_o=1.52$. The viewing angles are θ=0° for curves 1, θ=-45° and φ=0° for curves 2, θ=+45° and φ=0° for curves 3, θ=±45° and φ=+45° for curves 4, θ=-45° and φ=-45° for curves 5, θ=+45° and φ=-45° for curves 6.

Steeper slopes of the optical characteristics can also be obtained by using 2 frequency addressing schemes.[7,8,9,10] Curves calculated with the data of the Merck mixture 1085 are presented in Fig. 11. The simulation was done by increasing a low frequency voltage while a high frequency voltage was permanently applied.

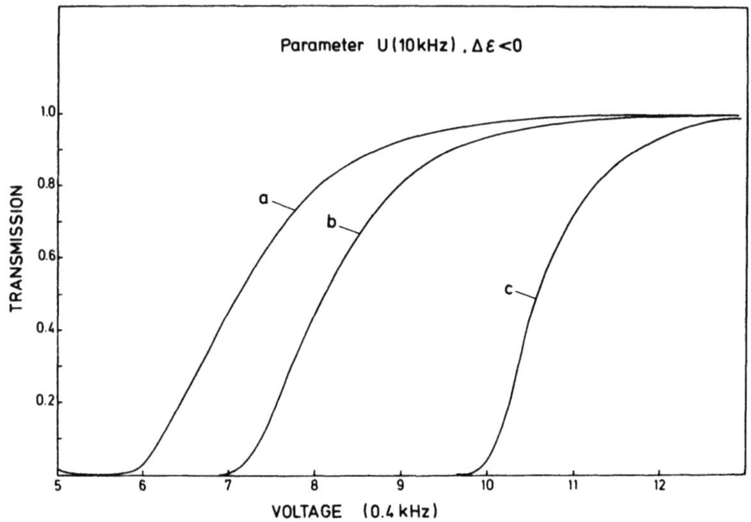


Fig. 11. Transmission of a TN-cell for normally incident light. Curves are calculated with the material data of ZLI 1085: $K_{11}=20.3\cdot10^{-7}$, $K_{22}=9.3\cdot10^{-7}$, $K_{33}=17.1\cdot10^{-7}$, d=18.2 µm, $n_e=1.6618$, $n_o=1.4974$, $\varepsilon_{\parallel}=5.68$ (0.4 kHz), $\varepsilon_{\perp}=4.86$ (0.4 kHz), $\Delta\varepsilon=-0.56$ (10 kHz) in cgs units. $\Delta\varepsilon$ changes sign at 6.5 kHz Parameter is the voltage for a frequency of 10 kHz: 0 Volts for a), 5 Volts for b), 10 Volts for c).

EXPERIMENTAL RESULTS

In the first part, some parameters relevant to the choice of suitable liquid crystal materials have been discussed. In this second part, an overview of material constants as well as a comparison of theoretical and experimental results are given. A similar comparison for MBBA with a positive dielectric additive was published earlier.[11] In Tables 2 and 3, elastic constants for some components and commercial mixtures are collected. Some of the data presented here were measured in our laboratory, others were collected from literature.

Table 2. Ratio of $K_{33}/K_{11}$ for various liquid crystals. 14 gives the reference from which the values were taken. * indicates measured in our laboratory.

| Material | $K_{33}/K_{11}$ | $T/T_c$ | $T_c$(°C) |
|---|---|---|---|
| PCH-3* | 2.27 | 0.95 | 45.2 |
| PCH-5* | 1.84 | 0.95 | 53.5 |
| PCH-7* | 1.58 | 0.95 | 57 |
| PCH-3/5/7* | 1.87 | 0.95 | 51.9 |
| CB 7* | 1.18 | 0.95 | 43 |
| ZLI 1085* | 0.97 | 0.90 | 83 |
| ZLI 1297* | 0.75 | 0.84 | 83.8 |
| ZLI 1341* | 0.86 | 0.90 | 90.5 |
| alkoxy-azoxybenzenes[14] | | | |
| n=3 | 1.95 | 0.95 | 167.7 |
| n=5 | 1.19 | 0.95 | 135.9 |
| n=7 | 0.96 | 0.95 | 129.3 |
| n=8 | 0.6 | 0.95 | 123.9 |
| alkyl-azoxybenzenes[14] | | | |
| n=5 | 0.91 | 0.95 | 67.5 |
| n=6 | 0.67 | 0.95 | 54.2 |

Table 3. Ratio of $K_{33}/K_{11}$ for some commercial mixtures. 15, 16 give the references from which the values are taken. * indicates measured in our laboratory.

| Material | $K_{33}/K_{11}$ | T(°C) |
|---|---|---|
| RO-TN-200 [15] | 2.16 | 22 |
| | 2.8 | 25 |
| RO-TN-201 [15] | 1.95 | 22 |
| RO-TN-211 [15] | 1.59 | 22 |
| RO-TN-103 [15] | 1.63 | 22 |
| RO-TN-132 [15] | 1.64 | 22 |
| RO-TN-403 [15] | 1.83 | 22 |
| RO-TN-430 [15] | 1.71 | 22 |
| E7 * | 1.43 | 24 |
| E7 [15] | 1.94 | 22 |
| E7 [16] | 1.20 | 25 |
| ZLI 1132 [15] | 2.41 | 22 |
| ZLI 1216 * | 1.31 | 25 |
| ZLI 1253 * | 1.4 | 25 |
| ZLI 1085 * | 0.85 | 25 |
| azoxy [16] | 1.0 | 25 |

In Table 2 values of $K_{33}/K_{11}$ for some components are given. The PCH and ZLI components have been measured in our laboratory.[12,13] The values for the alkoxy- and alkyl-azoxybenzenes are taken from de Jeu's paper.[14] For the ester mixture ZLI 1085 and the related mixtures ZLI 1297, ZLI 1341 as well as for some alkyl- and alkoxy-azoxybenzenes $K_{33}/K_{11}$ is less than 1.

In Table 3 a survey of commercial materials is given. Data from Hoffmann-LaRoche,[15] Hitachi[16] and our laboratory are put together. To our knowledge the lowest value of $K_{33}/K_{11}$ for a commercial material was found for ZLI 1085.

To confirm the calculations discussed before, calculated and experimental transmission curves are compared in Fig. 12. The solid line is the transmission curve calculated from material parameters. The dots are experimental values taken for an 18 μm cell prepared with ZLI 1085. The pretilt was less than 1°.

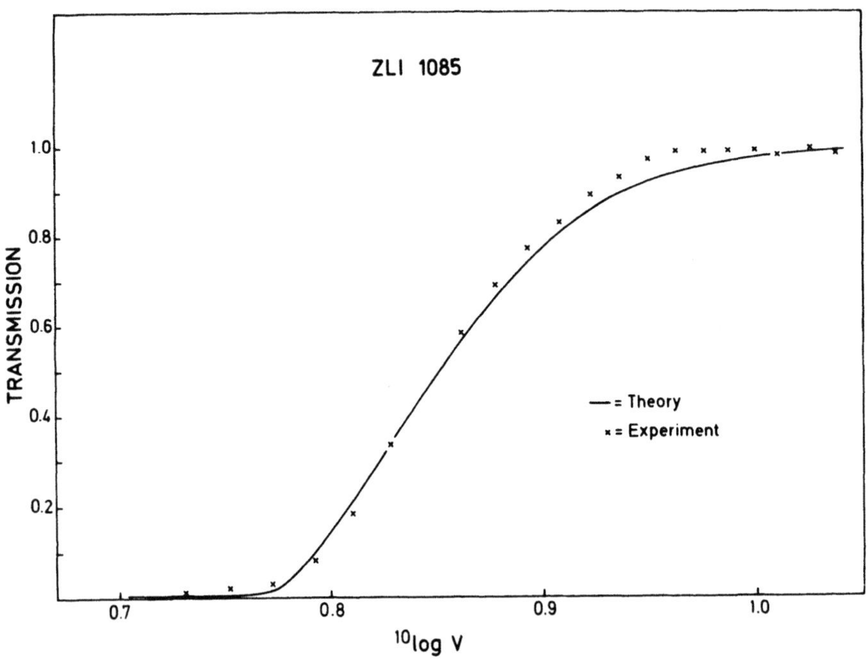


Fig. 12. Optical transmission for a TN-cell. The solid line is calculated with material data of ZLI 1085. The dots are experimental values.

THE FLUORESCENCE-ACTIVATED DISPLAY FLAD

The last part is concerned with a combination of a TN cell with a light collecting plate, the FLAD.[17] A schematic representation is shown in Fig. 13. The TN cells are operated in a transmissive mode. The thickness of the back glass plate should not be larger than 0.7 mm to avoid parallax effects.

The high brightness at the segments of the light collecting plate is achieved by absorption and spectral conversion of incident ambient light and subsequent trapping of fluorescent light by total internal reflection in the plate.

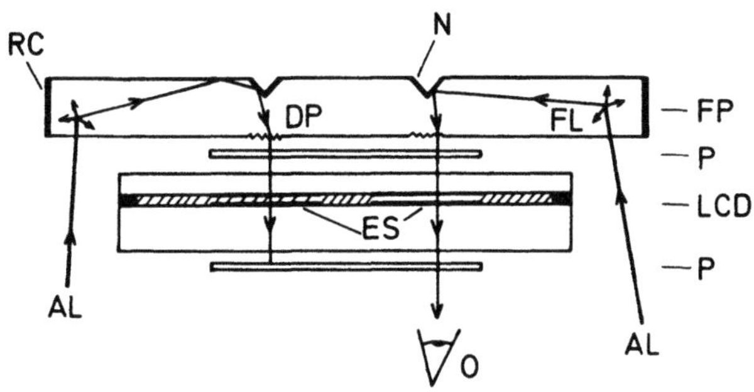


Fig. 13. Schematic diagram of a fluorescence activated LCD. The abbreviations are: Reflector RC, Notch N, Fluorescent plate FP, Dispersing pattern DP, Fluorescence light FL, Electrode segment ES, Polarizer P, Ambient light AL.

By virtue of the well-known conversion mechanism the short wavelength part of the light entering the plate is absorbed by the dyes and converted to fluorescent light, resulting in a high spectral intensity at wavelengths near the emission peak of the dyes. Depending on the color temperature of the ambient light, dyes with peak emission wavelengths ranging from green to red are efficient.

The fluorescent plate is a plastic sheet appropriately doped with fluorescent dyes and V-shaped notches at the backside. The notches represent the segments of a digit. The narrow sides of the fluorescent plate as well as the notches are coated with a reflective layer. Opposite to the notches, on the front side, a light dispersing pattern is attached.

Notches, dispersing pattern and electrode segments of the LCD coincide. According to the information to be displayed, only those segments of the display to which a voltage is applied are transparent for the fluorescent light. To the observer the displayed information appears as bright colored symbols on a dark background. From the mode of operation described, it is obvious that the brightness of the display is self-adapting to ambient light conditions.

The amount of ambient light absorbed by two different dyes is given in Fig. 14 for an AM 1.5 and an incandescent spectrum. For dye number I the emission peak occurs at the maximum of the sensitivity of the human eye for photopic vision.

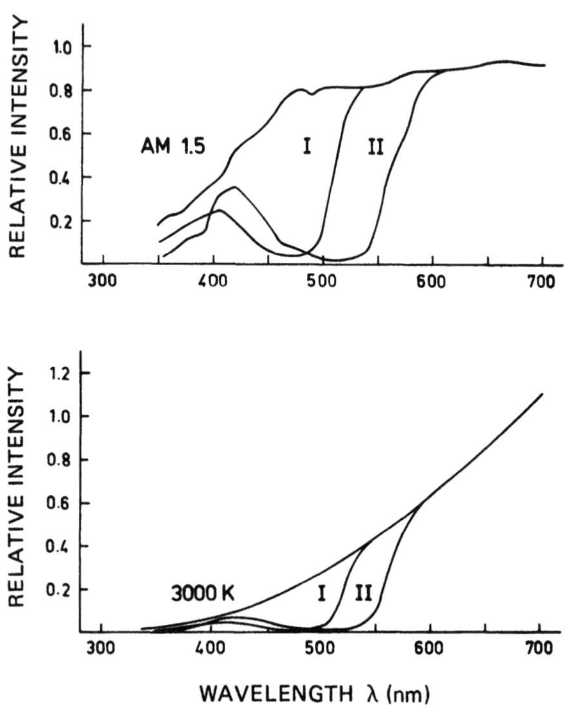


Fig. 14. Part of absorbed light for two different dyes, upper part in AM 1 ambient light, lower part in ambient light with a color temperature of 3000K.

If the brightness of the segments of the fluorescent plate is compared with the brightness of an ideal reflector, the concentration factor, needed to achieve the same brightness, is about 4 in AM 1.5 light and about 10 in incandescent light. The brightness is further diminished by the shift of the luminosity function to scotopic vision. Therefore, considering stable dye molecules available at present, a ratio of collecting area to segment area of at least 15 is recommended.

The use of complicated injection molds to fabricate plates with a special notch pattern is expensive and restricts the design flexibility. Therefore we tried to deflect the guided light by a printed pigment pattern. If thickness of the plate and width of the segments is adjusted the brightness achieved with pigment deflection is comparable to that achieved with notches.

This simplification of the production process will cut down the price of the plates to a level where the FLAD might be employed for many display applications.

## ACKNOWLEDGMENT

The author wishes to express his gratitude to Hp. Schad and B. Scheuble for measurements of material parameters, to W. Fehrenbach and V. Wittwer for performing calculations, R. Eidenschink from Merck for supplying us with LC-materials, G. Meier and A. Stieb for helpful discussions and D. W. Berreman for supplying us with a computer program and critically reading the manuscript.

## REFERENCES

1. M. Schadt and W. Helfrich, Appl. Phys. Lett. 18:127 (1971).
2. C. Z. van Doorn, J. Appl. Phys. 46:3738 (1975).
3. D. W. Berreman, J. Appl. Phys. 46:3746 (1975).
4. F. Gharadjedaghi and J. Robert, Rev. Phys. Appliquée 11:467 (1976).
5. Allan R. Kmetz, 1978 SID Int. Symposium, Digest p. 70.
6. P. Andrew Penz, 1978 SID Int. Symposium, Digest p. 68.
7. G. Baur, A. Stieb and G. Meier, Appl. Phys. 2:349 (1973).
8. B. Dargent and J. Robert, 1977 SID Int. Symposium Digest p. 60.
9. C. Z. van Doorn, J. M. J. de Klerk, 8. Freiburger Arbeitstagung Flüssigkristalle 1978, 1978 SID Int. Symposium Digest p. 100.
10. K.-H. Walter, M. Bechteler and H. Krüger, 8. Freiburger Arbeitstagung Flüssingkristalle 1978.
11. G. Baur, F. Windscheid and D. W. Berreman, Appl. Phys 8:101 (1975).
12. Hp. Schad, G. Baur and G. Meier, Appl. Phys. 17:177 (1978).
13. Hp. Schad, G. Baur and G. Meier, J. Chem. Phys. March 1979.
14. W. H. de Jeu and W. A. P. Claassen, J. Chem. Phys. 67:3705 (1977).

15. M. Schadt and F. Müller, IEEE Trans. on Electron Dev. ED-25/9:1125 (1978).
16. K. Suzuki, K. Toriyama, T, Ishibashi and K. Odawara, 7e Congrès Int. sur les Cristaux Liquides, July 1978.
17. G. Baur and W. Greubel, Appl. Phys. Lett. 31:4 (1977).

# MULTIPLEXING LIMITS OF TWISTED NEMATIC LIQUID CRYSTAL DISPLAYS AND IMPLICATIONS FOR THE FUTURE OF HIGH INFORMATION CONTENT LCDs

Frederic J. Kahn and Henryk Birecki

Solid State Laboratory
Hewlett-Packard Laboratories
Palo Alto, California 94304

## I. INTRODUCTION

"Why have liquid crystal displays (LCDs) become important in the decade of the 1970's?" The answer is that LCDs have satisfied a major need in the world of electronic displays (see Fig. 1), i.e., a need for compact displays with power dissipations on the order of microwatts per square centimeter of active display area. That such a need existed has been well confirmed by the commercial success of multiple generations of watches, calculators, electronic games and other portable instruments incorporating LCDs. Although there are many more dimensions to the display world that those accounted for in Fig. 1, the critical dimension resulting in the commercial proliferation of LCDs in the 1970's was power dissipation.

As we look to the 1980's a display gap still remains. We show in this paper that the device responsible for the commercial success of LCDs to date, the twisted nematic (TN) LCD, is limited to flat panel displays with less than about $10^4$ independently addressable picture elements (pels). Thus, there is no display technology available today that will provide a low power, flat panel display with high information content ($10^4$ to $10^6$ pels). We include for purposes of the present discussion only technologies for which one can extrapolate on the basis of published information manufacturing costs compatible with mass produced commercial instruments. This rules out systems in which the manufacturing cost of the display and its drive electronics will exceed $1,000. Order of magnitude lower costs would be preferred, and for some applications required.

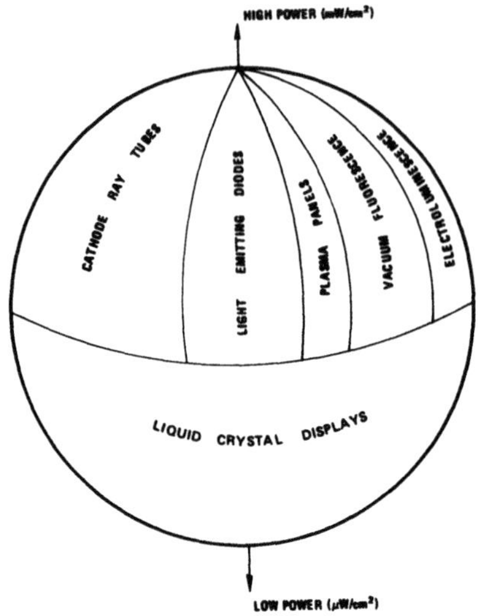


Fig. 1. The world of electronic displays. The upper or highpower hemisphere is crowded with displays dissipating milliwatts per $cm^2$ of active area while the lower or low power hemisphere is occupied solely by LCDs dissipating microwatts per $cm^2$.

High information content, low power displays are complementary to very large scale integrated circuits; to low cost, high density electronic memories; and to high bandwidth communication systems. It is axiomatic that high information content, low power displays will be required by projected advances in development of compact, low power, high performance electronic data and home entertainment products. Developments in military products should follow a parallel route.[1]

The purpose of this paper is to explore whether the high information content, low power display gap is likely to be filled by evolution of present TN-LCD technology based on use of the intrinsic electrooptical threshold of the LC for multiplexing or if a revolutionary solution is required. In the next section, projected limits to information content of multiplexed twisted nematic liquid crystal displays are computed.[2] This section should be of considerable interest not only to those interested in high information content displays but to those trying to make improved intermediate information content displays ($10^2$ to $10^4$ pels) where the limitations of the intrinsic liquid crystal threshold are very important. In the final section, alternative approaches to high information content flat panel LCDs are considered including approaches using threshold extrinsic to the liquid crystal such as can be obtained by use of semiconducting substrates.

## II. PERFORMANCE LIMITS FOR OPTIMIZED FAST SCAN MULTIPLEXED TN-LCDs

We consider here performance limits of optimized fast-scan multiplexed (FSM) twisted nematic liquid crystal displays. This multiplexing mode was originally reported theoretically by Alt and Pleshko[3] and Bigelow et al.[4] Operating eight character numerical displays with 64 independently addressed elements (pels) were described in 1975[5] while more recent progress has enabled displays with over 1,000 pels to be characterized.[6] Most recently a 2 line by 80 character display with 6,400 pels has been demonstrated by Hitachi.[7] Today virtually all commercially available twisted nematic liquid crystal displays with information contents greater than 40 or 50 independently addressable pels use the FSM TN-LCD technology.

In the FSM approach, an array of N×M pels is organized as a matrix addressed by N row electrodes and M column electrodes. The row scan voltage $AV_o$ is applied sequentially to the row electrodes with all the other row electrodes held at ground potential. Information voltages $+V_o$ or $-V_o$ are applied to the column electrodes. The instantaneous peak voltage across a single liquid crystal display element will then by $(A+1)\ V_o$ or $(A-1)\ V_o$ for selected ("ON") and nonselected ("OFF") elements respectively for 1/N of the frame time. For (N-1)/N of the frame time the LC elements will see $+V_o$ or $-V_o$. If the rms values of the voltages across the selected and nonselected LC elements are denoted by $V_S$ and $V_{NS}$ respectively then it can be shown that the ratio $V_{NS}/V_S$ will be a minimum for $A=N^{1/2}$, resulting in an optimized fast scan multiplexing voltage ratio given by

$$\frac{V_{NS}}{V_S} = \left(\frac{N^{1/2}-1}{N^{1/2}+1}\right)^{1/2} . \qquad (1)$$

Equation (1) is applicable to multiplexing of any material which responds to the rms value of the applied voltage and is therefore not limited to TN-LCDs. For very small values of N, e.g., 2 or 3, slightly lower $V_{NS}/V_S$ ratios can be obtained with recently reported alternative addressing schemes.[8] The ratio can also be reduced for displays in which each pel need not be independently addressable in all possible combinations.[9] However, Eq. (1) applies for the general case of intermediate ($10^2$ to $10^4$ pels) to high ($10^4$ to $10^6$ pels) information content displays being considered here. The figures of merit defined below enable us to relate display performance to the multiplexing voltage ratio specified by Eq. (1).

## III. FIGURES OF MERIT

We show here how to determine the number of rows N that can be multiplexed, i.e., the minimum multiplexing duty factor 1/N, from either measurements on actual displays or by computation of the static optical properties of TN-LCDs. We show that the figure of merit M can be decomposed into multiplicative factors $M_V$, $M_\Theta$ and $M_T$ specifying dependence on voltage, angle and temperature respectively. It is shown that these multiplicative factors permit customizing computations of N to match specific user requirements such as direct viewed or projection displays and temperature compensated or uncompensated drive voltage.

For simplicity, all the analysis described in this paper will be for transmission mode display cells. Therefore, the results presented for angular viewing dependence will be worse than those which actually would be obtained for reflection type cells where the superposition of contrast obtained from the 1st and 2nd passes of light through the LC layer will generally broaden the effectively usable viewing angle. The first pass of light through the display casts a "shadow image" on the reflector. This image has blurred edges, appears to be in the plane of the reflector and has a viewing angle which is essentially isotropic but modified by any directional properties of the reflector itself. Contrast of this image depends on the direction of the source as well as geometric properties of the cell. The second or reflected pass of light through the cell creates a "direct" or sharp image which appears to be in the plane of the liquid crystal and has the angular dependence of the transmission images discussed in this paper. The superposition of the "shadow" and "direct" images forms the composite "reflected" image viewed by the display user.

The tradeoff between angular viewing dependence and N is shown explicitly in Fig. 2 where measured contrast ratio is plotted as a function of the angle of incidence $\Theta$ in the principal viewing plane for a typical TN-LCD with E-7 biphenyl liquid crystal mixture between crossed polarizers. The principal viewing plane is defined by the sample normal and the liquid crystal director midway between the cell electrodes. Contrast ratio is defined as the ratio of cell transmission I with $V_{NS}$ applied to transmission with $V_S$ applied.

Typical computed[2] transmission versus voltage curves for E-7 LC between crossed polarizers are given in Fig. 3. These curves were computed for the principal viewing plane with $\Theta$ as a parameter. The strong angular dependence of electrooptical properties is readily apparent from Figs. 2, 3 and 4.

After careful visual inspection of TN-LCDs operated at various voltage levels we have concluded that the displayed image of a

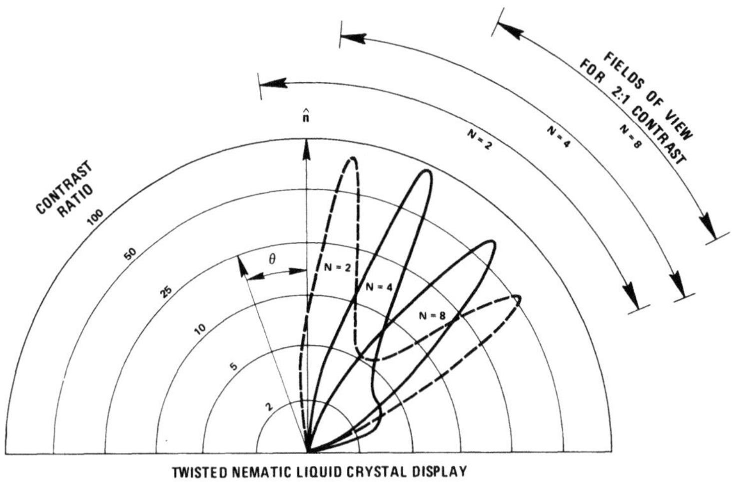


Fig. 2. Tradeoff between angular viewing dependence and number of multiplexed lines N. Plots are for contrast ratio versus angle of incidence Θ in principal viewing plane. Widest field of view is for low N. Polarizer alignment normal to the nematic director (n̂) at the adjacent LC surface emphasizes the double lobe for N=2 as discussed in the accompanying paper.[2]

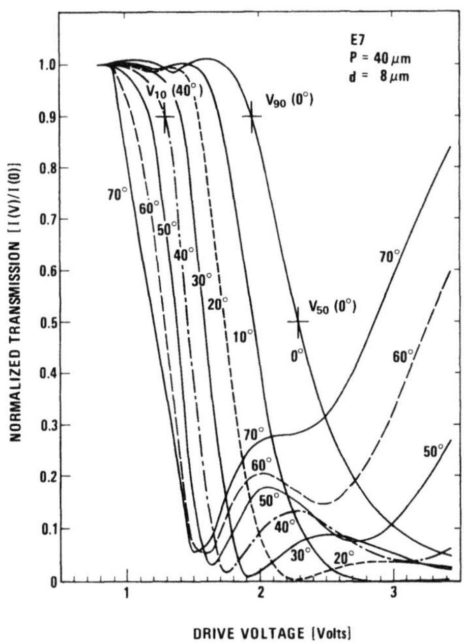


Fig. 3. Typical transmission versus voltage curves with angle of incidence in principal viewing plane as a parameter. "M"s are figures of merit.

typical font is usable in most viewing conditions if the selected elements have a contrast ratio greater than ~2:1. Similarly, if the nonselected elements have a contrast ratio less than ~1.1:1, the image quality is also acceptable. If $V_p(\Theta,T)$ is the voltage which results in p% transmission of the TN-LCD at a particular incidence angle $\Theta$ (in the principal viewing plane and at temperature T) then $V_{50}$ and $V_{90}$ correspond to 2:1 and ~1.1:1 contrast ratio respectively.

The dependence of $V_{50}$ and $V_{90}$ on viewing angle and temperature are shown explicitly in Fig. 4. Temperature dependence of threshold has been discussed extensively elsewhere[10] so we will not dwell on it here.

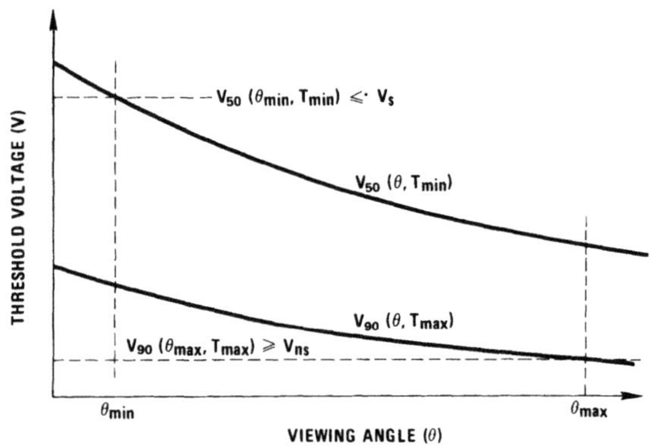


Fig. 4. Typical dependence of threshold voltages on temperature and viewing angle.

Satisfactory viewing quality will be obtained when

$$V_{NS} < V_{90\ min} = V_{90}(\Theta_{max}, T_{max}) \tag{2}$$

and when

$$V_S > V_{50\ max} = V_{50}(\Theta_{min}, T_{min}) \tag{3}$$

where $\Theta_{max}$ and $\Theta_{min}$ indicate the desired range of viewing angle in the principal viewing plane and $T_{max}$ and $T_{min}$ indicate the desired range of operating temperature. Thus, it is possible to define a multiplexing figure of merit M for a given TN-LCD display cell as

$$M = \frac{V_{90}(\Theta_{max}, T_{max})}{V_{50}(\Theta_{min}, T_{min})} . \tag{4}$$

For satisfactory visual quality we require

$$M \geq \frac{V_{NS}}{V_S} \tag{5}$$

where $V_{NS}/V_S$ depends only on the number of multiplexed lines as given by Eq. (10). The maximum number of multiplexed lines $N_{max}$ for a display with figure of merit M will then be given by

$$N_{max} = \left(\frac{1+M^2}{1-M^2}\right)^2 . \tag{6}$$

Furthermore by defining M in this manner, it can be factored into separate figures of merit explicitly quantifying the angular, temperature and voltage dependences as follows:

$$M = M_\Theta M_T M_V \tag{7}$$

where

$$M_\Theta = \frac{V_{90}(\Theta_{max}, T_{max})}{V_{90}(\Theta_{min}, T_{max})} \tag{8}$$

$$M_T = \frac{V_{90}(\Theta_{min}, T_{max})}{V_{90}(\Theta_{min}, T_{min})} \tag{9}$$

$$M_V = \frac{V_{90}(\Theta_{min}, T_{min})}{V_{50}(\Theta_{min}, T_{min})} . \tag{10}$$

For specific display requirements, it may be appropriate to set $M_\Theta$, $M_T$ or $M_V$ equal to unity. This use of the figure of merit is summarized in Table 1.

Unlike figures of merit which have been defined previously, the relative degradations in multiplexability due to angular viewing dependence, temperature dependence and lack of sharpness in the voltage threshold are immediately evident. Furthermore, these figures of merit can be obtained directly from experimental measurement or theoretical analysis. We now apply the figure of merit to determine the "ultimate" limit for $N_{max}$ for fast scan multiplexed TN-LCDs as well as the effects of the various cell and material parameters on $N_{max}$.

Table 1. Applying the Figure of Merit

| Display Type | Applicable Figure of Merit |
|---|---|
| Not temperature compensated | $M=M_\Theta M_T M_V$ |
| Temperature compensated | $M=M_\Theta M_V$ ($M_T=1$) |
| Collimated light | $M=M_V M_T$ ($M_\Theta=1$) |
| Collimated light, temperature compensated | $M=M_V$ ($M_\Theta=M_T=1$) |
| Ideal display | $M=1$ ($M_\Theta=M_T=M_V=1$) |

## VI. MULTIPLEXING LIMIT FOR FAST SCAN ADDRESSED TN-LCDs

Performance limits for fast scan multiplexed TN-LCDs are derived here. These limits were determined by means of computer calculations[2] referenced to current experimental results as a baseline. Results of the calculations are summarized in Table 2. The baseline for the computations was the cell and material parameters corresponding to low tilt display cells using E7 liquid crystal and are given in the accompanying paper in this volume.[2]

The cell parameters: thickness (d), twist ($\beta_o$), capacitive threshold voltage ($V_c$) and the liquid crystal material parameters: birefringence ($n_a$), dielectric anisotropy ($\varepsilon_a$), pitch (P), bend to splay elastic constant ratio ($K_3/K_1$) and twist elastic constant ($K_2$) were varied one at a time over a wide range of values. Transmission versus voltage curves were calculated at 0° and 40° incidence in the principal viewing plane and values for $M_\Theta$, $M_V$, $M_T$ and N determined. Parameter values giving the maximum value for N are listed in Table 2 along with the corresponding $N_{max}$ and maximum figures of merit. The range of parameter values was taken to be sufficiently wide to account not only for values that can be achieved today but for values which might be obtained as a result of future innovation. For $V_c$ we assumed the capacititance threshold could be varied without changing the slopes $\partial I/\partial V$ of the transmission versus voltage curves.

For purposes of these computations, the high transmission axes of the polarizers were assumed to be aligned parallel to the nematic directors ($\hat{n}$) at the nearest LC surfaces. A computation was also run for polarization normal to $\hat{n}$. The latter corresponds to the experimentally measured configuration denoted in Table 2 as the experimental baseline.

Table 2. Computed Improvements in Multiplexing by Varying Cell and Material Parameters

| Parameter | Range | Best Value | $M_\Theta$ | $M_V$ | $M_{max}$ | $N_{max}$ | $\Delta N$ |
|---|---|---|---|---|---|---|---|
| Computation Baseline (E7, pol. $\parallel\hat{n}$) | | | 0.65 | 0.84 | 0.54 | 3.4 | 0 |
| Computation (E7, pol. $\perp\hat{n}$) | | | 0.72 | 0.84 | 0.61 | 4.7 | 1.3 |
| Experimental Baseline (E7, pol. $\perp\hat{n}$) | | | 0.80 | 0.81 | 0.65 | 6.1 | 1.4 |
| | | | $\Delta_B = \Sigma(\Delta N)_{BASELINE} = 2.7$ | | | | |
| d | 6-15 μm | 6 μm | 0.71 | 0.86 | 0.61 | 4.8 | 1.4 |
| $\beta_o$ | 60-120° | ~100° | 0.70 | 0.87 | 0.61 | 4.8 | 1.4 |
| $V_c$ | 0.85/1.7 Volts | 1.7 | 0.77 | 0.89 | 0.69 | 7.9 | 4.5 |
| | | | | $\Delta_C = \Sigma(\Delta N)_{CELL} = 7.3$ | | | |
| $n_a$ | 0.1=0.3 | 0.1 | 0.76 | 0.86 | 0.66 | 6.5 | 3.1 |
| $\varepsilon_a$ | 0.5-50 | 50 | 0.67 | 0.86 | 0.58 | 4.1 | 0.7 |
| P | 40 μm/∞ | ∞ | 0.66 | 0.84 | 0.55 | 3.5 | 0.1 |
| $K_3/K_1^*$ | 0.2-20 | 0.2 | 0.82 | 0.86 | 0.70 | 8.7 | 5.2 |
| $K_2\times10^7$* | 2-50 | 2 | 0.72 | 0.86 | 0.62 | 5.2 | 1.7 |
| | | | | $\Delta_{LC} = \Sigma(\Delta N)_{LC} = 10.8$ | | | |
| $\Delta N_{TOTAL} = N_{\substack{COMPUTATION \\ BASELINE}} + \Delta_B + \Delta_C + \Delta_{LC} = 24$ | | | | | | | |

*P = ∞ For this computation.

Results of the computations and implications for multiplexing are discussed in this paper. Physical origins of the results are discussed in the accompanying paper in this volume.[2]

By subtracting the computation baseline value of N from each value of $N_{max}$ we find the increased number of multiplexed lines ΔN which can be obtained by varying that particular parameter. Although it is not strictly valid to do so, the total improvement in multiplexing which would be obtained by varying cell parameters simultaneously can be estimated by adding up all the "ΔNs." This provides an estimated upper limit for N which as shown in Table 3 turns out to be N=24 for temperature compensated direct view displays with $0 \le \Theta \le 40°$. This corresponds to a figure of merit M=0.81 or assuming $M_\Theta = M_V$ for a temperature compensated display, $M_\Theta = M_V = 0.9$. When all the best parameter values obtained by varying the parameters one at a time are substituted we obtain N=11. Further iterations involving variation of parameters one at a time would presumably increase N further.

The principal improvements required to achieve N=24 multiplexing are low d, high $\beta_o$, high $V_c$, low $n_a$, low $K_3/K_1$, low $K_2$ and high P. The significance of using long pitch, i.e., $P\approx\infty$ is not apparent from Table 3. However, low $K_3$ and low $K_2$ are most effective when the pitch is not stabilized by a cholesteric additive, i.e., $P=\infty$.[2]

For a display without temperature compensation if we assume $\partial V/\partial T/\text{ave}\approx 0.2\%/°C$ over the range 0°C to 50°C a value representative of LC mixtures with low temperature dependence of voltage threshold,[11] then $M_T=0.9$, M=0.73 and from Eq. (6) the multiplexing limit for a direct view display without temperature compensation will be N=11.

Table 3. Estimated Multiplexing Limits for Various Applications of FSM TN-LCDs

| Display Type | $M_V$ | $M_\Theta$ | $M_T$ | M | N |
|---|---|---|---|---|---|
| Direct view (0°≤Θ≤40°) | | | | | |
| T-compensated | 0.9 | 0.9 | 1 | 0.81 | 24 |
| Not T-compensated | 0.9 | 0.9 | 0.9 | 0.73 | 11 |
| Projection (Θ=0°)* | | | | | |
| T-compensated* | 0.9 | 1 | 1 | 0.9 | 91 |
| Not T-compensated | 0.9 | 1 | 0.9 | 0.81 | 24 |

*Higher values of N for projection displays could be obtained by tilting the cell and adjusting the polarizers to get higher $M_V$ such as discussed by Goscianski (M. Goscianski, J. Appl. Phys. 48:1426 (1977)). For example, if $M_V=0.95$, then N=381.

## V. ALTERNATIVE APPROACHES TO HIGH INFORMATION CONTENT

In the preceding sections, we learned that the ultimate multiplexing limit for direct view FSM TN-LCDs with reasonably broad viewing angles is going to be less than 25 rows. Assuming less than 200 columns, the maximum information content to be expected for FSM TN-LCDs will be $\sim 10^4$ picture elements (pels) even with column leads coming from both sides of the display. Thus, FSM TN-LCDs effectively span the intermediate information content range ($10^2$ to $10^4$ pels) but barely reach high information content ($10^4$ to $10^6$ pels).

Alternative approaches to obtaining high information content flat panels are listed in Table 4. These are divided into 2 classes - intrinsic and extrinsic threshold addressing. Intrinsic threshold means that a material property intrinsic to the liquid crystal itself is used for addressing. An example is the non-linearity in transmission versus voltage of TN-LCDs.

Table 4. Alternative Addressing Approaches for High Information Content Flat Panel LCDs

| Intrinsic Threshold | Some Problem Areas |
|---|---|
| 2-frequency addressed TN[12]<br>cholesteric-nematic[13]<br>dynamic scattering[14]<br>smectic storage[15] | relatively high voltage and/or power<br>slow transient response or long page refresh times<br>uniformity of appearance |
| **Extrinsic Threshold** | **Some Problem Areas** |
| transistor arrays (Si,[16] TFTs[17])<br>varistor control (ZnO)[18] | expensive patterning and/or fabrication |

Analysis of currently available data on intrinsic threshold approaches[12-15] indicates they are limited to less than about $5\times10^4$ pels due to relatively high voltage or power requirements, slow transient response times, or long page refresh times. Uniformity of appearance will also be a problem due to high sensitivity of the intrinsic LC threshold to variations in LC layer thickness or in the substrate LC interface.

Extrinsic threshold control[16-18] implies non-linear elements in series with each LC element. These series elements gate the voltage across the LC, thereby eliminating the need for a sharp intrinsic threshold in the LC itself. They do take advantage of the large electrooptical changes which can be obtained at low voltages in liquid crystals.

Due to the sharp voltage thresholds, which can be obtained with various semiconducting materials, extrinsic threshold devices of up to 1,000 or more rows should be possible. The principal limitation here is the yield of expense of patterning and processing the threshold control elements. The possibility of reducing interconnection of drive electronics by fabricating addressing circuitry directly on the threshold control substrate is an additional consideration in evaluating extrinsic threshold addressing.

Figure 5 shows the regions of information space, spanned by FSM TN-LCDs, other intrinsic threshold addressed LCDs, and extrinsic threshold addressed LCDs. Note that it doesn't seem appropriate to use extrinsic or other intrinsic threshold LCDs for less than about 10 rows of information due to the expected extra cost and complexity as compared to FSM TN-LCDs. It is also clear from Fig. 5 that if extrinsic threshold LCDs are available, then the alternative intrinsic threshold approaches will not be required unless they are of lower cost. Furthermore, the extrinsic threshold LCDs should have the best operational properties including lowest voltage, fastest transient response and shortest page refresh times.

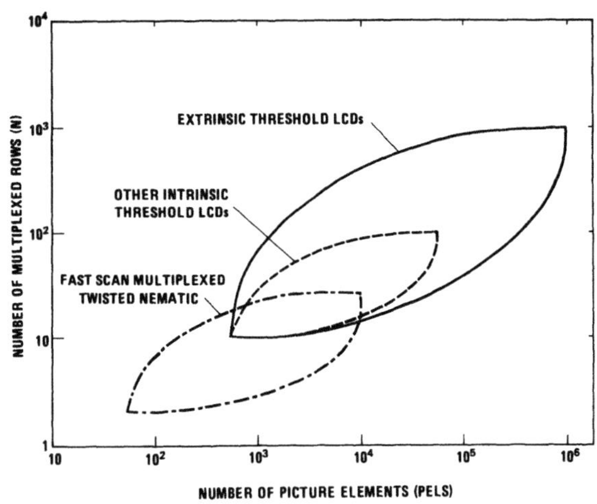


Fig. 5. Information space of LCD technologies.

## VI. CONCLUSIONS AND DISCUSSION

We have shown in this paper that single frequency fast scan multiplexed TN-LCDs with moderate viewing angles will be limited to less than about 25 multiplexed lines and about $10^4$ pixels. The principal improvements required to obtain N=25 multiplexing were determined by means of a computer calculation and are low thickness, high twist, high threshold voltage, low birefringence, low bend to splay elastic constant ratio, low twist elastic constant and low pitch.

Alternative intrinsic threshold LCDs will enable multiplexing up to ~100 lines and $\sim 5\times 10^4$ pixels. Extrinsic threshold technology will be required to multiplex high information content displays ($10^2$ to $10^3$ lines, $10^4$ to $10^6$ pixels) with greater than $\sim 5\times 10^4$ pixels. We therefore conclude that high information content flat panel LCDs will not be an evolutionary product of present day FSM TN-LCDs but will require development of a revolutionary generation of liquid crystal displays based on extrinsic threshold control.

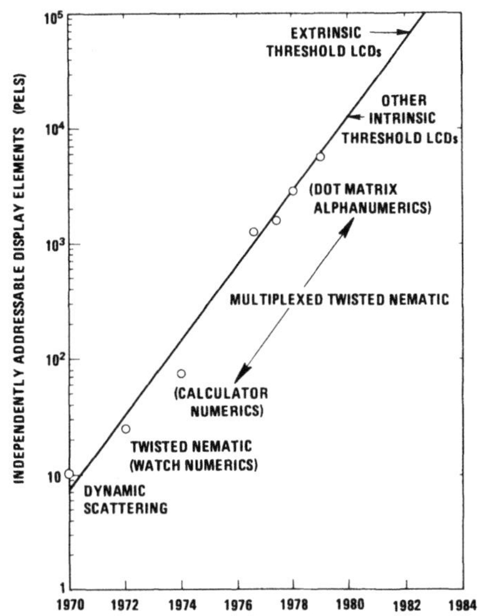


Fig. 6. Historical development of liquid crystal displays.

Since this revolutionary generation of extrinsic threshold LCDs will require technological breakthroughs in research and manufacturing development, it is difficult to predict an exact time frame for this to occur. The best we can do is note that the number of independently addressable LCD picture elements has increased by an order of magnitude (10×) every 3 1/2 years since 1970. This is shown graphically in Fig. 6. Assuming a linear extrapolation of this 9 year logarithic trend, high information content LCDs with extrinsic threshold addressing of greater than $10^5$ pels and suitable for commercial instrument applications should be reported in the technical literature by the mid 80's and commercially available by the end of this decade. The authors expect this to happen. Nevertheless, since there is no assurance that such trends will continue, the conservative assessor of the progress of LCD technology will from time to time prudently add new data points to Fig. 6 in order to determine if the trend continues to be valid.

## VII. ACKNOWLEDGMENTS

The authors wish to thank Robert Burmeister and Robert Archer for their continued support of this work and George LaBelle for performing the electrooptical measurements.

## VIII. REFERENCES

1. Ruth M. Davis, A Real Demand for Display Information, Keynote Address, 1979 SID International Symposium.
2. H. Birecki and F. J. Kahn, Effects of Cell and Material Properties on Multiplexing Levels of Twisted Nematic Liquid Crystal Displays, This volume.
3. P. M. Alt and P. Pleshko, IEEE Trans. Electron Devices ED-21:146 (1974).
4. J. E. Bigelow, R. A. Kashnow and C. R. Stein, IEEE Trans. Electron Devices ED-22:22 (1975).
5. F. J. Kahn and R. A. Burmeister, SID 1975 Digest, p. 86, L. Goodman and D. Meyerhofer, SID 1975 Digest, p. 76.
6. F. K. Kahn, SID 1978 Digest, p. 74; K. Odawara et al. SID 1978 Digest, p. 72.
7. K. Odawara et al., SID 1979 Digest, p. 120. Paper describes a 1 line by 80 character display. However a 2 line by 80 character display was also demonstrated at the meeting by K. Odawara et al.
8. A. Kmetz and J. Nehring, IEEE Trans. Electron Devices, ED-26:795 (1979).
9. I. Shanks, SID 1979 Digest, p. 110. Also p. 112.
10. F. J. Kahn and R. A. Burmeister in Nonemissive Electrooptic Displays, A. R. Kmetz and F. K. von Willisen eds., Plenum Press (1976) p. 289. For summary of more recent material results see K. Toriyama et al., this meeting, and Journal de Physique, 40:C3-317 (1979).
11. R. Eidenschink, E. Merck. See for example, data sheets on E. Merck ZLI-1216 and ZLI-1221.
12. M. Hosukawa, SID 1979 Digest, p. 116. C. Z. van Doorn and J. J. M. J. deKlerk, SID 1978 Digest, p. 100.
13. C. Tani et al., SID 1979 Digest, p. 114.
14. Z. Blank et al., Conference Record of 1978 Biennial Display Research Conference, p. 44.

15. Although intrinsic threshold devices using smectic LCs have not been as widely studied as devices using nematics and cholesterics, switching times are slower and voltages required are generally higher. Therefore, similar limitations are expected for applications where rapid page refresh times are required.
16. L. T. Lipton et al. SID 1978 Digest, p. 96.
17. R. C. Luo et al., SID 1978 Digest, p. 94.
18. D. E. Castleberry, Conference Record of 1978 Biennial Display Research Conference p. 42.

# INFLUENCE OF THE DEVICE PARAMETERS ON THE PERFORMANCE OF TWISTED-NEMATIC LIQUID-CRYSTAL MATRIX DISPLAYS

C. Z. van Doorn
C. J. Gerritsma
J. J. M. J. de Klerk

Philips Research Laboratories
Eindhoven, The Netherlands

## INTRODUCTION

It is well known that for a liquid-crystal matrix display without a static memory, and showing "root-mean-square (RMS) behavior," the maximum number of lines which can be scanned (the scanning capability) is a function of the static optical-response/voltage curve only.[1]

This paper presents the numerical calculation of the influence of several device and material parameters on the shape of the optical transition/voltage curve of a twisted nematic display. From this curve, the maximum number of scanned lines is obtained.

Parameters varied are the thickness of the layer, the wavelength of the light, the indices of refraction, the birefringence, the pretilt, the elastic constants and the dielectric anisotropy.

## THE TWISTED NEMATIC LIQUID-CRYSTAL DISPLAY CELL

Figure 1 shows typical transmission/voltage curves of a twisted nematic cell between parallel polarizers. We use this figure to define a number of characteristic voltages. At the Freederiksz threshold voltage $V_c$, the liquid crystal starts to deform and a sharp change occurs in the capacitance and the optical transmission if the polarizers are at angles of 45° with the rubbing directions. This voltage, in Gauss' units, is given by:[2,3]

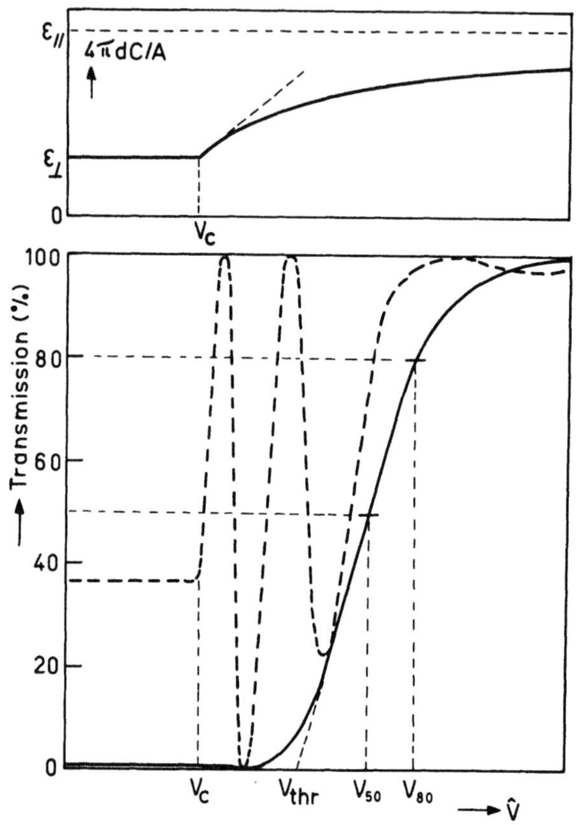


Fig. 1. Calculated capacitance C and optical transmission of a nematic liquid-crystal cell of thickness d and area A between parallel polarizers set parallel or perpendicular to the rubbing directions (solid curve) or at 45° to these directions (dashed curve) as a function of the applied RMS voltage $\hat{V}$. Definition of Freederiksz threshold $V_c$, optical threshold $V_{thr}$, $V_{50}$ and $V_{80}$ are indicated.

$$(\Delta\varepsilon/4\pi)V_c^2 = \pi^2 k_{11} + (k_{33}-2k_{22})\varphi_o^2 ,$$

where $\Delta\varepsilon=\varepsilon_{\parallel}-\varepsilon_{\perp}$ is the dielectric anisotropy, $k_{11}$, $k_{22}$ and $k_{33}$ are the Frank elastic constants and $\varphi_o$ is the total twist angle across the layer.

If there is a pretilt, i.e., the director tilt angle at the walls is nonzero, the changes near $V_c$ are gradual. We then define $V_c$ as the intersection of the extrapolated linear portion of the capacitance/voltage curve with the horizontal $C=C(\hat{V}=0)$ curve (see Fig. 1). We assume the surface coupling to be strong, so that the tilt angle at the surface is constant.

The optical threshold voltage $V_{thr}$ is obtained by extrapolating the linear portion of the transmission/voltage curve (with the polarizers parallel or perpendicular to the rubbing directions) as shown in Fig. 1.

The voltages where the transmission is 50 and 80% of the maximum define $V_{50}$ and $V_{80}$ respectively.

It should be noticed that, when crossed polarizers are used, as in reflective displays, the intensities are complementary to those shown in Fig. 1. We will, however, use the definitions of the parallel-polarizer case.

## SCANNING CAPABILITY OF TRANSMISSIVE AND REFLECTIVE DISPLAYS

The maximum number of scanned lines of a matrix display having RMS response is, according to Alt and Pleshko,[1] equal to

$$N_{max} = \left(\hat{V}_{on}^2 + \hat{V}_{off}^2\right)^2 \Big/ \left(\hat{V}_{on}^2 - \hat{V}_{off}^2\right)^2$$

where $\hat{V}_{on}$ and $\hat{V}_{off}$ are the RMS voltages in the ON and OFF states derived from the static optical-transmission/voltage curve. We have investigated three cases with the following definition of $\hat{V}_{on}$ and $\hat{V}_{off}$.

1. For a twisted nematic display between parallel polarizers used in transmission $\hat{V}_{on}$ was taken equal to $V_{80}$, and $V_{off}$ equal to $V_{thr}$.
2. For a display used in transmission like case 1 the number of lines can be increased if a lower intensity is accepted. We then put $\hat{V}_{on}=V_{50}$ and $\hat{V}_{off}=V_{thr}$.
3. For a display with crossed polarizers used in reflection we must take into account that the incident light comes from all directions. For some directions, the steepness of the response characteristic is strongly reduced and

we therefore put $\hat{V}_{off}$ equal to $V_c$. $V_{on}$ is put equal to $V_{80}$. It is not possible to put $\hat{V}_{on}=V_{50}$ to increase $N_{max}$ because this would cause an intolerable decrease of the contrast.

In using the above definitions of $\hat{V}_{off}$ and $\hat{V}_{on}$ we have chosen to consider two extreme cases of the angular variation of the optical transmission only. One case is that of a transmissive display where the viewing direction is restricted to normal incidence. The other case is that of a reflective display where a hemispherical illumination of the display is assumed; this is taken into account by using the definition $\hat{V}_{off}=\hat{V}_c$. We have not attempted to consider intermediate situations because the angular variation is very complicated and the definition of $\hat{V}_{off}$ and $\hat{V}_{on}$ becomes rather arbitrary.

## CALCULATION OF THE TRANSMISSION/VOLTAGE CURVE

In order to calculate the optical characteristic, it is necessary first to calculate the deformation of the liquid crystal. This calculation starts from the expression for the energy density U of a liquid crystal in terms of the direction $\vec{n}$ and its derivatives:[3,4]

$$U = \frac{1}{2}k_{11}(\mathrm{div}\ \vec{n})^2 + \frac{1}{2}k_{22}(\vec{n}.\mathrm{rot}\vec{n})^2 + \frac{1}{2}k_{33}(\vec{n}\times\mathrm{rot}\vec{n})^2 + \frac{1}{8\pi}\vec{D}.\vec{E}.$$

where $\vec{E}$ is the electric field strength and $\vec{D}$ the dielectric displacement. The director components are $n_x=\cos\varphi\cos\theta$, $n_y=\sin\varphi\cos\theta$ and $n_z=\sin\theta$. In a stratified layer $E_x=E_y=D_x=D_y=0$ and $D_z=\varepsilon\ E_z=$ constant, where $\varepsilon=\varepsilon_\perp\cos^2\theta+\varepsilon_\parallel\ \sin\theta$, $\varepsilon_\perp$ and $\varepsilon_\parallel$ being the dielectric constants for the field strength perpendicular and parallel to the director respectively. In the stationary state, with a constant charge on the electrodes, the total energy will be constant. From the corresponding Euler-Lagrange equations, we derive the differential equations[3]

$$\left(k_{11}\cos^2\theta+k_{33}\sin^2\theta\right)\frac{d^2\theta}{dz^2} = \sin\theta\ \cos\theta\left\{(k_{11}-k_{33})\left(\frac{d\theta}{dz}\right)^2 + \left[(k_{33}-2k_{22})\cos^2\theta - k_{33}\sin^2\theta\right]\left(\frac{d\theta}{dz}\right)^2 - \frac{\Delta\varepsilon}{4\pi}E_z^2\right\}, \tag{1}$$

$$\left(k_{22}\cos^2\theta+k_{33}\sin^2\theta\right)\cos^2\theta\ \frac{d\varphi}{dz} = k\ , \tag{2}$$

where k is an integration constant. Expressions for $d^2\theta/dz^2$ and $d\varphi/dz$ are obtained from (1) and (2) and are used for integration of these expressions in steps using a second order Taylor series. This integration usually starts at the middle of the layer and ends at the wall. Here, the boundary conditions for $\theta$ and $\varphi$ must be satisfied by adjusting the integration constants k and $\theta_m$ (the tilt angle in the middle of the layer). Also, $D_z$ must be adjusted so that the potential across the layer equals the applied voltage. This requires several iterations alternately for $\theta$ and $\varphi$.

After $\theta(z)$ and $\varphi(z)$ have been determined, the optical transmission for normally incident light has to be calculated. The method used is based on the division of the layer into a large number of mutually rotated birefringent plates. The electric field is repeatedly decomposed along the birefringent axes of the plates and the components are given phase retardations proporational to the refractive indices $n_o$ and $n=(\cos^2\theta/n_\ell^2+\sin^2\theta/n_o^2)^{-1/2}$ respectively, where n is the local effective index of refraction. Though this method is based on very simple assumptions, its results are in excellent agreement with those of the more exact methods of de Vries,[5,6] or Berreman.[7]

The capacitance is given by

$$C = (A/4\pi)\left[\int_o^d (\varepsilon_z)^{-1} dz\right]^{-1}$$

where A is the area, d the thickness and $\varepsilon_z=\varepsilon_\perp\cos^2\theta+\varepsilon_\parallel\sin^2\theta$ is the local dielectric constant relating $E_z$ and $D_z$.

## INFLUENCE OF PARAMETERS

We have investigated the influence of a number of parameters by varying each while keeping the others constant. The constant parameters were given values which are typical for liquid crystals. When not stated otherwise, these values are:

thickness of the layer d=15 µm,
ordinary index of refraction $n_o$=1.555,
extraordinary index of refraction $n_e$=1.779,
wavelength of the light $\lambda$=0.5461 µm,
director tilt angle at the walls $\theta_o$=0°,
ratio of elastic constants $k_{22}/k_{11}$=0.549,
ratio of elastic constants $k_{33}/k_{11}$=1.27,
dielectric constant parallel to the director $\varepsilon$ =17.3,
dielectric constant perpendicular to the director $\varepsilon_\perp$=5.7.

### Thickness of the Layers

Figure 2 shows the thickness dependence of $N_{max}$ for two values of the dielectric anisotropy. The scattering of the calculated points in the lower figure is caused by a thickness dependent optical interference which is superimposed on the

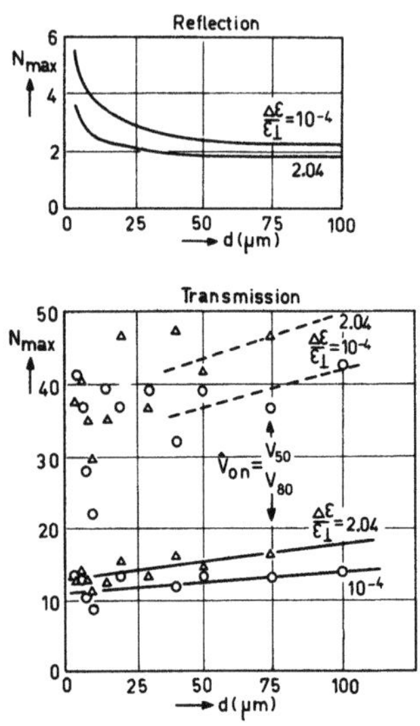


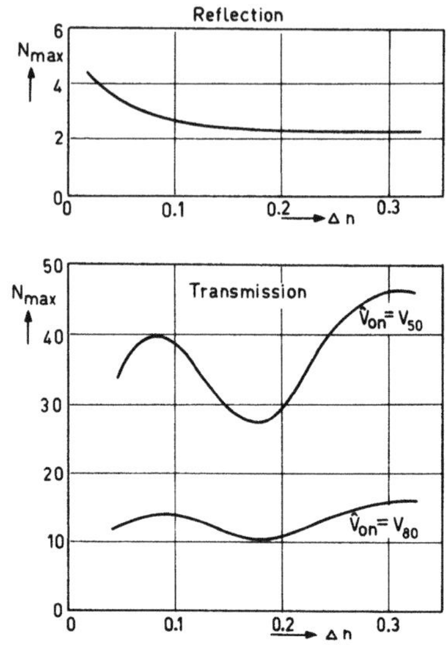


Fig. 2. $N_{max}$ as a function of cell thickness d for two values of the dielectric anisotropy. In transmission the result is obscured by optical interference.

Fig. 3. $N_{max}$ as a function of birefringence Δn.

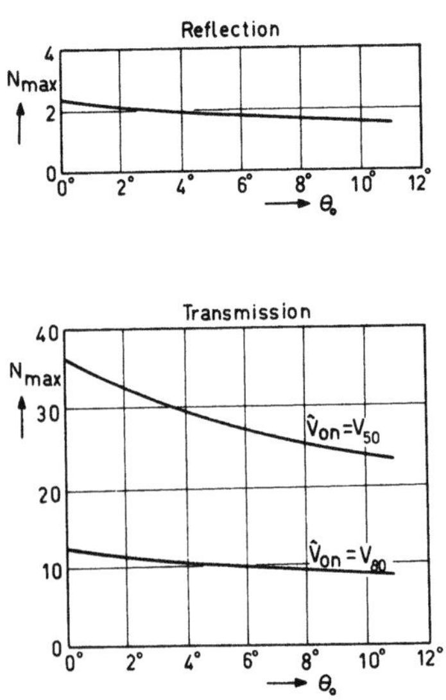


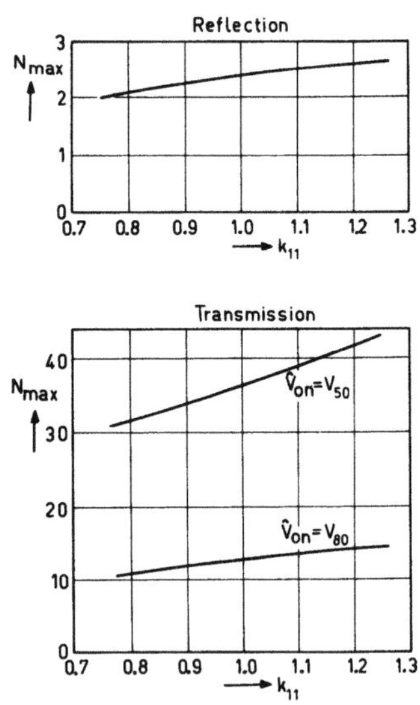


Fig. 4. $N_{max}$ as a function of pretilt $\Theta_o$.

Fig. 5. $N_{max}$ as a function of $k_{11}$.

transmission/voltage curve and influences $V_{thr}$ and $\hat{V}_{off}$ of the transmission cell. The origin of this interference is the fact that plane-polarized light becomes elliptically polarized as it passes through the twisted layer.[8] This causes the transmission T to be an oscillatory function of $u=2d\Delta n/\lambda$, where $\Delta n=n_e-n_o$ and $\lambda$ is the wavelength of the light in vacuum. The minimum and maximum values of this function are given by $T_{min}=0$ and $T_{max}=(1+u^2)^{-1}$.

At low applied voltage near $\hat{V}_{off}$ the transmission is still approximately given by the function T(u), but with $\Delta n$ replaced by the effective birefringence n' (see next section) which is a function of V.

Figure 2 suggests that a small thickness is favorable for a reflectance cell whereas the opposite is true for a transmission cell.

## Refractive Indices and Birefringence

Figure 3 illustrates the effect of the variation of $\Delta n$ with $n_o=1.5$. This is similar to that of the thickness variation. Changing $n_o$ while keeping $\Delta n$ constant has a negligible influence. This can be understood by expanding the effective birefringence $\Delta n'=(\cos^2\theta/n_e^2+\sin^2\theta/n_o^2)^{-1/2}-n_o$ as a power series in the small quantity $\Delta n/n_o$. It then appears that $\Delta n'\approx\Delta n\cos^2\theta$, independent of $n_o$.

## Wavelength of the Light

The wavelength dependence of $V_{thr}$ appeared to be so strong and fluctuating that no further attempt was made to calculate $N_{max}$.

## Pretilt

A small pretilt at the walls is usually introduced in order to suppress areas of reverse tilt and twist. The value of $V_c$ used in the calculation was obtained from the capacitance/voltage curves. The pretilt appears to lower $N_{max}$ (see Fig. 4) and should therefore be kept to a minimum; d=10 μm, $\lambda=0.6353$ μm, $n_o=1.549$, $n_e=1.764$.

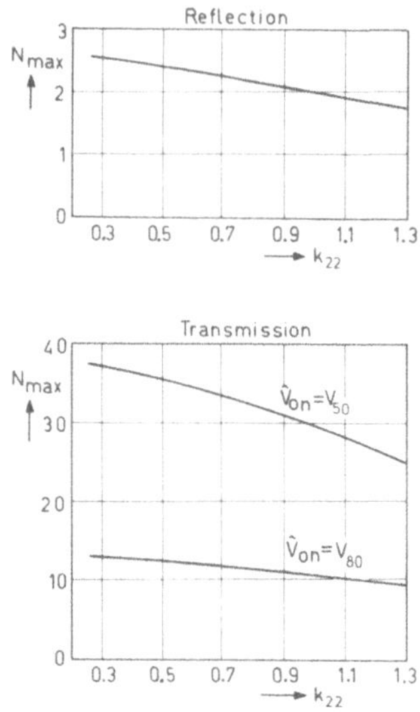


Fig. 6. $N_{max}$ as a function of $k_{22}$.

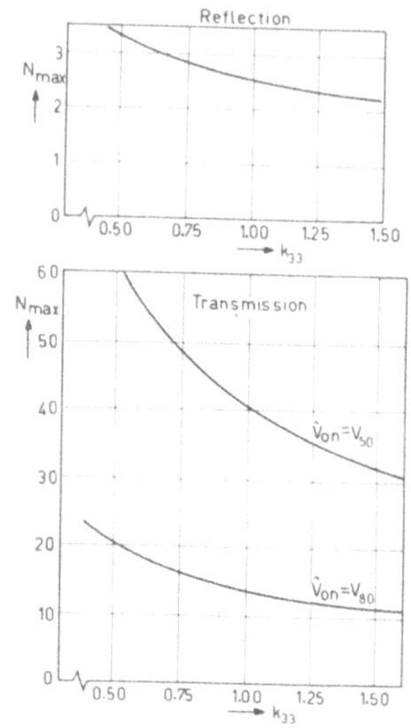


Fig. 7. $N_{max}$ as a function of $k_{33}$.

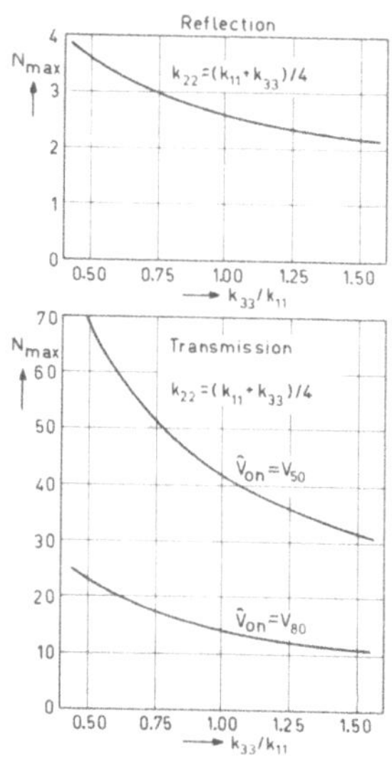


Fig. 8. $N_{max}$ as a function of $k_{33}/k_{11}$ with $k_{22}$ equal to half the mean value of $k_{11}$ and $k_{33}$.

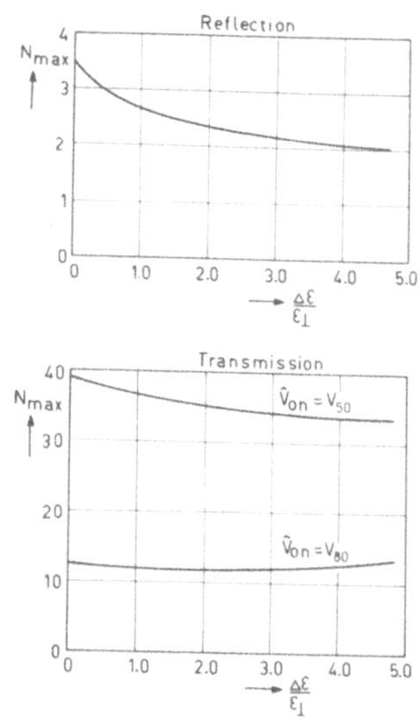


Fig. 9. $N_{max}$ as a function of the dielectric anisotropy $\Delta\varepsilon$.

Elastic Constants

Though we have varied $k_{11}$, $k_{22}$ and $k_{33}$, it must be born in mind that $N_{max}$ is influenced in fact by two independent variables only: the ratios of the constants. Figures 5, 6 and 7 give the influence of $k_{11}$, $k_{22}$ and $k_{33}$. In Fig. 8, $k_{33}/k_{11}$ has been varied while keeping $k_{22}$ equal to $(k_{11}+k_{33})/4$. This situation is typical for many liquid crystals. It is seen from Figs. 7 and 8 that a low value of $k_{33}$ is favorable though the gain obtainable is small because we have varied the elastic constants to a much greater extent then is attainable in practice.

Dielectric Anisotropy

Figure 9 shows the effect of the variation of $\Delta\varepsilon$ from zero to twenty while keeping $\varepsilon_{\perp}$ at five. For transmissive displays, the dependence is weak. For reflective displays, it is stronger and a low value of $\Delta\varepsilon$ is to be preferred.

## DISCUSSION

The values of $N_{max}$ calculated should not be considered as maximum values, because simple viewing conditions have been assumed. The main purpose of the analysis is the prediction of trends in the value of $N_{max}$ if several parameters are varied. Actually, much higher values can be obtained. For example, if $V_{off}$ and $V_{on}$ of a transmissive display are chosen equal to the voltages at 1% and 15% transmission, respectively, $N_{max}=100$ can be obtained. Oblique illumination may further increase this value. The optical interference which disturbs the calculation of the influence of the thickness, the birefringence and the wavelength of the light on $N_{max}$ can be turned to advantage by steepening a part of the transmission/voltage curve. A slight rotation of one of the polarizers may have the same effect.

In reflection, $N_{max}\approx 6$ is possible if the viewing angle is restricted to values <30°.

The analysis applies to the display of stationary pictures only so that response times do not come into play.

## CONCLUSION

The influence of the variation of several parameters on the maximum number of scanned lines of a reflective of transmissive twisted nematic liquid-crystal matrix display has been investigated by numerical calculation.

Increasing the layer thickness and the birefringence lower $N_{max}$ in reflection. Increasing the pretilt, the ratio $k_{33}/k_{11}$ and the ratio $k_{22}/k_{11}$ lower $N_{max}$. Increasing the dielectric anisotropy lowers $N_{max}$ in reflection and has only a small effect in transmission.

## REFERENCES


1. P. M. Alt and P. Pleshko, IEEE Transact. on Electron Devices ED21:146 (1974).
2. F. M. Leslie, Mol. Cryst. Liq. Cryst. 12:57 (1970).
3. H. J. Deuling, Mol. Cryst. Liq. Cryst. 27:81 (1975).
4. F. C. Frank, Disc. Faraday Soc. 25:19 (1958).
5. H. de Vries, Acta Cryst. 4:219 (1951).
6. C. Z. van Doorn, J. Appl. Phys. 46:3738 (1975).
7. D. W. Berreman, J. Opt. Soc. Am. 62:502 (1972).
8. C. H. Gooch and H. A. Tarry, J. Phys. D: Appl. Phys. 8:1575 (1975).

# ULTIMATE LIMITS FOR RMS MATRIX ADDRESSING

A. R. Kmetz

Bell Laboratories
Murray Hill, New Jersey

J. Nehring

Brown Boveri Research Center
CH-5405 Baden-Dattwil
Switzerland

## I. INTRODUCTION

In digital wristwatches where the number of digits is small each display segment can be individually addressed. For more complex displays, the number of driven connections required for direct addressing becomes prohibitive. The number of external connections can be substantially reduced by interconnecting the individual display segments into a matrix, but the infinite selection ratio of direct addressing is usually sacrificed. How well a liquid crystal display performs with the limited selection ratio which prevails under matrix addressing conditions is determined by the parameters of the liquid crystal material, the cell technology and the addressing scheme. In this paper, we concentrate on the last factor.

Liquid crystal displays typically respond to the root-mean-square (rms) value of an applied voltage waveform. Applying this fact to the multiplexed addressing schemes which are conventionally used, Alt and Pleshko[1] showed that the ratio of the voltages across selected and unselected display segments cannot exceed $[(\sqrt{N}+1)/(\sqrt{N}-1)]^{1/2}$ where N is the number of rows in the matrix. We now ask whether some unconventional addressing scheme exists which would allow us to surpass this limit.

## II. GENERALIZED DESCRIPTION OF MULTIPLEXING

Consider a matrix consisting of N rows and an arbitrary number of columns. We label the N elements in a given column by $i=1,2,\ldots,N$ and describe their collective information state by a quantity $\lambda$ which we define by

$$\lambda = \sum_{i=1}^{N} \eta_i \cdot 2^{i-1} \tag{1}$$

where

$$\eta_i = \begin{cases} 1 \text{ if } i \text{ is in the on-state} \\ 0 \text{ if } i \text{ is in the off-state} . \end{cases} \tag{2}$$

Thus, a given pattern of on- and off-elements in a column is identified with the integer $\lambda$ whose binary form has the same pattern of ones and zeroes. It follows that $\lambda$ can assume the $2^N$ values 0 through $(2^N-1)$ corresponding to the $2^N$ possible combinations of on- and off-states in the column.

Let us assume that periodic potentials $F_i(t)$ $(i=1,\ldots,N)$ are applied respectively to the N rows of the matrix. In order to generate a certain information state $\lambda$ in a given column, a potential $G_\lambda(t)$ is applied to that column. The voltage across the element i is then

$$U_i = F_i(t) - G_\lambda(t) . \tag{3}$$

The rms value of $U_i$ may be written as

$$\bar{U}_i = \left[(F_i \cdot F_i) + (G_\lambda \cdot G_\lambda) - 2(F_i \cdot G_\lambda)\right]^{1/2} \tag{4}$$

where the expressions in the parentheses are the "inner products"

$$(F_i \cdot F_i) = \frac{1}{T}\int_0^T F_i^2 dt$$

$$(G_\lambda \cdot G_\lambda) = \frac{1}{T}\int_0^T G_\lambda^2 dt$$

$$(F_i \cdot G_\lambda) = \frac{1}{T}\int_0^T F_i \cdot G_\lambda dt \tag{5}$$

and T is a period in which all signals repeat. The first two expressions (5) are identical to the squares of the rms values, $\bar{F}_i$ and $\bar{G}_\lambda$, of $F_i$ and $G_\lambda$ and the third one is the overlap integral of $F_i$ and $G_\lambda$. For convenience, we will normalize all voltages by the electrooptic threshold voltage $V_t$. It is important to note that the rms voltage across i (Eq. (4)) depends only on the inner products (5) and not upon the detailed waveforms of $F_i$ and $G_\lambda$.

In the multiplexing scheme considered by Alt and Pleshko, rectangular voltage pulses with duty ratio 1/N and a common amplitude are applied to each of the N rows in succession. These row signals $F_i$, which may be called strobe signals, are mutually orthogonal and have identical rms values $\bar{F}$, i.e.,

$$(F_i \cdot F_j) = \bar{F}^2 \, \delta_{ij} \; . \tag{6}$$

Data are applied to a given column by choosing the polarity of a fixed amplitude column signal according to the desired state of the display element at the intersection with the row being strobed at any given instant. This data waveform can be written as

$$G_\lambda = \sum_{i=1}^{N} a_i^{(\lambda)} \, F_i \; , \tag{7}$$

where

$$a_i^{(\lambda)} = \begin{cases} +b \text{ if the } i^{th} \text{ element is OFF} \\ -b \text{ if the } i^{th} \text{ element is ON} \; . \end{cases} \tag{8}$$

By substituting Eqs. (6), (7) and (8) into (4), and solving for the maximum voltage across an on-element subject to the constraint that the normalized voltage across an off-element does not exceed unity, the Alt and Pleshko result is obtained:

$$\bar{U}_{on} = \left[\frac{\sqrt{N}+1}{\sqrt{N}-1}\right]^{1/2} , \quad N > 1 \; . \tag{9}$$

Equations (6), (7) and (8) comprise a generalized definition of multiplexing: each information source i exerts independent control over one specific component $F_i$ of the data signal $G_\lambda$. The on-voltage given by Eq. (9) constitutes the upper limit, not only for the pulsed waveforms considered by Alt and Pleshko, but also for any multiplexed addressing scheme so defined. Thus, no other set of orthogonal strobe functions, such as sinusoids of different frequencies or Walsh functions of different sequencies, can lead to better performance. We will show next that this limit can be exceeded if the functional dependence of $a_i^{(\lambda)}$ is more general than in Eq. (8).

III. NON-MULTIPLEXED ADDRESSING

We assume that the signals $F_i$ which we continue to call "strobe signals" are orthogonal and of equal rms value $\bar{F}$ and that the "data signals" $G_\lambda$ are linear combinations of the $F_i$, i.e., that Eqs. (6) and (7) hold. In contrast to Eq. (8), we will generalize the coefficients $a_i^{(\lambda)}$ by allowing them to depend on the collective information state λ. It can be shown that for all λ with a given total number of on-elements

$$n = \sum_{i=1}^{N} \eta_i , \tag{10}$$

the coefficients $a_i^{(\lambda)}$ may, without loss of of generality, be assumed to be identical for all on-elements and for all off-elements. Hence, we may express the dependence on λ by writing

$$a_i^{(\lambda)} = \begin{cases} b_n & \text{if } i \text{ is in the off-state} \\ c_n & \text{if } i \text{ is in the on-state} . \end{cases} \tag{11}$$

The rms voltages across off- and on-elements in a column addressed by $G_\lambda$ are then calculated from Eq. (4) as

$$\bar{U}_{off}^{(n)} = \bar{F} \cdot \left[ 1 + \left(c_n\sqrt{n}\right)^2 + \left\{ b_n\sqrt{N-n} - \frac{1}{\sqrt{N-n}} \right\}^2 - \frac{1}{N-n} \right]^{1/2}$$

$$\bar{U}_{on}^{(n)} = \bar{F} \cdot \left[ 1 + \left(b_n\sqrt{N-n}\right)^2 + \left\{ c_n\sqrt{n} - \frac{1}{\sqrt{n}} \right\}^2 - \frac{1}{n} \right]^{1/2} . \tag{12}$$

In the first Eq. (12), n runs from 0 to (N-1) and in the second one from 1 to N. Using the method of Lagrange multipliers we find the following expression for $\bar{U}_{on,max}^{(n)}$ under the constraint $U_{off}^{(n)}=1$:

$$\bar{U}_{on,max}^{(n)} = \left[ 1 + \frac{N}{\sqrt{n(N-n)(N-1)}-n} \right]^{1/2} , \quad (0<n<N-1) . \tag{13}$$

Equation (13) is plotted in Fig. 1 as a function of n for several values of N as a parameter. It is clear that $\bar{U}_{on,max}^{(n)}$ diverges at n=0, where none of the elements is selected, and at n=N-1, where every element in the column is turned on except one. The first case is trivial; the second case is the basis of several special purpose displays such as electrically alterable reticles,[2] bargraphs,[3] pointers[4] and oscilloscopes[5,6] where only patterns with n=N-1 are allowed. In all of these devices, the voltage applied to the single selected element in each column is zero

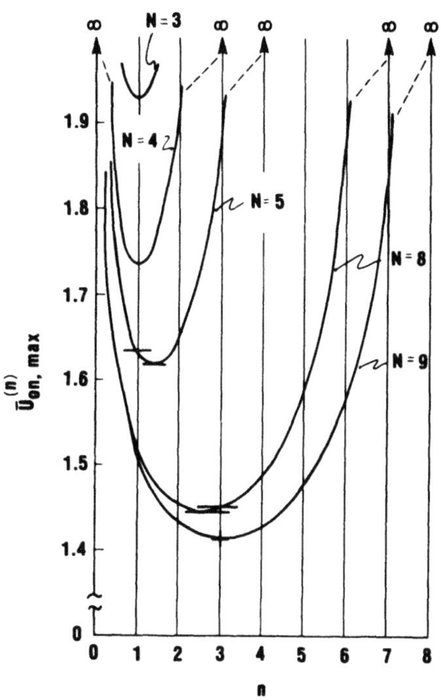


Fig. 1. Maximum voltage across a selected element normalized to threshold versus the number n of selected elements in the given column, with the number N of matrix rows as a parameter.

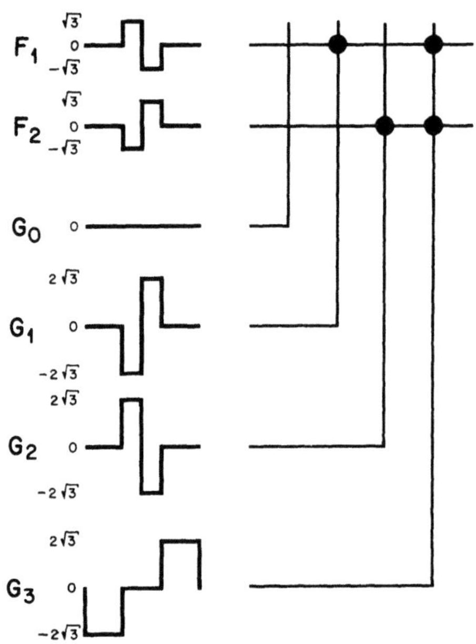


Fig. 2. Practical scheme for optimum addressing of matrix with two rows and any number of columns. Voltage ratio between selected and unselected elements is 3.0.

while the voltage on the other elements can be as large as desired.

For intermediate values of n the maximum selection voltage ratio is finite, with a minimum at

$$n_o = \frac{1}{2}(N-\sqrt{N}) \ . \tag{14}$$

The minimum value of $\bar{U}^{(n)}_{on,max}$ is identically the Alt and Pleshko limit given by Eq. (9). We see that multiplexed addressing, characterized by Eqs. (6), (7) and especially (8), could be represented as a horizontal straight line passing through the minimum of the curve in Fig. 1 for the appropriate N: $\bar{U}_{on}$ for multiplexing is constant for all n and equal to the worst-case $\bar{U}^{(n)}_{on,max}$.

Evidently, patterns near a minimum in Fig. 2 are much harder to display than others, so attainable performance could be significantly improved if these "worst-case" patterns could be avoided. For instance, eliminating the n=1 pattern from an N=4 matrix can improve its performance as shown by Fig. 1 to the level obtained by multiplexing an N=3 matrix. The seven-segment numeric format provides considerable latitude in this regard because only 10 out of $2^7$ possible patterns are actually used to display numerals. It is possible[7] to connect the seven segments into a matrix in such a way that the worst-case patterns do not occur.

Even when all possible patterns must be displayed, it is often possible in principle to improve on the multiplexing limit by recognizing that n is a discrete variable. Equation (13) and Fig. 1 have physical significance only at integer values of n. Therefore, as long as the minimum of the curve in Fig. 1 does not fall at an integer value of n, the theoretical worst case is never encountered in practice and addressing schemes exist which are better than "optimum" multiplexing.

The amount of improvement thus obtainable is seen in Fig. 1 as the difference between $\bar{U}^{(n)}_{on,max}$ at $n_o$ and at the nearest integer value of n. These values are compared in Table 1 for N=2 through N=9. No improvement is possible when N is the square of an integer, because $n_o$ from Eq. (14) is then also an integer. The improvement is significant only for N=2. It is shown elsewhere[8] that these results are not restricted to orthogonal strobe functions but are completely general. No addressing scheme exists which can display all possible patterns on a matrix with larger $\bar{U}_{on}$.

A practical implementation of optimized addressing for a matrix with N=2 is illustrated in Fig. 2. The two strobe signals and the corresponding column signals which produce all four possible patterns are shown. Computing the rms potential difference across each display element reveals that every unselected element has a voltage normalized to threshold of 1.0, and every selected element has a normalized voltage of 3.0. This is a 24% improvement over multiplexing and, compared to direct addressing, entails a minimal sacrifice in performance to approach a 50% reduction in lead count.

## IV. SUMMARY

For the uniform display of all possible patterns on an rms-responding matrix, the ultimate limits for selection voltage ratio as a function of matrix complexity have been found (Table 1). No addressing scheme exists which improves substantially on Alt and Pleshko multiplexing except in a two-line matrix. For this special case, a practical scheme is given which provides a 24% improvement. Enhanced performance in more complex matrices must be sought instead through optimization of display materials, device parameters and optical design.

For displays of restricted patterns, the addressing scheme remains a fruitful object of ingenuity. Figure 1 suggests, however, that success is likely only with very simple patterns or simple matrices.

Table 1. Ratio of RMS Voltages Across Selected and Unselected Elements

| Number of Rows | 2 | 3 | 4 | 5 | 6 | 7 | 8 | 9 |
|---|---|---|---|---|---|---|---|---|
| RMS Multiplex | 2.414 | 1.932 | $\sqrt{3}$ | 1.618 | 1.543 | 1.488 | 1.447 | $\sqrt{2}$ |
| Ultimate Limit | 3.0 | 2.0 | $\sqrt{3}$ | 1.633 | 1.545 | 1.489 | 1.450 | $\sqrt{2}$ |

## REFERENCES

1. P. M. Alt and P. Pleshko, Scanning Limitations of Liquid Crystal Displays, IEEE Trans. Electron Dev. ED-21:146 (1974).
2. C. H. Gooch, R. Bottomley, J. J. Law and H. A. Tarry, J. Phys. E6:485 (1973).
3. S. Sherr, A Liquid Crystal Bar Graph Meter, SID Symp. Digest 7:42 (1976).

4. W. G. Freer, Current Liquid Crystal Display Technology, Microelectronics and Reliability, Pergamon, UD, 15:315 (1976).
5. F. Ueda, H. Arai and H. Amagasaki, German Patent Appl. OS 2414 608 (1974).
6. I. A. Shanks and P. A. Holland, Addressing Methods for Non-Multiplexed Liquid Crystal Oscilloscope Displays, SID Symp. Digest 10:112 (1979).
7. A. Tsuzuki, H. Fujita and S. Morokawa, German Patent Appl. OS 2727 010 (1977).
8. J. Nehring and A. R. Kmetz, Ultimate Limits for Matrix Addressing of RMS-Responding Liquid-Crystal Displays, IEEE Trans. Electron Dev. ED-27:795 (1979).

# ACCURATE OPTICAL MEASUREMENT OF SMALL TILT ANGLES IN THIN TWISTED NEMATIC LAYERS

H. Birecki and F. J. Kahn

Hewlett-Packard Laboratories
Palo Alto, California 94304

## INTRODUCTION

Liquid crystal (LC) molecules have a preferred orientation on a bounding surface. The angle, $\alpha_o$, that they form with the surface in the absence of applied fields is called the initial tilt angle and is an important measure of the liquid crystal-surface interaction.

Tilt angle has also an important influence on the performance of twisted nematic liquid crystal displays (TN-LCDs) because the electrooptical threshold and angular viewing uniformity decrease as the tilt angle is increased.[1,2] Thus it is important to be able to measure tilt accurately. Low initial tilt angles ($\lesssim 10°$) are of particular interest for multiplexed TN-LCDs.

It is also advantageous to have a simple technique which is highly independent of specific material properties of the liquid crystal, which requires neither electric nor magnetic fields, and which is applicable to the thin liquid crystal layers ($\lesssim 10$ μm) commonly used in display devices. The method described in this paper requires neither electric nor magnetic fields and permits accurate measurements of low initial molecular tilts in both twisted and uniformly oriented LC layers as thin as 3 μm.

## PREVIOUS TILT MEASUREMENTS

Several techniques have been developed for measuring tilt.[3-8] They can be divided into four major groups according to whether electric or magnetic fields are required, and whether electrical or optical properties are used to probe the orientation of the liquid crystal layer.

Only surface probing methods such as reflective ellipsometry acutally measure tilt at the boundaries of the liquid crystal layer. All other techniques measure a tilt averaged over the liquid crystal layer thickness,[9] and thus yield surface tilt only in case of uniform tilt through the LC layer.

Tilt varies throughout the cell thickness in twisted cells due to a competition between the splay and bend elastic energies and the twist elastic energy.[9] For small tilts and twists, ≤90°, this variation is calculated to be small. Doping the liquid crystal with a cholesteric compound, as is commonly done in TN-LCDs,[10] stabilizes the twisted structure, thereby decreasing tilt variation.

Magnetocapacitive measurements[4,5] of tilt have been particularly attractive because they do not require the liquid crystal's material constants or even the LC layer thickness to be known quantitatively. However, as with other magnetic field methods, the relatively high magnetic fields required result in geometries limiting the size and shape of cells which can be studied.

G. Baur, W. Wittwer and D. W. Berreman[11] proposed an optical measurement method which, being based on finding the geometry for which optical phase retardation is extremal, should not be very sensitive to actual values of indices of refraction. No detailed analysis of the retardation extrema method nor factors influencing its accuracy has been published. However, the original report on this method indicated that its use was limited to LC layers with thicknesses greater than about 10 μm. This would greatly reduce its usefulness for measuring properties of many display devices.

We believed the phase extrema method to be sufficiently simple and accurate to merit a detailed evaluation of its advantages and limitations. We have carried out this evaluation and find that by varying the wavelength of the incident light we can make accurate measurements of small tilt angles even in liquid crystal layers thinner than 10 μm.

## RETARDATION EXTREMA METHOD ANALYSIS

The retardation extrema method can be understood by considering a twisted nematic layer in which the nematic pitch is long compared to the wavelength of light. The polarization of the light will then be quided by the nematic twist.[12] In that case the incidence angle $\Psi$ for which the phase difference between light propagating in the ordinary and extraordinary modes is extremal (symmetry offset angle) is given by

$$0 = \int_0^d \left\{ \left(n_e^2 - n_o^2\right) \sin\alpha\cos\alpha\cos\beta - \frac{n_o}{n_e} \cdot \frac{\left[n_e^2 - \left(n_e^2 - n_o^2\right)\cos^2\alpha\sin^2\beta\right]\sin\Psi}{\sqrt{n_o^2\cos^2\alpha + n_e^2\sin^2\alpha - (\sin^2\Psi)\left[1 - \left(1 - \frac{n_o^2}{n_e^2}\right)\cos^2\alpha\sin^2\beta\right]}} + \frac{\left(n_o^2\cos^2\alpha + n_e^2\sin^2\alpha\right)\sin\Psi}{\sqrt{n_o^2 - \sin^2\Psi}} \right\} dz \ , \qquad (1)$$

where $\alpha(z)$ is the tilt angle, $\beta(z)$ is the azimuth of liquid crystal molecules measured relative to the plane of incidence, $n_e$ is the extraordinary and $n_o$ the ordinary index of refraction, and d is the layer thickness. In all further discussion we will assume constant tilt and uniform twist throughout the liquid crystal layer.

For twist symmetrical around the incidence plane, Eq. (1) can be solved in the limit of small tilts to give

$$\sin\alpha_o = \frac{\beta_o \frac{n_e - n_o}{n_e + n_o} + \sin\beta_o}{4 n_e \sin(\beta_o/2)} \sin\Psi \ , \qquad (2)$$

where $\beta_o$ is the total twist. This further reduces to $\alpha_o \approx (2n_e\sqrt{2})^{-1}\sin\Psi$ for $\beta_o = \pi/2$, $(n_e - n_o)/(n_e + n_o) \ll 1$ and $\alpha_o < 10°$.

The symmetry offset angle $\Psi$ can be determined experimentally from a plot of transmission versus angle of incidence. The experimental setup and geometry are shown in Fig. 1. The incidence plane bisects the twist ($\phi_o = 0$) and the sample is positioned between crossed polarizers with the polarizer axes in the incidence plane or perpendicular to it. Typical experimental transmission curves are shown in Fig. 2.

Each transmission curve has the following characteristics at the symmetry offset angle $\Psi$: (1) it has an extremum at this point which is the only extremum with value intermediate between maximum

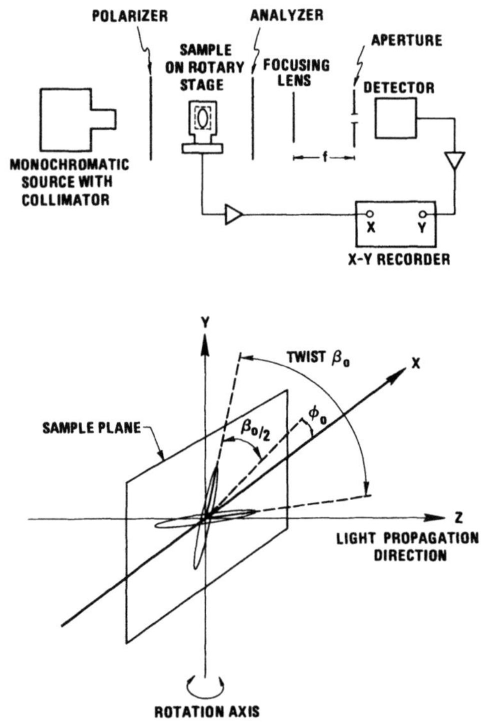


Fig. 1. Experimental setup and geometry. For details see text.

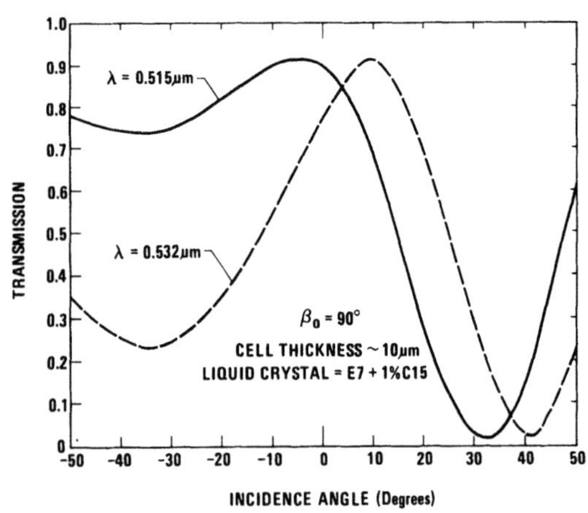


Fig. 2. Typical experimental transmission curves.

and minimum transmission; (2) for small tilts the curve is roughly symmetrical around that point; (3) the angle $\Psi$ is independent of the layer thickness and is insensitive to changes in wavelength of the incident light. This last characteristic makes it possible to optimize the measurement by changing the wavelength of light and thus making the extremum sharper. It also enables unambiguous identification of the transmission extremum corresponding to the extremal phase difference.

The numerical solutions of Eq. (1) are shown in Fig. 3 for three different twists and typical liquid crystal refractive indices ($n_o$=1.52, $n_e$=1.75). Figures 4, 5 and 6 show that the tilt measurement is relatively insensitive to the liquid crystals optical constants, and to small alignment errors. The tilt measurement is most sensitive to the actual value of the ordinary refractive index and to the errors ($\phi_o$) in alignment of the sample with plane of incidence bisecting the twist. The refractive index can be measured and the alignment error is easy to minimize.

The relative insensitivity of this tilt measurement to the liquid crystal material constants enables the use of universal curves such as those in Fig. 3 to determine the tilt of liquid crystal layers from transmission curves. We have used this technique to determine tilt in several LC layers of about 10 μm thickness and with tilts in the range of 0 to 7° and compared the results with those obtained using magnetocapacitive null technique of Ref. 4. The results agreed within experimental errors ($\lesssim$1°) arising principally from the magnetocapacitive null method.

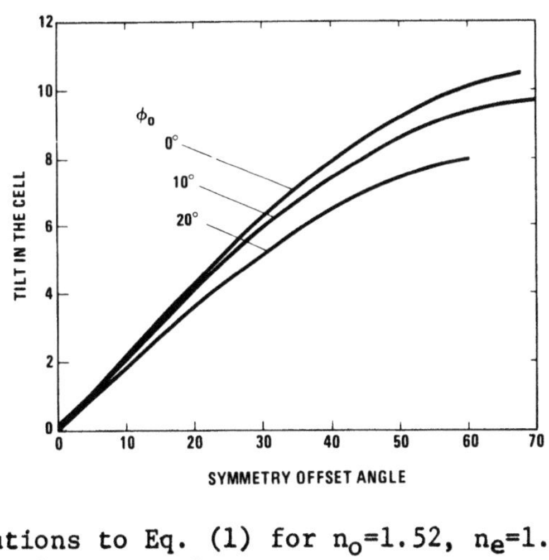


Fig. 3. Solutions to Eq. (1) for $n_o$=1.52, $n_e$=1.75 ($\phi_o$=0) for several twist values.

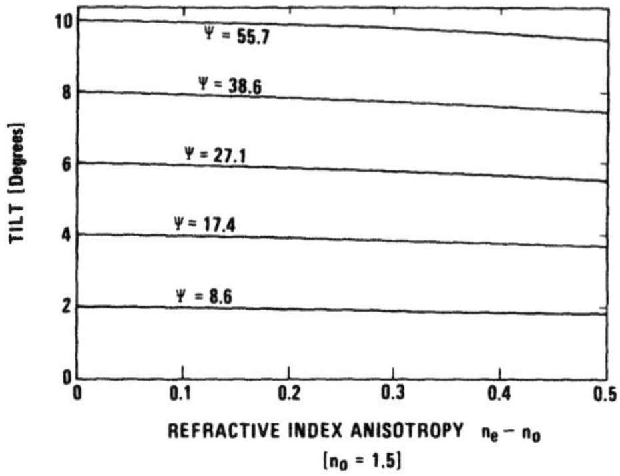


Fig. 4. Dependence of tilt determined from Eq. (1) on the refractive index anisotropy for several values of symmetry offset angle Ψ and twist=90°.

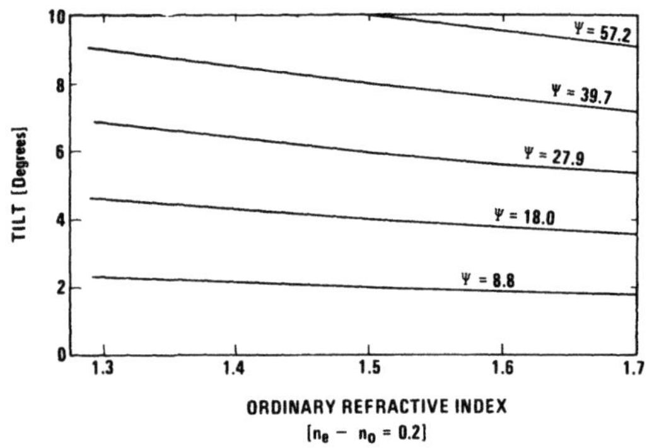


Fig. 5. Dependence of tilt determined from Eq. (1) on the ordinary refractive index for several values of symmetry offset angle (twist=90°).

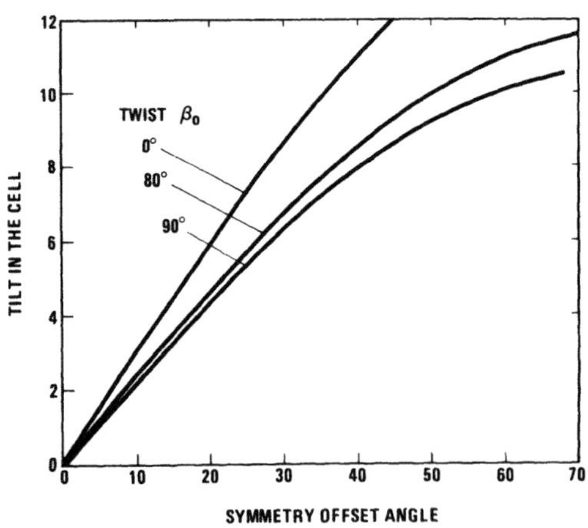


Fig. 6. Effects of misalignment of the sample on tilt determination from Eq. (1).

ACCURACY FOR THIN LC LAYERS

The question arises as to the accuracy of the retardation extrema method for measuring tilts in thin liquid crystal layers. It could be expected that, as the nematic pitch decreases due to decreasing thickness of a TN layer, the polarization guidance is lost and formula (1) becomes invalid. By numerical solution of Maxwell's equations for a twisted nematic structure, we have found that the error in tilt determination due to the loss of polarization guidance is less than 0.3° for a 1 μm thick LC layer with tilt $\alpha_o$=9° and twist $\beta_o$=90°.

The major source of error for thin LC layers is due to depolarization from reflection at glass-liquid crystal interfaces and to intensity modulation caused by the reflection at glass-air interfaces. The latter source of error can be corrected relatively easily, while correcting for the former is impractical. Measurements done on thin layers are most severely affected since the thinner the layer, the weaker the variation of the transmission versus angle curve.

Figure 7 shows the errors in tilt determined by applying formula (1) to calculated transmission curves for various LC layer

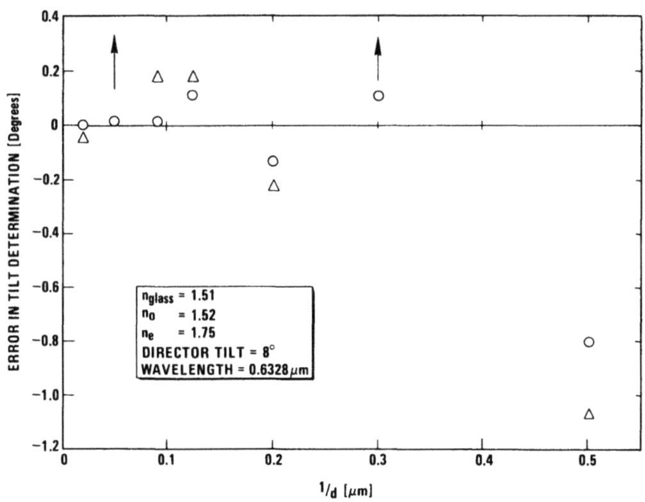


Fig. 7. Numerically computed errors in tilt determination resulting from Eq. (1). Results plotted are for 8° tilt and 90° twist as a function of liquid crystal layer thickness. Triangles are the errors without correction for air-glass reflection, circles are the errors after correcting and arrows correspond to cases where an extrenum is not observed at the symmetry offset angle. For details, see text.

thicknesses. The polarization of light for this calculation was in the plane of incidence, and therefore the Brewster peak in transmission strongly influenced the results. The extremum in the transmission curve for the 3.3 μm thick layers was shifted almost to the Brewster's angle and a weak extremum for a 20 μm cell disappeared completely with only one of the two neighboring extrema in the transmission curve remaining. Figure 7 shows that a simple correction for the glass air interface reflection results in measurement accuracies of better than ±0.2° for 8° of tilt and LC layers thicker than 3 μm. For lower tilts the errors are smaller.

## CONCLUSION

We have shown that the retardation extrema method of measuring molecular tilt is applicable to small tilt angles (this method can also be used for very large tilt angles (small deviations from homeotropic alignment)) (less than 11° for 90° twist) and that, contrary to a previous report,[11] this method can yield accurate measurements of tilt even in thin (≥3 μm) liquid layers. Measurement conditions can be optimized by changing the wavelength of the incident light, thereby obviating the need for detailed calculation of conoscopic patterns.

Advantages of the retardation extrema method are that neither electric nor magnetic fields are necessary and that the results do not depend strongly on liquid crystal material parameters, thereby enabling use of universal curves for data analysis. The accuracy of the method can be further improved by a simple correction for the reflections at the glass-air interfaces.

## ACKNOWLEDGMENTS

The authors wish to thank the Hewlett-Packard Corvallis Division for providing several of the cells used in this work.

## REFERENCES

1. L. A. Goodman, D. Meyerhofer and S. Digiovanni, IEEE Trans. Electr.
2. F. J. Kahn, 1978 SID International Symposium, Digest of Technical Paper, p. 74.
3. W. A. Crossland, J. H. Morrissy and B. Needham, J. Phys. D9:2001 (1976).
4. K. Toriyama and T. Ishibashi in: "Nonemissive Electrooptic Display," edited by A. R. Kmetz and F. K. von Willisen (Plenum N.Y., 1976).

5. F. J. Kahn, Mol. Cryst. Liq. Cryst. 38:109 (1977).
6. R. W. Gurtler and J. W. Casey, Mol. Cryst. Liq. Cryst. 35:275 (1976).
7. T. Y. Scheffer and J. Nehring, J. Appl. Phys. 48:1983 (1977).
8. K. Suzuki and K. Toriyama, Appl. Phys. Lett. 33:561 (1978).
9. T. J. Scheffer, J. Appl. Phys. 49:5835 (1978).
10. E. P. Raynes, Electron. Lett. 10:141 (1974).
11. G. Baur, V. Wittwer and D. W. Berreman, Phys. Lett. 56A:142 (1976).
12. P. G. deGennes, "The Physics of Liquid Crystals," (Claredon Press, Oxford, 1974) p. 220.

# EFFECTS OF CELL AND MATERIAL PROPERTIES ON MULTIPLEXING LEVELS OF TWISTED NEMATIC LIQUID CRYSTAL DISPLAYS

H. Birecki and F. J. Kahn

Solid State Laboratory
HP Laboratories
Palo Alto, California 94304

## INTRODUCTION

Multiplexed twisted nematic liquid crystal devices (TN-LCDs) are the mainstay of low power, medium information content displays. At present, the voltage threshold and the nonlinearity of the electrooptic response intrinsic to the twisted nematic structure are used to achieve multiplexing. The multiplexing level achievable using this technology is limited by the sharpness of the electrooptic response and by the variations of this response with viewing angle.[1]

In order to attain high multiplexing levels, one can vary several liquid crystal material and cell design parameters. In practice, it is very difficult to vary one material parameter without affecting all others. Thus, experimental work provides little insight into the physical origins of the dependences of multiplexing levels on these parameters. However, the electrooptic response of a twisted nematic structure can be numerically calculated.[2-5] One can thus study the effects of these parameters by performing computer calculations. The drawback of such an approach is that numerical calculations generally do not provide a physical insight into the reasons behind observed trends. Thus, one tends to question the validity of such studies as the results may not be widely applicable.

In this paper, we attempt to provide a physical insight into the dependence of multiplexing levels on the material and design parameters. This insight is tested against a computer calculation of the transmission-voltage characteristic of a typical TN-LCD where one parameter at a time is varied.

The calculations of the electrooptic response of TN-LCDs are usually conducted on high speed computers with large memories. By reducing the complexity of optical calculations and by solving all equations in terms of the natural variable $\lambda$ (see Appendix) of the twisted nematic structure, we were able to develop a computer program for calculation of the static electrooptic response of TN-LCDs which can run on a desktop computer (HP 9845) without sacrificing the accuracy of computations. The program was checked against the calculations of Ref. 5 and gave identical results. For completeness, we give in the Appendix the theoretical basis for this program.

## PARAMETERS AND RESULTS OF COMPUTATION

The study reported here considers only the effects in the principal viewing plane of a transmissive liquid crystal display and neglects the effects of temperature. In this case, the multiplexing level N is determined[1] by the merit figure

$$M = V_{90}(40°)/V_{50}(0°) \tag{1}$$

through the equation

$$N = \left(\frac{1+M^2}{1-M^2}\right)^2 \tag{2}$$

where $V_x(\theta)$ is defined as the voltage at which the transmission of light at incidence angle $\theta$ is x% of the transmission with no voltage applied. In order to describe the constraints imposed by the viewing angle requirements, we use the angular merit figure

$$M_\theta = V_{90}(40°)/V_{90}(0°) \ . \tag{3}$$

And in order to describe sharpness of the transmission versus voltage curve at normal incidence, we use the voltage merit figure

$$M_V = V_{90}(0°)/V_{50}(0°) \ . \tag{4}$$

As the baseline for our calculations described in the Appendix, we take the typical LCD cell configuration of thickness d=8 μm filled with E7 liquid crystal mixture with a cholesteric additive giving a natural cholesteric pitch P=40 μm. The boundary conditions imposed on the liquid crystal are: total twist $\beta_0$=90°, pretilt $\alpha_0$=3° and strong surface anchoring. The polarizers are crossed and parallel to the nematic director at the neighboring liquid crystal surfaces. In calculations where the total twist

$\beta_0$ was varied, polarizers remained crossed at 90° and were oriented symmetrically with respect to the principal viewing plane (see Fig. 1).

In the available literature, we find the following set of material constants for E7:

splay elastic constant[6] $K_1 = 10.7 \times 10^{-7}$ dynes[7]
twist elastic constant $K_2 = 10 \times 10^{-7}$ dynes[7]
bend elastic constant $K_3 = 20.7 \times 10^{-7}$ dynes[7]
dielectric constant normal to the director $\varepsilon_0 = 5.65$[7,8]
dielectric anisotropy $\varepsilon_a = 13.5$[7,8]
ordinary refractive index $n_0 = 1.5222$[7]
refractive index anisotropy $n_a = 0.224.$[7]

The light wavelength used was 0.55 μm. Since it lies in the middle of the visible range, it should be representative of white light illumination, especially as the calculations were designed to eliminate interference due to reflections. The index of refraction of glass was taken to be 1.51.

The liquid crystal parameters: $K_2$, $K_3/K_1$, $\varepsilon_a$, $n_a$ and cell design parameters: $\beta_0$, $\alpha_0$, d and the orientation of polarizers were varied one at a time. The ratio $K_3/K_1$ was chosen as a parameter in order to keep the Freedericksz transition threshold voltage[9] constant while varying the elastic constants. Variations in the threshold voltage due to the natural cholesteric pitch were not compensated for. Several calculations were also conducted for the case of infinite cholesteric pitch (no cholesteric compound in the liquid crystal mixture).

The resulting figures of merit and the multiplexing levels are tabulated in Tables 1-8. Typical calculated transmission curves are shown in Fig. 2.

## PHYSICAL ORIGINS OF TRENDS IN MERIT FIGURES

In order to understand the trends in merit figure dependences on the elastic constants and on the total twist of the cell, one has to realize that the transmission versus voltage curve at normal incidence depends crucially on the azimuthal distribution of the director within the nematic layer. For normal incidence, the minimal tranmission occurs at high voltages in 90° twist cells. At these voltages, all molecules are in two mutually orthogonal planes,[2] and the birefringence of one-half of the twisted nematic layer compensates the birefringence of the other half.[10] For higher $\beta_0$, this compensation occurs at lower voltages when the molecules of the two half-layers are on the average in orthogonal planes. For $\beta_0$ less than 90°, the birefringence

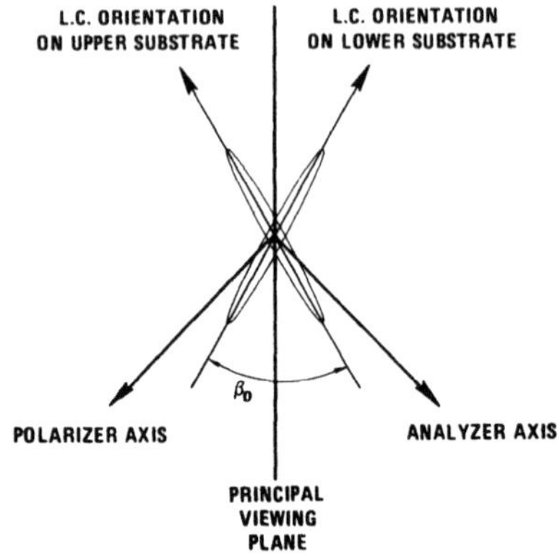


Fig. 1. Projection of the TN-LCD geometry on its substrate plane showing the total director twist.

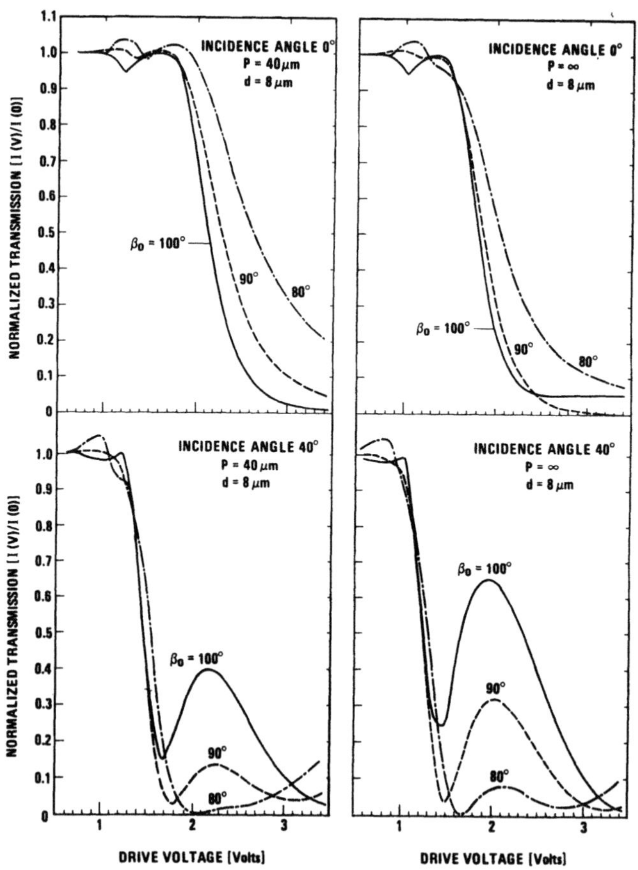


Fig. 2. Effects of total twist $\beta_0$ and natural cholesteric pitch P on the transmission-voltage characteristic of a typical TN-LCD.

Table 1. Dependence of Figures of Merit on the Total Twist $\beta_o$.

| P[μm] | Total Twist [degrees] | $M_\Theta$ | $M_V$ | M | M |
|---|---|---|---|---|---|
| 40 | 110 | 0.74 | 0.89 | 0.66 | 6.5 |
| | 100 | 0.70 | 0.87 | 0.61 | 4.8 |
| | 90 | 0.65 | 0.84 | 0.54 | 3.3 |
| | 80 | 0.61 | 0.80 | 0.49 | 2.7 |
| | 70 | 0.57 | 0.77 | 0.44 | 2.2 |
| | 60 | 0.56 | 0.74 | 0.41 | 2.0 |
| ∞ | 110** | 0.72 | 0.86 | 0.62 | 5.0 |
| | 100* | 0.70 | 0.87 | 0.61 | 4.8 |
| | 90 | 0.68 | 0.84 | 0.57 | 3.9 |
| | 80 | 0.63 | 0.78 | 0.49 | 2.7 |
| | 70 | 0.62 | 0.72 | 0.45 | 2.3 |
| | 60 | 0.57 | 0.72 | 0.41 | 2.0 |

*Excessive transmitted intensity at 40° incidence angle at voltage equal to $V_{50}(0°)$.

**As * also there is a substantial increase in the transmission of light at high voltages at normal incidence which increases $V_{50}(0°)$ and thus decreases $M_V$.

Table 2. Dependence of figures of merit on the ratio of bend to splay elastic constants $K_3/K_1$.

| P[μm] | $K_3/K_1$ | $M_\Theta$ | $M_V$ | M | N | $K_1 \times 10^7$ [dynes] |
|---|---|---|---|---|---|---|
| 40 | 2 | 0.65 | 0.83 | 0.54 | 3.3 | 10.6 |
| | 1 | 0.67 | 0.84 | 0.56 | 3.7 | 12.7 |
| | 0.5 | 0.70 | 0.82 | 0.57 | 3.9 | 14.1 |
| | 0.2 | 0.73 | 0.83 | 0.61 | 4.8 | 15.1 |
| ∞ | 2 | 0.67 | 0.85 | 0.57 | 3.9 | 10.6 |
| | 1 | 0.73 | 0.84 | 0.62 | 5.2 | 12.7 |
| | 0.5 | 0.74 | 0.86 | 0.64 | 5.7 | 14.1 |
| | 0.2 | 0.82 | 0.86 | 0.70 | 8.7 | 15.1 |

Table 3. Dependence of figures of merit on the twist elastic constant $K_2$.

| P[μm] | $K_2 \times 10^7$ [dynes] | $M_\Theta$ | $M_V$ | M | N | $K_3$, $K_1 \times 10^7$ [dynes] |
|---|---|---|---|---|---|---|
| 40 | 50 | 0.63 | 0.83 | 0.52 | 3.0 | 46.8, 24.2 |
| | 20 | 0.64 | 0.83 | 0.53 | 3.2 | 27.2, 14.1 |
| | 10 | 0.65 | 0.84 | 0.54 | 3.3 | 20.7, 10.7 |
| | 5 | 0.67 | 0.84 | 0.56 | 3.7 | 17.4, 9.0 |
| | 2 | 0.71 | 0.86 | 0.61 | 4.8 | 15.5, 8.0 |
| ∞ | 50 | 0.55 | 0.78 | 0.43 | 2.1 | 46.8, 24.2 |
| | 20 | 0.64 | 0.82 | 0.52 | 3.0 | 27.2, 14.1 |
| | 10 | 0.66 | 0.84 | 0.55 | 3.5 | 20.7, 10.7 |
| | 5 | 0.69 | 0.85 | 0.59 | 4.3 | 17.4, 9.0 |
| | 2 | 0.72 | 0.86 | 0.62 | 5.2 | 15.5, 8.0 |

Table 4. Dependence of figures of merit on the director pretilt.

| P[μm] | Director Pretilt [degrees] | $M_\Theta$ | $M_V$ | M | N |
|---|---|---|---|---|---|
| 40 | 0 | 0.66 | 0.84 | 0.55 | 3.5 |
| | 3 | 0.65 | 0.84 | 0.54 | 3.3 |
| | 7 | 0.67 | 0.83 | 0.56 | 3.7 |
| | 10 | 0.68 | 0.84 | 0.57 | 3.9 |
| ∞ | 0 | 0.69 | 0.83 | 0.57 | 3.9 |
| | 3 | 0.66 | 0.84 | 0.55 | 3.5 |
| | 7 | 0.68 | 0.83 | 0.56 | 3.7 |
| | 10 | 0.67 | 0.83 | 0.55 | 3.5 |

Table 5. Dependence of figures of merit on the dielectric anisotropy.

| P[μm] | $\varepsilon_a$ | $M_\Theta$ | $M_V$ | M | N |
|---|---|---|---|---|---|
| 40 | 0.5 | 0.67 | 0.80 | 0.54 | 3.3 |
| | 2.0 | 0.69 | 0.81 | 0.56 | 3.7 |
| | 13.5 | 0.65 | 0.84 | 0.54 | 3.3 |
| | 50.0 | 0.67 | 0.86 | 0.58 | 4.1 |

Table 6. Dependence of figures of merit on the refractive index anisotropy.

| P[μm] | $n_a$ | $M_\Theta$ | $M_V$ | M | N |
|---|---|---|---|---|---|
| 40 | 0.30 | 0.65 | 0.82 | 0.53 | 3.2 |
| | 0.22 | 0.65 | 0.84 | 0.54 | 3.3 |
| | 0.15 | 0.69 | 0.85 | 0.59 | 4.3 |
| | 0.1 | 0.76 | 0.86 | 0.66 | 6.5 |

Table 7. Dependence of figures of merit on the liquid crystal layer thickness.

| P[μm] | d[μm] | $M_\Theta$ | $M_V$ | M | N |
|---|---|---|---|---|---|
| 40 | 6 | 0.71 | 0.86 | 0.61 | 4.8 |
| | 9 | 0.65 | 0.84 | 0.54 | 3.3 |
| | 12 | 0.66 | 0.79 | 0.53 | 3.1 |
| | 15 | 0.70 | 0.71 | 0.50 | 2.8 |
| ∞ | 6 | 0.70 | 0.86 | 0.60 | 4.6 |
| | 9 | 0.67 | 0.83 | 0.56 | 3.7 |
| | 12 | 0.71 | 0.82 | 0.58 | 4.1 |
| | 15* | 0.71 | 0.84 | 0.60 | 4.6 |

*Excessive transmitted intensity at 40° incidence angle at $V_{50}(0°)$.

Table 8. Dependence of figures of merit on the orientation of polarizers axes with respect to the nematic director at neighboring interfaces.

| Polarizer Orientation | $M_\Theta$ | $M_V$ | M | N |
|---|---|---|---|---|
| ∥ to director | 0.65 | 0.84 | 0.54 | 3.3 |
| ⊥ to director | 0.72 | 0.84 | 0.61 | 4.7 |

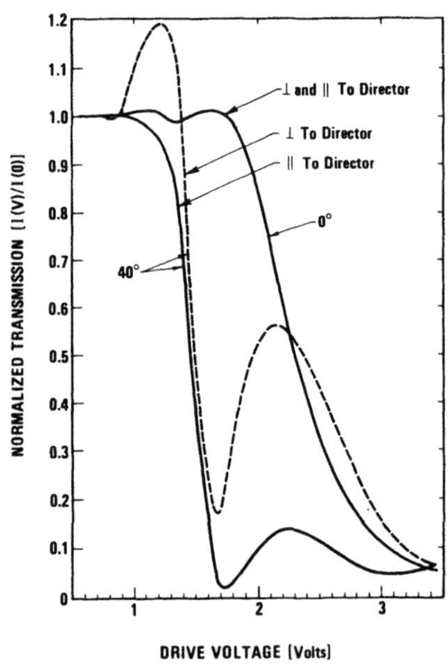


Fig. 3. Effects of polarizers orientation with respect to nematic directors at neighboring interfaces on the transmission-voltage characteristic.

compensation can never be achieved. In such cells, the decrease in transmitted intensity is associated with the formation of a Maltese cross conoscopic image,[10,11] and occurs at much higher voltages. Thus, $V_{50}(0°)$ is highly dependent on the total twist of the nematic director and increases with decreasing $\beta_0$. $V_{90}(0°)$ is expected to behave in the same fashion, but as it is more sensitive to the intial rearrangement of the director twist and tilt in the middle of the nematic layer, it varies slower than $V_{50}(0°)$ with change in $\beta_0$. Moreover, it changes faster than $V_{90}(40°)$ as the latter is more sensitive to the director tilt in the center of the cell than to total twist. Thus, both $M_\theta$ and $M_V$ should increase with increasing twists as is the case (Table 1). In practice, improvement in the figures of merit by increasing $\beta_0$ is limited by the fact that at off normal incidence, there is a "bump" in the transmission versus voltage curves which becomes large at large $\beta_0$ (see Fig. 2).

The dependence of transmission versus voltage characteristics of TN-LCDs on the cell twist $\beta_0$ was studied experimentally.[12] However, the definition of cell twist of that reference appears to be complementary to our $\beta_0$, i.e.,

$$\text{cell twist}_{[12]} = \frac{\pi}{2} - \beta_0 \ .$$

Otherwise, the changes in general appearance of the transmission-voltage curves with varying $\beta_0$ look identical to those calculated here (Fig. 2).

As the director twist distribution through the cell thickness $\beta(z)$ depends predominantly on the twist elastic constant and the natural cholesteric pitch, it is understandable that the sharpness of the transmission versus voltage curve for normal incidence does not depend significantly on the $K_3/K_1$ ratio when $K_2$ and p are constant. The initial deformation of the twisted nematic layer is that of splay. Thus, increased splay elastic constant decreases the initial deformation and shifts $V_{90}(40°)$ to higher voltages. As $V_{90}(40°)$ is more sensitive to small deformations than $V_{90}(0°)$, the angular merit figure $M_\theta$ increases with decreasing $K_3/K_1$ ratio which explains the trend observed in Table 2.

The dependence of merit figures on the twist elastic constant (Table 3) is analogous to the dependence of total twist. As $K_2$ increases, the director twist distribution is stabilized since small rearrangements lead to large changes in energy. Decreasing $K_2$ is thus equivalent to increasing the total twist and the observed increase in figures of merit is expected.

We do not observe clearly defined trends in the dependence of figures of merit on the director pretilt $\alpha_0$ (Table 4). At the relatively small pretilt values studied, the pretilt provides a minor bias to the nematic structure which results in a small, almost parallel shift of the transmission versus voltage curves. This shift is too small to significantly affect the figures of merit we chose to discuss. However, even a small pretilt can have an effect on merit figures when incidence angles larger than 40° are considered. At higher pretilts the merit figures decrease.

It is usually expected[13,14] that the merit figure decreases with increasing dielectric anisotropy due to expulsion of electric field from the center of the nematic layer when the molecules reorient in the applied field. However, the expelled field concentrates in the outer parts of the layer driving them harder and thus compensates for decreased field in the center. The angular merit figure $M_\theta$ depends to a large extent on the rate of change of tilt with voltage. Its apparent independence of $\varepsilon_a$ (Table 5) could be attributed to the above compensation.

The sharpness of the transmission versus voltage curve for normal incidence actually increases slightly with increasing dielectric anisotropy. This increased sharpness is due to the fact that the rearrangement of the director twist distribution takes place at relatively high voltages.[10] At these voltages, the director in the center of the nematic layer is nearly normal to the boundaries. Thus, the twist energy is concentrated in the boundary layers. As the field in these layers is increased, due to expulsion of the field from the center of the cell, the rearrangement of the twisted structure toward the structure with molecules in two orthogonal planes is easier to achieve, and higher $M_V$ results.

The dependence of the merit figure on the dielectric anisotropy is a rather sensitive function of the choice of the threshold voltage. Had we selected the Freedericksz transition (capacitance) threshold voltage instead of an optical threshold $V_{90}(40°)$, we would observe a decrease in figure of merit with increasing $\varepsilon_a$, such as reported elsewhere.[15] However, for an optical reponse, the latter seems more relevant.

For higher refractive index anisotropy $n_a$, the geometrical requirements on the nematic structure which need to be satisfied in order to achieve compensation of birefringence become more stringent. Thus, the transmission versus voltage curve for normal incidence shifts toward higher voltages and becomes less steep. At off normal incidence, however, the curves are limited in their shift toward higher voltages because the positions of their minima are determined by the geometrical configuration of the director tilt and twist independently of the refractive indices.[10,16] The net effect is that of decreased $M_\theta$ and $M_V$ as observed in Table 6.

One would expect the above argument to explain also the dependence of the figures of merit on the twisted nematic layer thickness (Table 7). In fact, there is a general trend for $M_V$ to increase with decreasing d. The one exception at infinite P and d=15 μm may be due to interference effects.[15] The angular merit figure $M_\theta$ shows no significant dependence of the layer thickness. This and the above interference is at present not understood and deserves further study.

It is usually accepted that it does not matter whether the polarizers are parallel or perpendicular to the nematic director at bounding surfaces. As Fig. 3 and Table 8 show, this is the case only for normal incidence. In the case of off normal incidence and the polarizers perpendicular to the director, light is being admitted in both the ordinary and extraordinary modes. These modes are being guided separately,[17] and interference occurs at the output. This interference results in our case in the increased angular merit figure $M_\theta$. As our calculation has been performed for monochromatic light and this is an interference effect, it may be averaged out in white light.

In general, the figures of merit are higher when no cholesteric compound is added. This is due to the stabilizing effect of the natural cholesteric pitch on the twist structure. The notable exceptions to this rule are the cases of high twist elastic constant $K_2$ and low $\beta_0$. For large $K_2$, the Freedericksz transition threshold voltage is substantially increased,[18] due to a term which is proportional to $\beta_0 K_2/P$. This increase of the threshold voltage compensates for the decreased sharpness of the transmission versus voltage curve resulting in a higher figure of merit as seen in Table 3. For high $\beta_0$ this increase is masked by the increase in the $\beta_0^2(K_3-2K_2)$ term in the Freedericksz threshold voltage.

The addition of cholesteric compound can be beneficial as it decreases the height of the "bump" in the transmission versus voltage curves at off normal incidence (see Fig. 2). The explanation of this effect is beyond the scope of this paper, however, it follows logically from a new model for operation of TN-LCDs reported elsewhere.[10]

## DISCUSSION AND CONCLUSIONS

The important factors determining the number of lines which can be multiplexed are the birefringence of the liquid crystal and the ability to easily distort its twist structure into two compensating half-layers. The strong dependence of figures of merit on total twist $\beta_0$ and twist distribution $\beta(z)$ explained in the present work is a logical extension of a new model of operation of TN-LCDs reported elsewhere,[10] which provides a better description of transmission patterns than the usually accepted model.[16] The latter stresses the importance of director tilt in the middle of the cell.

In order to obtain a high level of multiplexing, small birefringence, small twist elastic constant, small concentration of cholesteric additives and total twists greater than 90° are desirable. In addition, the higher the $K_3/K_1$ ratio, the wider the expected viewing angle. The director pretilt is relatively unimportant for incidence angles less than 40° and pretilts less than 10°. Contrary to reports of others,[13,14] we find the effects of dielectric anisotropy to be small and that large $\varepsilon_a$ may be desirable because it increases $M_V$ and would allow operation at low voltages.

We were able to explain the observed trends on the basis of general arguments which do not depend on the specific values of parameters used. Thus, we believe the above conclusions to be generally valid.

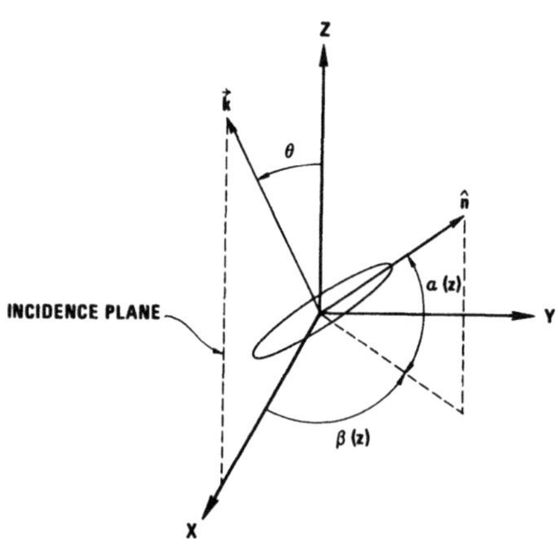


Fig. 4. Coordinate system used in the calculations.

APPENDIX

## Calculation of Static Electrooptic Properties of TN-LCDs

In order to calculate transmission of light through a twisted nematic layer, one needs to know the orientation of molecules at each point. Since we are interested only in static properties of the layer, we can obtain this information by minimizing the free energy.[6]

$$F = \frac{1}{2} K_1 (\nabla\cdot\hat{n})^2 + \frac{1}{2} K_2 \left[\frac{2\pi}{P} + \hat{n}\cdot(\nabla\times\hat{n})\right]^2$$

$$+ \frac{1}{2} K_3 (\hat{n}\times\nabla\times\hat{n})^2 + \frac{1}{8\pi} \vec{\mathscr{E}} \cdot \overleftrightarrow{\varepsilon}(\hat{n}) \cdot \vec{\mathscr{E}} \qquad (A.1)$$

where $\overleftrightarrow{\varepsilon}$ is the dielectric tensor, $\vec{\mathscr{E}}$ is the electric field and $\hat{n}$ is the nematic director. The minimization has to be conducted subject to boundary conditions. In our case, the director tilt angle $\alpha(z)$ is symmetric with respect to the mid-plane of the nematic layer and the director twist azimuth angle $\beta(z)$ satisfies

$$\beta(d-z) = \beta_0 - \beta(z) \ . \qquad (A.2)$$

It is therefore sufficient to find the orientation of molecules only for $z \le d/2$.

Following Leslie[19] we arrive at the following set of equations

$$\partial_z \beta = \frac{\kappa + 2\pi K_2 \cos^2\alpha / P}{g(\alpha)} \qquad (A.3)$$

$$\partial_z \alpha = f(\alpha)^{-1/2} \left[ \frac{D_\perp^2}{4\pi} \left( \frac{1}{\varepsilon_o + \varepsilon_a \sin^2\alpha} - \frac{1}{\varepsilon_o + \varepsilon_a \sin^2\alpha_m} \right) - \frac{(\kappa + 2\pi K_2 \cos^2\alpha / P)^2}{g(\alpha)} + \frac{(\kappa + 2\pi K_2 \cos^2\alpha_m / P)}{g(\alpha_m)} \right]^{1/2} \qquad (A.4)$$

where

$g(\alpha) = [K_2 + (K_3 - K_2) \sin^2\alpha] \cos^2\alpha$
$f(\alpha) = K_1 + (K_3 - K_1) \sin^2\alpha$
$D_\perp$ is the normal component of the displacement vector
$\alpha_m$ is the tilt in the middle of the cell and
$\kappa$ is the azimuthal torque exerted on the bounding surface by the nematic layer.

$D_\perp$, $\alpha_m$ and $\kappa$ are constants of integration and can be found by self-consistent solution of the following set of boundary conditions

$$V = 2D_\perp \int_{\alpha_o}^{\alpha_m} \frac{d\alpha}{[\varepsilon_o + \varepsilon_a \sin^2\alpha]\partial_z\alpha} \tag{A.5}$$

$$d = 2\int_{\alpha_o}^{\alpha_m} \frac{d\alpha}{\partial_z\alpha} \tag{A.6}$$

$$\beta_o = 2\int_{\alpha_o}^{\alpha_m} \frac{\partial_z\beta}{\partial_z\alpha}\, d\alpha \tag{A.7}$$

where V is the applied voltage.

The integrands are divergent at $\alpha_m$ but a substitution of a new variable $\lambda$ such that

$$\sin\lambda = \frac{\sin\alpha}{\sin\alpha_m} \tag{A.8}$$

eliminates the divergences and the above equations can be solved numerically by a nonlinear least squares fit coupled with Gauss-Legendre integration. Variable $\lambda$ is a natural variable for the nematic structure and we shall take advantage of it later.

To calculate the propagation of light through the twisted nematic structure, we start with the 4×4 matrix linear differential equation approach of Berreman[2,20] and Teitler and Henvis.[21] As we were not interested in the interference effects and reflections internal to the twisted nematic layer are negligible, we were able to reduce the calculations to a 2×2 matrix linear differential equation. Further reduction in the amount of computation was achieved by solving this equation in terms of the natural variable $\lambda$ of the twisted nematic structure, rather than in terms of distance.

For completeness, we give here a brief summary of the 4×4 matrix approach:

For light propagating in the free space with wavevector $\vec{k}$, we choose a coordinate system such that $k_y=0$ (see Fig. 4.). In this coordinate system, the electromagnetic vectors in the liquid crystal layer can be written as:

$$\vec{\mathscr{E}} = \vec{E}(z)\ \exp[i(kx\sin\Theta-\omega t)] \tag{A.9}$$

$$\vec{\mathscr{H}} = \vec{H}(z)\ \exp[i(kx\sin\Theta-\omega t)] \tag{A.10}$$

where $\Theta$ is the incidence angle in air.

Vectors $\vec{\mathscr{E}}$ and $\vec{\mathscr{H}}$ satisfy Maxwell's equations:

$$\nabla\cdot\overleftrightarrow{\varepsilon}\cdot\vec{\mathscr{E}} = 0 \tag{A.11}$$

$$\nabla\cdot\vec{\mathscr{H}} = 0 \tag{A.12}$$

$$-i\partial_z\frac{c}{\omega}\begin{pmatrix}E_x\\E_y\\H_x\\H_y\end{pmatrix} = \begin{bmatrix} K\dfrac{\varepsilon_{xz}}{\varepsilon_{zz}} & , K\dfrac{\varepsilon_{yz}}{\varepsilon_{zz}} & , 0 , & \dfrac{K^2}{\varepsilon_{zz}} - 1 \\ 0 & , 0 & , 1 , & 0 \\ \varepsilon_{xy} - \dfrac{\varepsilon_{yz}\varepsilon_{xz}}{\varepsilon_{zz}} & , \varepsilon_{yy} - \dfrac{\varepsilon_{yz}^2}{\varepsilon_{zz}} - K^2 & , 0 , & -K\dfrac{\varepsilon_{yz}}{\varepsilon_{zz}} \\ -\varepsilon_{xx} + \dfrac{\varepsilon_{xz}^2}{\varepsilon_{zz}} & , -\varepsilon_{xy} + \dfrac{\varepsilon_{xy}\varepsilon_{yz}}{\varepsilon_{zz}} & , 0 , & k\dfrac{\varepsilon_{xy}}{\varepsilon_{zz}} \end{bmatrix}\begin{pmatrix}E_x\\E_y\\H_x\\H_y\end{pmatrix}$$

$$\equiv \overleftrightarrow{A}\begin{pmatrix}E_x\\E_y\\H_x\\H_y\end{pmatrix} \tag{A.13}$$

where

$$K = \sin\Theta$$
$$\varepsilon_{zz} = n_o^2 + \left(n_e^2-n_o^2\right)\ \sin^2\alpha$$
$$\varepsilon_{xx} = n_o^2 + \left(n_e^2-n_o^2\right)\ \cos^2\alpha\ \cos^2\beta$$
$$\varepsilon_{yy} = n_o^2 + \left(n_e^2-n_o^2\right)\ \cos^2\alpha\ \sin^2\beta$$
$$\varepsilon_{xz} = \left(n_e^2-n_o^2\right)\ \sin\alpha\ \cos\alpha\ \cos\beta$$
$$\varepsilon_{yz} = \left(n_e^2-n_o^2\right)\ \sin\alpha\ \cos\alpha\ \sin\beta$$
$$\varepsilon_{xy} = \left(n_e^2-n_o^2\right)\ \cos^2\alpha\ \sin\beta\ \cos\beta$$
$$n_e = n_o + n_a$$

and the director twist $\beta$ is measured from the plane of incidence. If the derivative matrix $\overleftrightarrow{A}$ is a constant then Eq. (A.13) has a solution of the form

$$\begin{pmatrix} E_x(z) \\ E_y(z) \\ H_x(z) \\ H_y(z) \end{pmatrix} = \exp\left[i \frac{\omega}{c} \overleftrightarrow{A} z\right] \begin{pmatrix} E_x(0) \\ E_y(0) \\ H_x(0) \\ H_y(0) \end{pmatrix} \tag{A.14}$$

Equation (A.13) can be solved by a propagator method[2] wherein the LC layer is divided into several slabs, each with a constant derivative matrix. Then we find

$$\begin{pmatrix} E_x(z) \\ E_y(z) \\ H_x(z) \\ H_y(z) \end{pmatrix} = \exp\left[i \frac{\omega}{c} \overleftrightarrow{A}(d-h)h\right] \cdot \ldots \cdot \exp\left[i \frac{\omega}{c} \overleftrightarrow{A}(0)h\right] \begin{pmatrix} E_x(0) \\ E_y(0) \\ H_x(0) \\ H_y(0) \end{pmatrix} \tag{A.15}$$

where h is the slab thickness. The division of the liquid crystal layer into slabs of equal thickness would be inefficient as the derivative matrix in Eq. (A.13) depends on angles rather than distance. The computation of angles as function of z is also time consuming as they are given by integrals. It is thus better to change the variable in Eq. (A.13) from z to $\lambda$. In this case, the director tilt angle $\alpha$ is given directly by relation (A.8) and only the twist angle $\beta$ has to be found by integration. In order to minimize the amount of computation, we made the step size in $\lambda$ variable. The step size is determined by the conditions which assure small variation of the derivative matrix at each step. We found that limiting steps in $\lambda$ so that the resulting steps in $\alpha$ and $\beta$ are less than 2° was sufficient. The linear errors in calculation of the propagator matrix were removed by computing the derivative matrices at the middle of each step in $\lambda$. Thus, Eq. (A.15) was transformed into

$$\begin{pmatrix} E_x \\ E_y \\ H_x \\ H_y \end{pmatrix} = \begin{matrix} P'\left(\frac{\lambda_1}{2}, \lambda_1\right) \cdot P'\left(\lambda_1 + \frac{\lambda_2}{2}, \lambda_2\right) \cdots P'\left(\frac{\pi}{2} - \frac{\lambda_n}{2}, \lambda_n\right) \\ \cdot P\left(\frac{\pi}{2} - \frac{\lambda_n}{2}, \lambda_n\right) \cdots P\left(\lambda_1 + \frac{\lambda_2}{2}, \lambda_2\right) \cdot P\left(\frac{\lambda_1}{2}, \lambda_1\right) \cdot \end{matrix} \begin{pmatrix} E_x \\ E_y \\ H_x \\ H_y \end{pmatrix} \tag{A.16}$$

where

$$P(\lambda, \lambda_m) = \exp\left[i \frac{\omega}{c} \overleftrightarrow{A}(\alpha(\lambda), \beta(\lambda)) \frac{\partial z}{\partial \lambda} \lambda_m\right]$$

and

$$P'(\lambda, \lambda_m) = \exp\left[i \frac{\omega}{c} \overleftrightarrow{A} (\alpha(\lambda), \beta_o - \beta(\lambda)) \frac{\partial z}{\partial \lambda} \lambda_m\right] .$$

The symmetry of the twisted nematic layer about its midplane was explicitly taken into account in the above equation. The amount of computation is minimized if $P'(\lambda,\lambda_m)$ and $P(\lambda,\lambda_m)$ are calculated simultaneously.

A further simplification that we made was to reduce the order of Eq. (A.13) from 4 to 2. The derivative matrix of Eq. (A.13) has four eigenvalues.

$$\varepsilon_{1,2} = \pm\sqrt{n_o^2 - \sin^2\Theta} \qquad \text{(A.17)}$$

$$\varepsilon_{3,4} = \frac{\left(n_e^2-n_o^2\right)\sin\alpha\cos\alpha\cos\beta\sin\Theta}{n_o^2+\left(n_e^2-n_o^2\right)\sin^2\alpha} \pm \frac{n_o n_e}{n_o^2+\left(n_e^2-n_o^2\right)\sin^2\alpha}$$

$$\sqrt{n_o^2+\left(n_e^2-n_o^2\right)\sin^2\alpha-\sin^2\Theta\left[1-\left(1-\frac{n_o^2}{n_e^2}\right)\cos^2\alpha\sin^2\beta\right]} \qquad \text{(A.18)}$$

corresponding to the ordinary and extraordinary transmitted and reflected waves.

Since at the voltages with which we are concerned there are neither abrupt changes in the indices of refraction (except at the boundaries) nor any periodicity in the structure, we can safely neglect the reflected waves and solve Eq. (A.13) for transmitted waves only. This transforms Eq. (A.13) into a 2×2 differential matrix equation. The 4×4 matrix formulation is necessary only when considering the reflections at surfaces. Neglecting the reflection eigenmodes also eliminates the effects of interference as the surface reflections can be treated independently.

The above approach has the disadvantage of having to compute two eigenvectors and a transformation matrix into the vector space spanned by them. However, all the necessary algebra is in 2×2 matrices rather than 4×4 and the computation of the exponents of matrices is now replaced by computation of exponents of scalars, as the propagator in the space spanned by the eigenvectors is given by

$$P(\lambda,\lambda_m) = \begin{bmatrix} \exp i\left(e_1 \frac{\omega}{c}\frac{\partial z}{\partial\lambda}\lambda_m\right), & 0 \\ 0, & \exp i\left(e_3 \frac{\omega}{c}\frac{\partial z}{\partial\lambda}\lambda_m\right) \end{bmatrix}. \qquad \text{(A.19)}$$

The approach described above allowed us to do the computations on a desktop computer (HP 9845). The program is written in BASIC and takes up 45K bytes of memory. The number of transmission vs. voltage curves that can be computed simultaneously depends on the numbers of incidence angles, azimuthal viewing angles and wavelengths required but in general is greater than 45 for a machine for 64K bytes of memory. Typical time needed to compute two curves simultaneously (such as in Figs. 2 and 3) is 35 minutes with voltage steps of 0.15 volts.

## REFERENCES


1. F. J. Kahn, H. Birecki, "Multiplexing Limits of Twisted Nematic Liquid Crystal Displays and Implications for the Future High Information Content LCDs," this volume.
2. D. W. Berreman, J. Opt. Soc. Am. 63:1374 (1973).
3. D. W. Berreman, J. Appl. Phys. 46:3746 (1975).
4. C. Z. van Doorn, J. Appl. Phys. 46:3738 (1975).
5. G. Baur, F. Windscheid and D. W. Berreman, Appl. Phys. 8:101 (1975).
6. F. C. Frank, Disc. Farad. Soc. 25:19 (1958).
7. M. Schadt and F. Muller, IEEE Trans. Electr. Dev. 25:1125 (1978).
8. M. Schadt and F. Muller, 7th Intern. Liq. Cryst. Conf. Bordeaux, (1978).
9. M. Schadt and W. Helfrich, Appl. Phys. Lett. 18:127 (1971).
10. H. Birecki and F. J. Kahn, submitted to J. Appl. Phys.
11. P. A. Penz, SID Int'l Symposium 1978, Digest Tech. Papers, p. 68.
12. E. P. Raynes, "Recent Advances in Liquid Crystal Materials and Display Devices," Biennial Display Research Conference Record, 1978 p. 8.
13. K. Toriyama, K. Suzuki, T. Nakagomi, T. Ishibashi and K. Odawara, J. Phys. Colloq. 40-C3:317 (1978).
14. A. R. Kmetz, SID Int'l Symposium, 1978, Digest of Tech. Papers, p. 70.
15. C. Z. van Doorn, C. J. Gerritsma and J. J. M. J. de Klerk, IBM Symposium on Liq. Cryst., Feb. 1979.
16. C. Z. van Doorn and J. L. A. M. Heldens, Phys. Lett. 47A:135 (1974).
17. P. G. De Gennes, "Physics of Liquid Crystals," Clarendon Press, Oxford (1974), Section 6.1.
18. D. W. Berreman, Appl. Phys. Lett. 25:12 (1979).
19. F. M. Leslie, Mol. Crystl. Liq. Cryst. 12:57 (1970).
20. D. W. Berreman, J. Opt. Soc. Am. 62:502 (1972).
21. S. Teitler and B. W. Henvis, J. Opt. Soc. Am. 60:830 (1970).

# SENSITIVE MEASUREMENT OF TWIST IN LCD's

Wilbur Kaye

Beckman Instruments, Inc.
Irvine, California 92713

## INTRODUCTION

The contrast between on and off states of twisted liquid crystal displays (LCD's) can exceed $10^5$ under certain circumstances.[1] This contrast is far in excess of that usually encountered in LCD's and an understanding of the controlling parameters has obvious technological significance. It suggests nondisplay applications such as light modulation. Numerous papers have appeared discussing the parameters controlling contrast. This paper is primarily concerned with the significance and measurement of cell twist.

Contrast is here defined as the ratio of transmittances, $T_{on}/T_{off}$, where "on" refers to a cell with voltage applied. The discussion is simplified by considering only the passage of collimated rays at normal incidence through uniformly twisted liquid crystal (LC) cells between crossed polars. With this limitation, the factors controlling contrast are: (1) Dichroism of the polars. (2) Orientation of the polars relative to the surface directors. (3) Voltage applied to the cell. (4) Cell twist angle.

In the limiting case of high voltage applied to a "perfect" cell between crossed polars, the transmittance should be dictated by the dichroism of the polars. Defining dichroism of a polar as the difference between absorbances of the polar in crossed and parallel orientations to a perfectly polarized beam, it should be apparent that polar dichroism would have to be somewhat greater than 5 to achieve a contrast ratio of $10^5$ under the above conditions. Literature on commercially available polars does

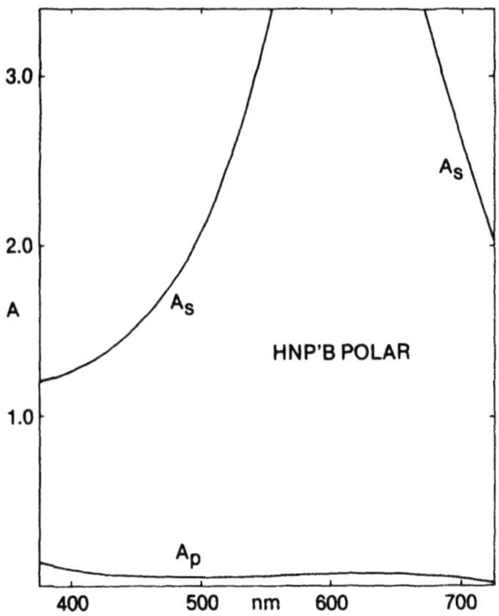


Fig. 1. Absorbance spectra of a single film unmounted polaroid type HNP'B polar.

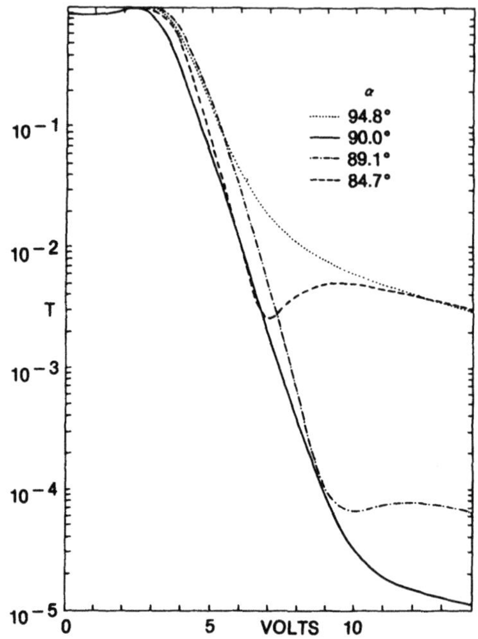


Fig. 2. Log transmittance vs. voltage at 300 Hz for LCD's of different twist.

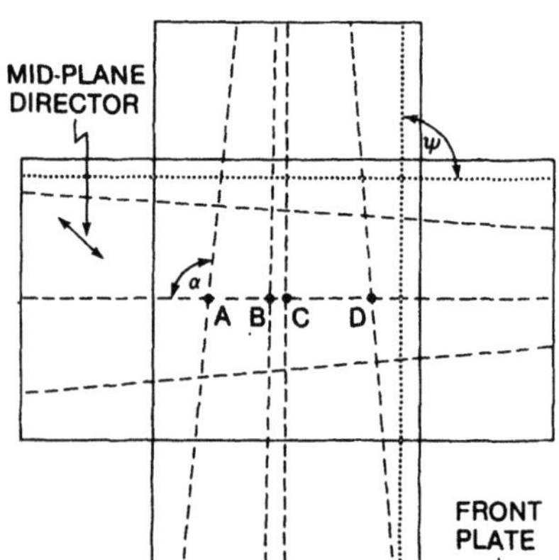


Fig. 3. Geometry of a demountable LC cell.

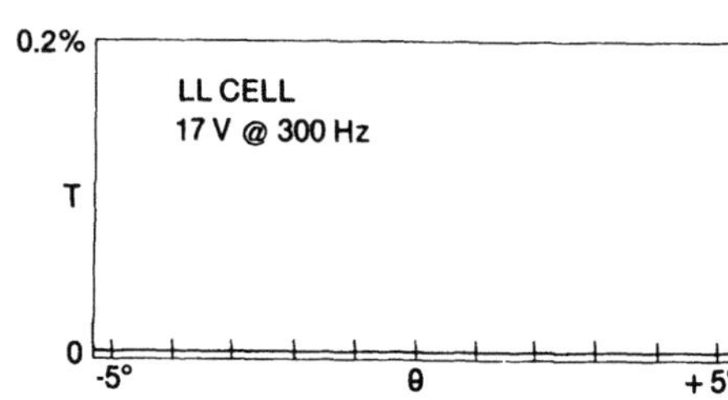


Fig. 4. T vs. Θ for a symmetric tilt, 90° twist LC cell between crossed polars with 17v at 300 Hz applied.

not indicate the ready availability of polars with this high a dichroism; however, this is believed to be a limitation of the instrumentation used.

Figure 1 shows absorption spectra (corrected for surface reflections) of a single unsupported film of Polaroid type HNP'B material oriented parallel and perpendicular to the plane of polarization in a spectrophotometer. Commercial HNP'B polars usually consist of two such films sandwiched between isotropic cellulose acetate films. Absorbances in the perpendicular ($A_s$) orientation were not measurable between 550 and 670 nm because of instrumental difficulties (primarily depolarization at optical surfaces in the spectrophotometer). However, in measurements with the HeNe instrumentation described previously[2] the dichroism was seen to exceed 5 at 6328Å. Obviously several layers of this material would be required in each polar if a dichroism of 5 were required throughout the visible region.

In order to show the significance of twist on the transmittance of LC cells between crossed polars, Fig. 2 was obtained using the polarized HeNe instrument. It shows semilog plots of the transmittance vs. voltage for 4 different production LCD's. Only cells with a twist of exactly 90° are capable of a polar-limited contrast. Preparation of plots such as those shown in Fig. 2 is very difficult. A major difficulty is in aligning the polars relative to the surface directors of the cell. Even more difficult is the measurement of the twist in a sealed cell.

Factors (surface uniformity, splay, etc.) other than twist can limit contrast, hence an independent measurement of twist is important. It might be surmised that twist could be easily measured by determining the angular rotation of the plane of polarization of a beam passing through the cell. In practice, this is not possible without a knowledge of the orientation of the surface directors relative to the polars.

All the controlling variables can be studied using demountable cells. Figure 3 shows the geometry of a demountable LC cell fabricated with two spring clamped glass plates. Straight lines (indicated by dotted lines) were first scribed on these plates, and the plates were aligned in the SiO evaporator relative to these scribed lines. Dashed lines identify the intersection of the evaporating SiO with the plates. The surface directors of the LC molecules assume orientations defined by this evaporated material.[3] Twist angle $\alpha$ is dictated by the angles between juxtaposed intersection paths. Only along a particular line will $\alpha$ be exactly 90°. Knowing the distance between the source of evaporating material and the plates in the evaporator, the orientation of the scribed lines in the evaporator and the angle $\Psi$ between the scribed lines in the assembled cell, one can predict the exact twist angle at any position on the cell. A tool-makers microscope was used to determine $\Psi$. A knowledge of the mid-plane director orientation permits one to determine the direction of twist, hence whether the twist is acute or obtuse.

In manufacturing LCD's with evaporated orienting layers, the distance between the cell plates and the evaporating source may be large hence the "fanning" of intersection paths of evaporated material with the plates may be small, but the edges of the cell plates are not necessarily parallel to a line between the center of the plate and the source of evaporating material. Under extreme conditions twist angles may differ from 90° by more than ±5°. Furthermore, the plate edges cannot be unambiguously used to orient the polars.

The uniform variation of transmittance with increasing voltage in a properly oriented cell of exactly 90° twist (Fig. 2) can only indicate a uniform variation of director tilt while retaining twist. However, most of the twist is associated with nearly homeotropic molecules.[4] This does not mean that all the molecules are homeotropic even at relatively high voltages. A sizeable fraction of the LC molecules may have a homogeneous component, but most of these molecules have little twist component and can be considered to be tilted within planes defined by the two surface directors and field lines. These molecules can induce a retardation when the planes defined above are neither parallel nor perpendicular to the plane of polarization. If the two planes defined by the surface directors and the field lines are not orthogonal (i.e., $\alpha \neq 90°$), the LC molecules in one or both planes have to induce retardation. This retardation is largely responsible for the loss in contrast of cells with a twist other than 90°.

When a beam of plane polarized light passes through a uniaxial crystal, the beam behaves as though it were split into components of orthogonal polarization propagating at velocities varying with the refractive indices in these orthogonal planes. The phase difference, $\delta$, between these rays is:

$$\delta = 2\pi \, d(n_e - n_o)/\lambda \,, \tag{1}$$

where $n_e$ and $n_o$ are the refractive indices along the beam propagation paths and d is the path length in this medium.

In an untwisted LC cell under a field the LC molecules will exist with a distribution of tilt within the plane defined by the surface director and the field lines. Consequently, the propagation velocity of the ray whose E vector is in this plane varies as it passes through the cell. However, one can still assume an integrated $dn_e$. Upon passing through a second polar the components of the above two rays in the plane allowed by the second polar will interfere and the emerging light intensity follows the equation:

$$I = K\sin^2\Theta\cos^2\Theta\sin^2\delta/2 \ , \tag{2}$$

where $\Theta$ is the angle between the incident plane of polarization and the plane defined by the surface director and the field lines.

A twisted LC cell subjected to a high field can be considered as two separate uniaxial crystals whose optic axes are in planes subtending the twist angle. Each of the two rays emerging from the first half of the cell are subsequently split into two rays by the second half of the cell. Four rays emerge. The components of these four rays allowed by the second polar then interfere and the intensity of the emerging ray is given by the equation:

$$T = X^2 + Y^2 \ . \tag{3}$$

Where:

$$X = -\sin\Theta\cos\alpha\cos(\alpha+\Theta) - \sin\Theta\sin\alpha\sin(\alpha+\Theta)\cos\delta_2 - \cos\Theta\sin\alpha\cos(\alpha+\Theta)\cos\delta_1 + \cos\Theta\cos\alpha\sin(\alpha+\Theta)\cos(\delta_1+\delta_2), \tag{4}$$

and

$$Y = -\sin\Theta\sin\alpha\sin(\alpha+\Theta)\sin\delta_2 - \cos\Theta\sin\alpha\cos(\alpha+\Theta)\sin\delta_1 + \cos\Theta\cos\alpha\sin(\alpha+\Theta)\sin(\delta_1+\delta_2) \ , \tag{5}$$

$\delta_1$ is the phase angle determined by the front half of the cell and $\delta_2$ is that determined by the rear half of the cell.

Equation (3) reduces to:

$$I = K(\sin^2 w\Theta)\ (1-\cos(\delta_1-\delta_2))/2 \ , \tag{6}$$

when

$$\alpha = 90° \ .$$

Consider a 90° twist cell with symmetric surface director tilt between crossed polars after application of a relatively high AC field. Then $\delta_1=\delta_2$ and Eq. (6) reduces to zero. Light intensity transmitted through this system is essentially zero for all values of $\Theta$. Figure 4 shows the experimental verification of this situation. The cell was known to have a twist of 90°. It was fabricated using SiO alignment layers evaporated at a high angle of incidence thus producing equal to surface director tilts

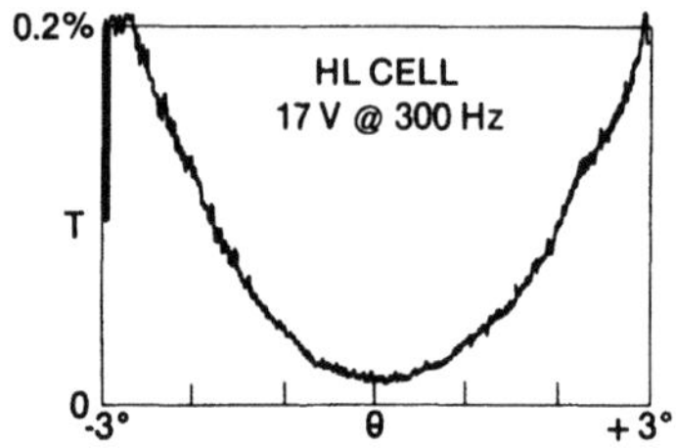


Fig. 5. T vs. Θ for an unsymmetrical tilt, 90° twist LC Cll between crossed polars with 17v at 300 Hz applied.

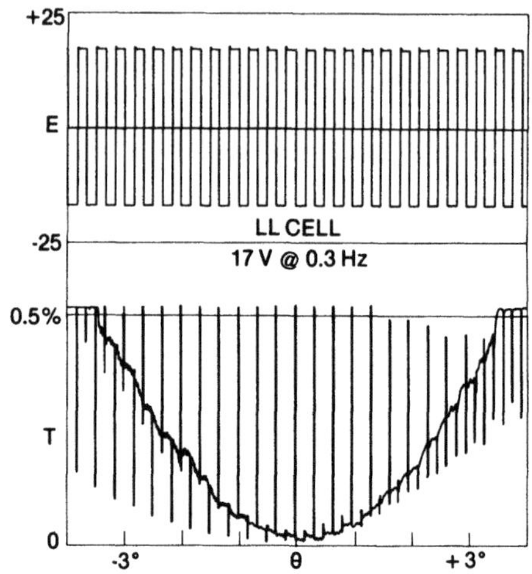


Fig. 6. T vs. Θ for a symmetric tilt, 90° twist LC cell between crossed polars with 17v at 0.3 Hz applied.

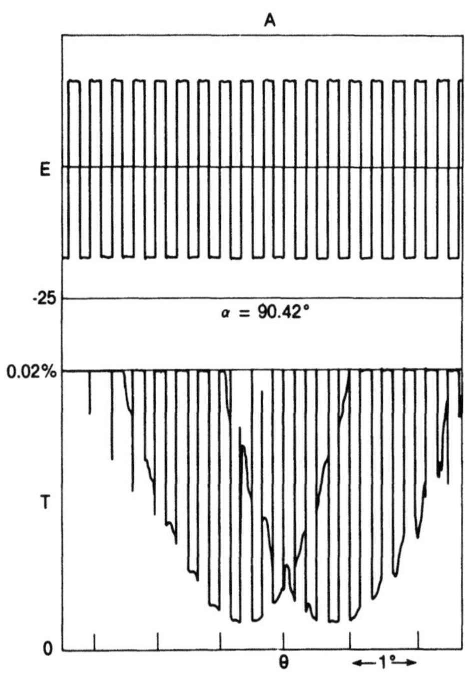


Fig. 7. T vs. Θ for a demountable symmetric tilt, twist α=90.42° LC cell between grossed polars with 17v at 0.3 Hz applied.

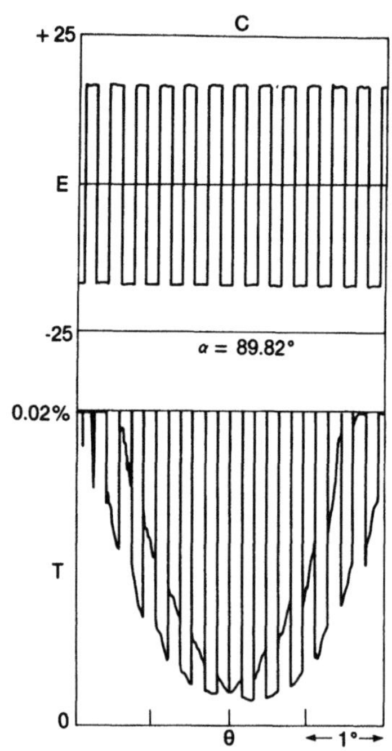


Fig. 8. T vs. Θ for a demountable symmetric tilt, twist α=89.82° LC cell between crossed polars with 17v at 0.3 Hz applied.

of about 30°.[5] The cell was rotated about an axis parallel to the beam while 17v at 300 Hz square wave were applied to the cell.

In contrast, consider a 90° twist cell of unsymmetric surface director tilts. Such dissymmetry can be introduced by depositing SiO at different angles of incidence on the two plates.[5] Under this circumstance $\delta_1 \neq \delta_2$. Light intensity falls to zero only when $\Theta=0°$, i.e., when the front surface director lies in the incident plane of polarization. Light intensity increases with the absolute value of $\Theta$. Figure 5 shows the behavior of such a cell.

When DC or low frequency AC fields are applied to LC cells, negative ions migrate from the negative to the positive electrode.[6,7] In doing so they increase the average tilt at the negative electrode and decrease the average tilt at the positive electrode.[8] A low frequency field then renders unsymmetric a cell having identical orienting layers. Figure 6 shows how light transmission then varies with $\Theta$. The same cell was used here as for Fig. 4. Only the frequency has been changed from 300 Hz to 0.3 Hz. A dual trace oscillograph was used to record the signals. The upper trace is a recording of the applied voltage while the lower trace is a recording of transmittance. Transients accompanying polarity reversals are believed to be caused by turbulence in the LC when the ion flow reverses. The cell used for Figs. 4 and 6 differed from that of Fig. 5 only in the character of one of the alignment layers. Note that T varies at a greater rate with changing $\Theta$ in Fig. 6 than in Fig. 5. This indicates that ionic flow is quite effective in tilting the LC molecules.

When $\alpha \neq 90°$ light transmittance follows Eq. (3), a distinction can be made between front and rear halves of the cell on the basis of their electrical polarity and surprising small deviations of $\alpha$ from 90° can be measured. To demonstrate this, the cell whose geometry is described by Fig. 3 was used. Recordings of transmittance vs. $\Theta$ were made with the beam passing through the four positions denoted by A, B, C and D. The record at position B was similar to Fig. 6. These recordings are found in Figs. 7-9. No effort was made to determine the absolute $\Theta$ values. The cell was of the symmetric type with surface director tilts of approximately 30°. The ion-flow induced director tilt changes on each plate as polarity changes and two T vs. $\Theta$ envelopes can be identified. The recording at position B where $\alpha=90°$ can be applied to determine $\delta_1-\delta_2$ using Eq. (6). Knowing $\alpha$ from the cell geometry (Fig. 3), it is possible to estimate $\delta_1$ and $\delta_2$ by an iterative match of Eq. (3) to one of the experimental recordings where $\alpha \neq 90°$. Figure 10 is a computer simulated match of Eq. (3) to Fig. 9. Note that the two curves were obtained by interchanging $\delta_1$ and $\delta_2$ (the consequence of polarity reversal).

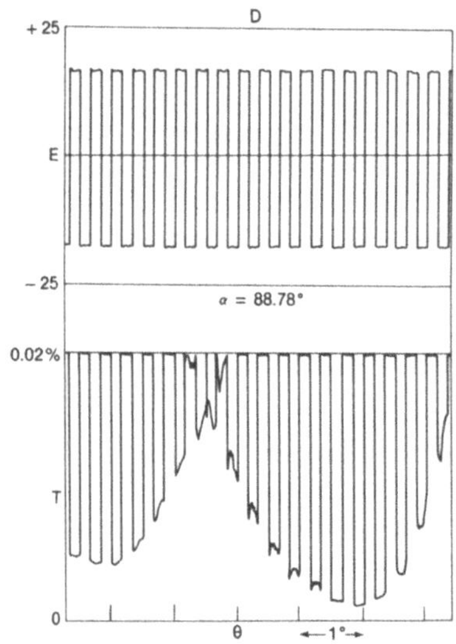


Fig. 9. T vs. Θ for a demountable symmetric tilt, twist α=88.78° LC cell between crossed polars with 17v at 0.3 Hz applied.

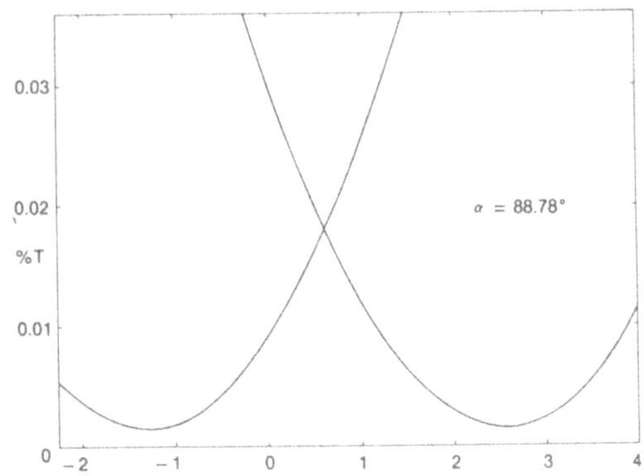


Fig. 10. Theoretical T vs. Θ, twist α=88.78°, $\delta_1$=25°, $\delta_2$=46° and $\delta_2$=25°, $\delta_1$=46°.

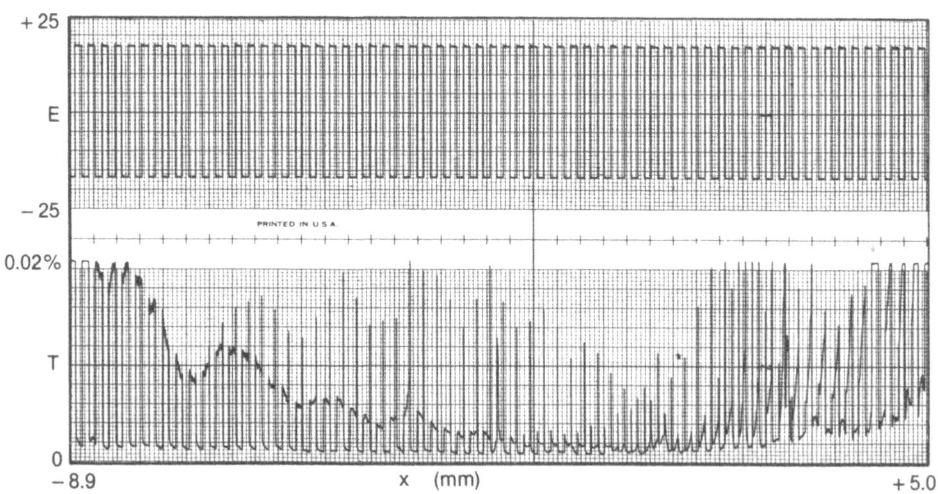


Fig. 11. T vs. translation position for a demountable symmetric tilt LC cell of approximately 90° twist with 17v at 0.3 Hz applied.

Table 1 summarizes the relationship between the angular ($\Theta$) separation of the two envelope minima and the deviation of twist angle from 90° for Figs. 6-9. So long as the cells contain the same LC material, the same alignment layers, the same low frequency voltage is used and the cells are of approximately the same thickness, the twist angle can be determined from the angular separation of envelope minima. If any of these conditions change, a recalibration is required.

Table 1. Relation Between Twist and Angular Separation of Transmittance Minima

| Figure | $\alpha$ | $\Theta_1-\Theta_2$ | $(\alpha-90°)/(\Theta_1-\Theta_2)$ |
|---|---|---|---|
| 7 | 90.42° | 1.3° | 0.32 |
| - | 90.00° | 0.0 | - |
| 8 | 89.82° | -0.6 | 0.30 |
| 9 | 88.78° | -4.0° | 0.31 |

Secondary effects can introduce small errors in this method of measuring twist. Dust and surface scatches can, of course, seriously affect local alignment. More subtle are perturbations apparently produced by variations in the flow pattern (hydrodynamic domains) of neutral molecules. In the cells used here hydrodynamic instabilities become quite pronounced at field frequencies around 5 Hz at 17v and resemble the patterns described by Sussman.[9] They were not visible at 0.3 Hz and 17v but Fig. 11 shows evidence of their presence. In this case, the $\Theta$ orientation of the cell was fixed and the cell was translated relative to the beam while applying 17v at 0.3 Hz. A slow variation in polarity modulation amplitude is expected because of the effects discussed in connection with Fig. 3. Superimposed are local oscillations of modulation amplitude possibly attributable to hydrodynamic domains. It may be surprising that neutral molecule flow would not exhibit a more pronounced effect; however, the flow adjacent to the cell walls must be parallel to the walls and exert a minor orienting force on the thin layer of molecules with significant homogeneous component at 17v.

ACKNOWLEDGMENT

The author is indebted to Dr. W. S. Gallaway for assistance in the development of Eq. (3).

REFERENCES

1. W. Kaye, Mol. Cryst. Liq. Cryst. 37:147 (1976).
2. W. Kaye, ibid. p. 137.
3. M. Schadt and W. Helfrich, Appl. Phys. Lett. 18:127 (1971).
4. W. W. Berreman, J. Opt. Soc. Amer. 63:1374 (1973).
5. E. Guyon, P. Pieranski and M. Boix, Lett. Appl. Eng. Sci. 1:19 (1973).
6. G. H. Heilmeyer, L. A. Zanoni and L. A. Barton, Proc. IEEE 56:1162 (1968).
7. W. Helfrich, J. Chem. Phys. 51:4092 (1969).
8. W. Kaye, "Electrochemical Polarization in Field Effective LCD's," presented at the 7th International Congress on Liquid Crystals, Bordeaux, July 1978.
9. A. Sussman, Appl. Phys. Lett. 29:633 (1976).

# LIQUID CRYSTAL FOR MULTIPLEXED TWISTED NEMATIC DISPLAYS - ITS PHILOSOPHY AND PRACTICE

Kazuhisa Toriyama, Kenkichi Suzuki, Tamihito Nakagomi
Tadashi Ishibashi and Kōzō Odawara

Electron Tube Division, Hitachi, Ltd.
3300 Hayano
Mobara, Chiba, Japan

## INTRODUCTION

To extend the applicability of twisted nematic liquid crystal displays (TN-LCDs), it is very important to develop liquid crystals (LCs) themselves. A few papers have been published[1] concerning the design of LCs for TN-LCD application, discussing the relationship between material properties and display performance. But, most of those papers have neglected problems of multiplexing capability of displays, including factors of practical importance, such as angular dependence of viewing characteristics.

The purpose of this paper is to report on efforts for the systematic design of liquid crystals from the chemical and physical viewpoints.

The first part of this paper discusses factors that govern the multiplexing capability of TN-LCD, and introduces some definitions of figure of merit. For various chemical classes of LCs such as Schiff's base, esters, biphenyls, azoxy and cyclohexanes, their physical properties and their relationship to display performance are discussed. We classify those LCs into three categories, $N_p^s$ (weakly positive dielectric anisotropy nematic LC) and $N_n$+$N_p$ (binary mixed system of negative and positive dielectric anisotropy nematic LC). An evaluation of each class is given.

In the last part of the paper, it is proposed that the binary mixed system $N_n$+$N_p$ and its variations are the most promising of the three classes, as low threshold LC materials for multiplexed TN-LCD. The roles and importance of the $N_n$ in the binary mixed system $N_n+N_p$ are discussed in detail and also difficulties in design of better binary mixed systems are pointed out.

## DEFINITION OF DISPLAY PERFORMANCE PARAMETERS FOR MULTIPLEXED TN-LCD

In the multiplexing drive of a TN-LCD, the variation of optical response of the threshold voltage is critical. In addition to temperature dependence of the threshold voltage, the factors limiting operating voltage tolerance are threshold sharpness and viewing angle dependence of the threshold voltage.

The operating voltage tolerance in a TN-LCD, for example, in the case of the 1:3 multiplexing drive as shown in Fig. 1 is given by

$$V_{thL} + \Delta V_{thL} \leqq V_o \leqq V_{thH} - \Delta V_{thH} \qquad (1)$$

where $\Delta V_{thL}$ and $\Delta V_{thH}$ are variation values of threshold voltage due to mass production and degradated life. For an evaluation of LC materials, we give definitions of performance parameters for multiplexed TN-LCDs. The relations between threshold voltage and viewing angle in a TN-LCD are shown in Fig. 2. The operating region of the device is enclosed within the solid lines in Fig. 2 due to the following limitations:

1) There are restrictions on drive voltage due to crosstalk problems in multiplexing drives, as illustrated in Fig. 1.

2) An operating temperature range as narrow as 0-40°C may be acceptable. Therefore, the operating voltage region is limited between $V_{thL}$ at 0°C and $V_{thH}$ at 40°C.

3) A viewing angle as narrow as 10-40° may be acceptable.

Thus, $\Delta V_{th1}$ is the error margin of $V_{thL}$ at 0°C temperature and 10° viewing angle. $\Delta V_{th2}$ is the error margin of $V_{thH}$ at 40°C temperature and 40° viewing angle.

Therefore, the figure of merit for a liquid crystal material under a given molecular alignment technique can be defined as follows:

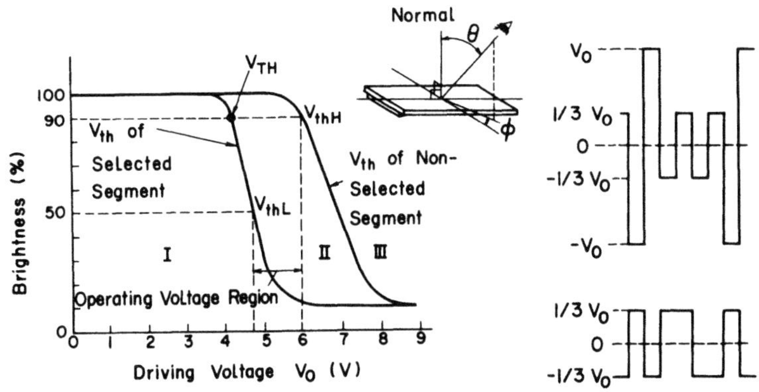


Fig. 1. Brightness curves of a TN-LCD in the case of the 1:3 selection multiplexing drive. The applied voltage signals are given on the right side. ($\theta$=0°, $\phi$=0°, T=25°C, tilt angle $\alpha$=3°), Region I: "OFF" state, Region II: "ON" state on selected segment, Region III: "ON" state on nonselected segment due to crosstalk.

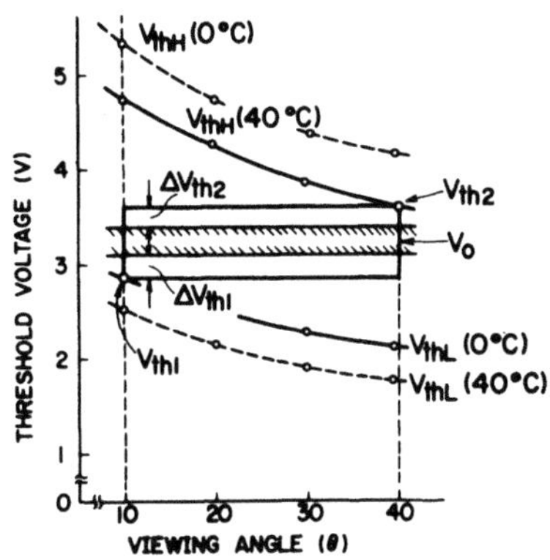


Fig. 2. Definition of figure of merit.

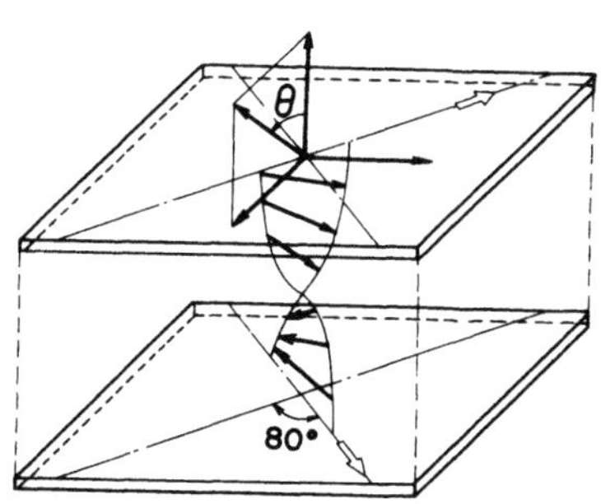


Fig. 3. Geometrical configuration of TN-LCD.

Figure of merit:

$$M = \frac{V_{thH}(40°C,\theta=40°) - V_{thL}(0°C,\theta=10°)}{V_{thH}(40°C,\theta=40°) + V_{thL}(0°C,\theta=10°)}$$

$$= \frac{1 - (\gamma/\Delta V^{\theta})(1/(\alpha\cdot\Delta V^{T}))}{1 + (\gamma/\Delta V^{\theta})(1/(\alpha\cdot\Delta V^{T}))} \qquad (2)$$

where

$$\gamma = \frac{V_{thL}(0°C,\theta=10°)}{V_{TH}\ (0°C,\theta=10°)} \qquad (3)$$

$$\alpha = \frac{V_{thH}(0°C,\theta=40°)}{V_{thL}(0°C,\theta=40°)} \qquad (4)$$

$$\Delta V^{\theta} = \frac{V_{thH}(0°C,\theta=40°)}{V_{thH}(0°C,\theta=10°)} \qquad (5)$$

$$\Delta V^{T} = \frac{V_{thH}(40°C,\theta=40°)}{V_{thH}(0°C,\ \theta=40°)} \qquad (6)$$

Here we also define $M_T$:

$$M_T = \frac{V_{thH}(0°C,\theta=40°) - V_{thH}(40°C,\theta=40°)}{V_{thH}(0°C,\theta=40°) + V_{thH}(40°C,\theta=40°)} \qquad (7)$$

## BOUNDARY CONDITION OR CELL CONFIGURATION FOR LC MATERIAL

The technology of molecular alignment has much influence on multiplexed display performance. A technique applicable to any LC material and suitable for mass production is required which reliably provides uniform quiescent alignment with low tilt angle and is free from induced domain of reverse tilt.

The authors and another research group at Hitachi, have investigated[2,3] alignment techniques for accurate detection of low tilt angle of LC molecules in the device. Using these techniques, we have constructed LC devices in which the LC molecules' directors are all tilted 3±1 degrees in the same direction over the patterned electrodes in the OFF state, and this ensures stable alignment and freedom from induced domain of reverse tilt in the ON state. The device's configuration is illustrated in Fig. 3.

## CLASSIFICATION AND SEARCH OF LCs FOR MULTIPLEXED TN-LCDs

In the development of LCs for multiplexed TN-LCD, we have encountered the following problems (cf. Table 1):

1) Temperature dependence of the threshold voltage

2) Threshold sharpness and viewing angle dependence of the threshold voltage

3) Dull response at low temperature.

We have investigated various LC materials such as Schiff's base, ester, biphenyl, azoxy, and cyclohexane-ring-containing LCs, focusing on the correlation between their display performance factors and material constants, such as viscosity, refractive index, dielectric constant, and elastic constant.

Considering data on the correlation between display performance parameters and material constants, we have classified those LCs into three categories, $N_p^S$ (strongly positive dielectric anisotropy nematic LC), $N_p^W$ (weakly positive dielectric anisotropy nematic LC), and $N_n+N_p$ (binary mixed system of negative and positive dielectric anisotropy nematic LC). The definitions of symbols are tabulated in Table 2.

### Comparisons of $N_p$ and $N_n+N_p$ in Various Chemical Classes of LC

To clarify the characteristics of the three classes, $N_p^S$, $N_p^W$, and $N_n+N_p$, we have diagrammed the correlations between threshold voltage ($V_{th}$), its temperature dependence ($M_T$), and viscosity ($\psi$). The correlations between threshold voltage and viscosity in each $N_p$($N_p^S$ and $N_p^W$), $N_n+N_p$ are given in Figs. 4 and 6, respectively. The cross hatched region in the figures is the acceptable region for 3 volt, 1:3 multiplexing drive, in relaxation to factors and $V_{th}$ of LC.

Similar diagrams of the correlations between $M_T$ and $\psi$ for $N_p$ and $N_n+N_p$ are given in Figs. 5 and 7, respectively.

From these diagrams, we can easily conclude that:

1) Any $N_p$ LC is unacceptable for its high viscosity ($N_p^S$) or high threshold voltage property ($N_p^W$)

## Table 1 Design Outline of TN - LCD for Multiplexed Drive

| | Temperature Dependence of Threshold Voltage | Threshold Sharpness, Viewing Angle Dependence of $V_{th}$ | Viewing Angle ( Contrast ) | Response Time |
|---|---|---|---|---|
| LC Material | • Binary Mixture of $N_n + N_p$,<br>• Selection of $N_n$ | Optical and Dielectrical Properties of LC | Optical Properties of LC | • Viscosity<br>• Temperature Dependence of $V_{th}$ |
| Cell Construction | | Control of Low Tilt Angle | | Gap Control |

* $N_n$ and $N_p$ : Nematic Liquid Crystal of Dielectrically Negative and Positive Material Respectively

## Table 2 Notations

| | | Definition | Example |
|---|---|---|---|
| $N_p$ | $N_p^s$ | Dielectrically Positive, $\Delta\epsilon \approx 20$ | 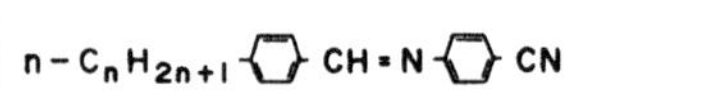 |
| | $N_p^w$ | Dielectrically Positive, $\Delta\epsilon \approx 10$ | 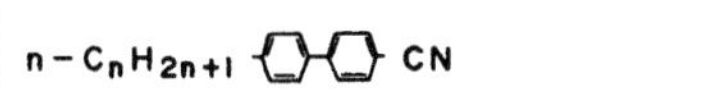 |
| $N_n$ | $N_n^-$ | Dielectrically Negative, $\Delta\epsilon \lesssim 0$ | 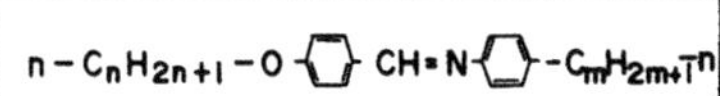 |
| | $N_n^+$ | Positive but Very Small, $\Delta\epsilon \gtrsim 0$ | 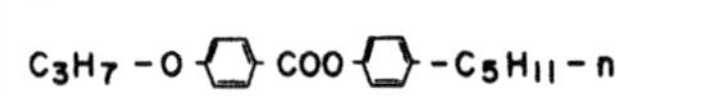 |
| | $N_n^p$ | Dielectrically Negative with Polar Group | 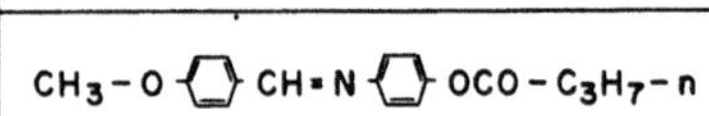 |

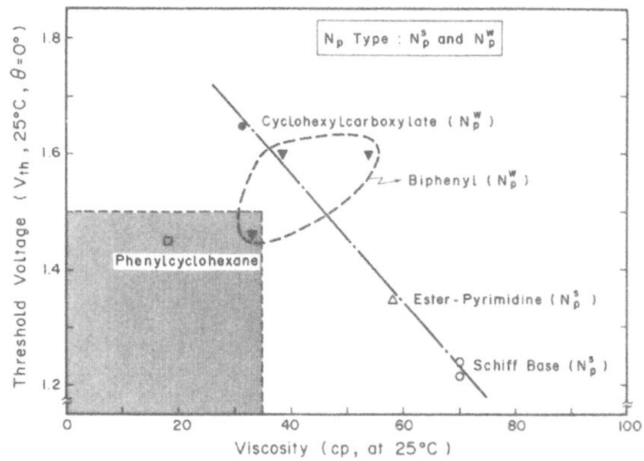


Fig. 4. Correlation between threshold voltage ($V_{th}$, 25°C, θ=0°) and viscosity in $N_p$ type LCs.

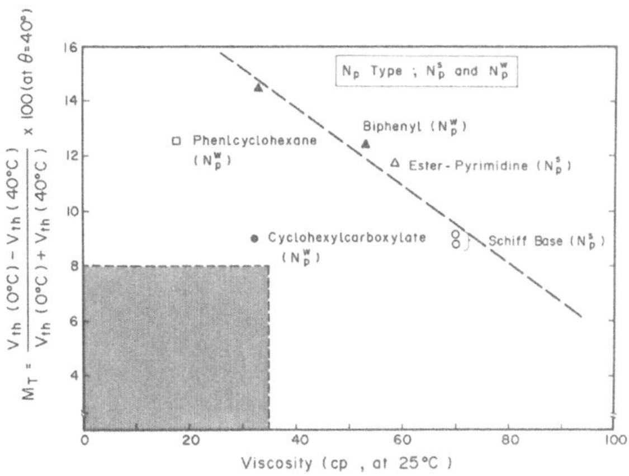


Fig. 5. Correlation between temperature dependence of $V_{th}$ and viscosity in $N_p$ type LCs.

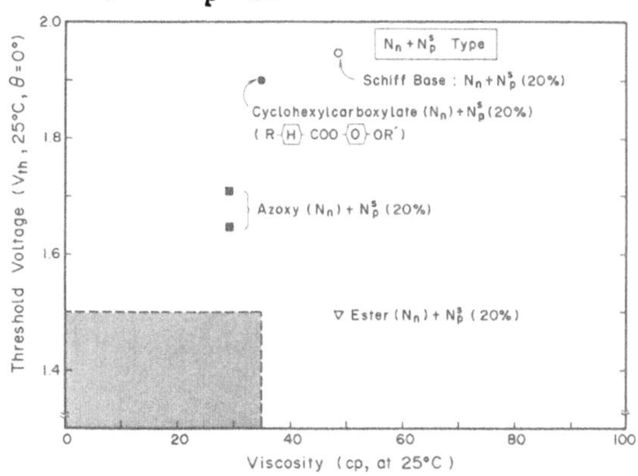


Fig. 6. Correlation between threshold voltage ($V_{th}$, 25°C, θ=0°) and viscosity in $N_n$+$N_p^s$ type LCs.

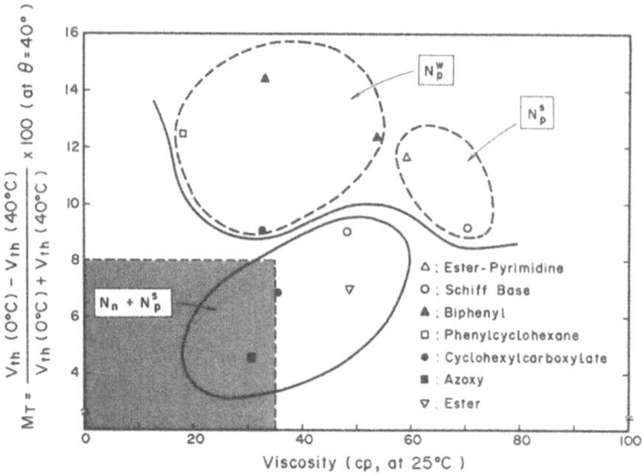


Fig. 7. Correlation between temperature dependence of $V_{th}$ and viscosity in $N_n$+$N_p^S$ type LCs.

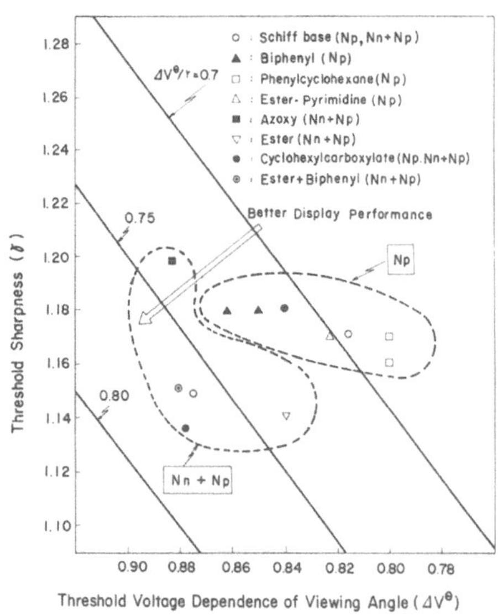


Fig. 8. Properties γ, $\Delta V^{\theta}$ in various LC materials.

2) $N_n+N_p$ systems give a performance better than $N_p$'s, but still far from requirement for multiplexing drive: $N_n+N_p$ systems have a small temperature dependence of $V_{th}$, but a high threshold voltage

3) We can see that the cyclohexane-ring-containing LCs form a group with better display performance than other LCs, viz., LCs with two benzene rings.

Other important factors that govern the margin of LC are threshold sharpness ($\gamma$) and viewing angle dependence of $V_{th}$ ($\Delta V^{\theta}$). We have studied various LC materials focusing on $\Delta V^{\theta}$ and $\gamma$. In our experience, LCs consisting of dielectrically positive ($N_p$) type molecules have poor $\Delta V^{\theta}$ values, while LCs consisting of mixtures of dielectrically negative material ($N_n$) and dielectrically positive material ($N_p$) give good $\Delta V^{\theta}$ values as shown in Fig. 8.

## BINARY MIXED SYSTEM OF $N_n+N_p$ LC

In the course of studies of binary $N_n+N_p$ systems, we have encountered the following difficulties:

1) Maintenance of a wide mesomorphic range (MR)

2) Reduction of threshold voltage without sacrificing the figure of merit (M): The figure of merit tends to decrease with the lowering of the LC threshold voltage.[4]

### Phase Diagram of $N_n+N_p$ System

A common feature of a binary mixed system of $N_n+N_p$ is that the melting point rises steeply as either of the two components increases.[5,6] An example for an $N_n+N_p$ binary system is given in Fig. 9. A mixed binary system of cyclohexanecarboxylate (ECH) as the dielectrically negative type LC and biphenyl as the dielectrically positive type LC gives a poor alloying system. A poor alloying tendency is also seen in a binary system in which the components $N_n$ and $N_p$ have similar chemical structure. Figure 10 is a phase diagram of a binary mixture of negative Schiff's base and positive Schiff's base.

A clue to overcoming poor alloying in a binary mixed system of $N_n+N_p$ will be found in the following experiment conducted for the purpose of improving the alloying in the $N_n+N_p$ system. In the experiment, we have studied a "third dopant." Doping an $N_n+N_p$ Schiff's base system with 4-n-alkanoyloxy-N-(4-n-alkyloxy-benzylidene)-aniline as "third dopant" increases the alloying

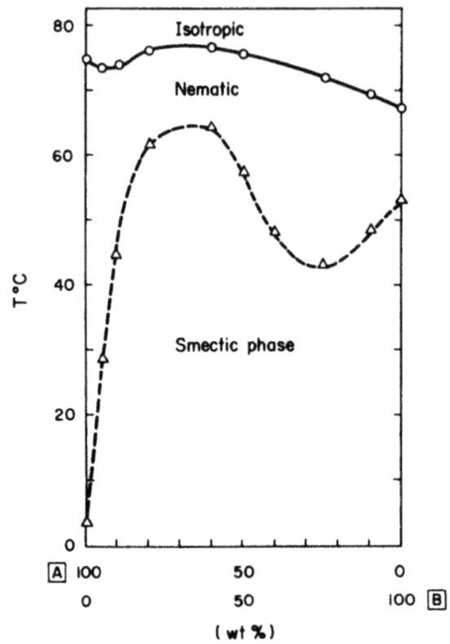


Fig. 9. Phase diagram of mixture of ECH(A) and Biphenyl(B)

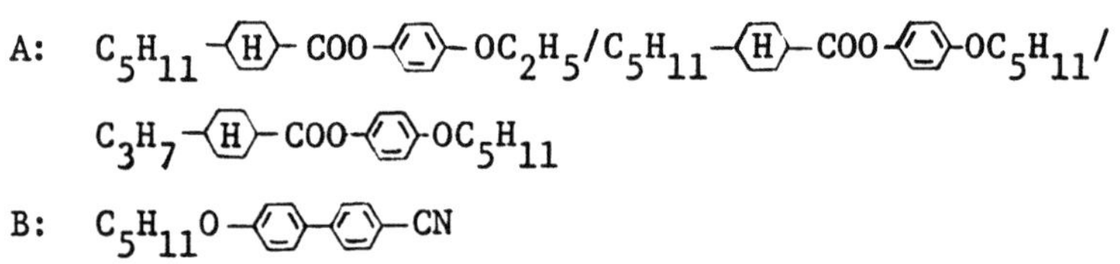


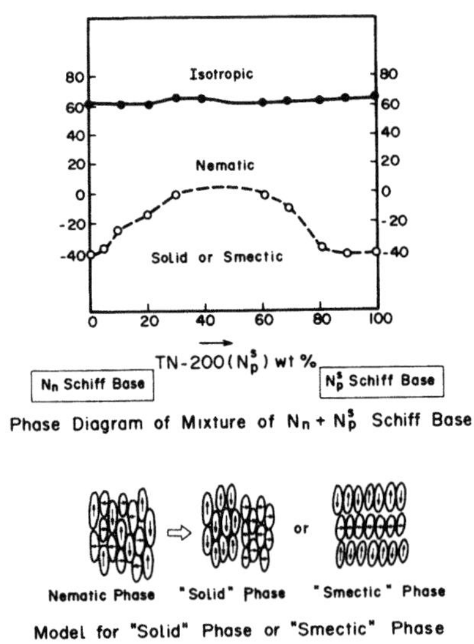


Fig. 10. Binary mixture of $N_n$+$N_p^S$ in Schiff's base.

property of the system and widens the mesomorphic range as shown in Fig. 11. We regard this dopant as a "mediator" between $N_n$ and $N_p^s$. From the results of this and other similar experiments on dopants in our laboratory, we inferred such dopants as effective $N_n^p$ type mediator of which symbol means that the dopant is dielectrically negative but contains polar group in the molecule. These mediators (third dopants) are extremely important in overcoming poor alloying in $N_n+N_p$ systems. This problem will be considered in more detail in the following sections.

## Selection of $N_n$ for Binary Mixed $N_n+N_p$ System

At present, we have four important $N_n$ type LCs, ester, Schiff's base, azoxy and cyclohexanecarboxylate (ECH). Ester and Schiff's base are unsuitable because of their high viscosity (cf. Fig. 6). Azoxy, although it has a good figure of merit, must be discarded because of its photochemical instability. Cyclohexanecarboxylate (ECH) was introduced in 1975 by Demus[1a] in the field of LCD applications, but no one, at least to the authors' knowledge, has applied it to multiplexed TN-LCD. We investigated ECH as a useful $N_n$ type LC because of its low viscosity and unique properties (small temperature dependence of $V_{th}$, small birefringence ($\Delta n$) value), both of which make it suitable for display performance for multiplexed TN-LCDs. Another major advantage of ECH is that it has many homologs[7] some of which possess wide mesomorphic ranges, so that it is easy to make up mixtures of $N_n$ with wide mesomorphic range. The mesomorphic ranges of ECH and their analogs, viz., ester with two benzene rings, are tabulated in Table 3.

## Modification of $N_n(ECH)+N_p^S$ Binary System

Another problem in simple $N_n(ECH)+N_p^s$ binary mixed system other than the MR problem is its somewhat high threshold voltage (cf. Fig. 6).

We studied the ternary system $N_n(ECH)+N_p^w+N_p^s$, adding $N_p^w$ type molecules as a mediator between $N_n(ECH)$ and $N_p^s$. The correlations between threshold voltage and viscosity in $N_n(ECH)+N_p^w+N_p^s$ in which $N_p^w$ is the 4-n-alkyl-4'-cyanophenylcyclohexane (PCH) and $N_p^s$ is ester, are shown in Fig. 12. It must be pointed out that $N_p^w$ as a mediator is effective in lowering of $V_{th}$ as well as in widening MR.

In general, adding of $N_p^w$, increases $M_T$ as shown in Fig. 13. In most cases, it is compensated by a technique that might be generally effective. The technique is to dope the ternary system with LC of a long molecular structure that normally consists of

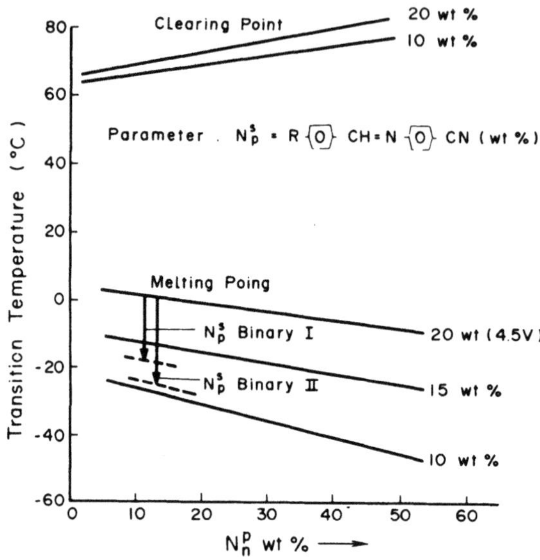


Fig. 11. Mesomorphic range of a ternary system: $N_n$-$N_n^p$-$N_p^s$, $N_n^p$ means dielectrically negative but contains polar group in the molecule.

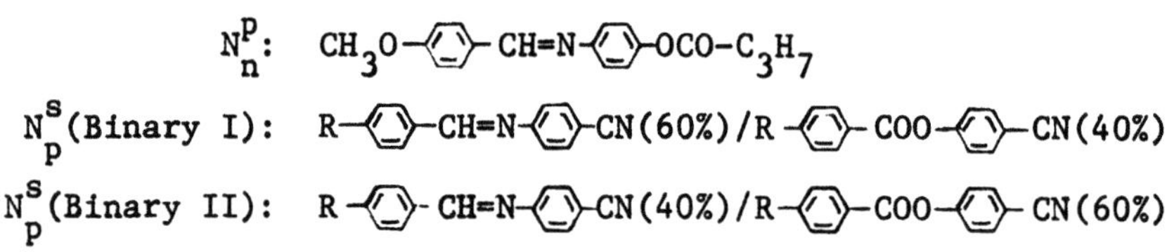


Table 3. Mesomorphic Range of $N_n$ Type LCs, Cyclohexylcarboxylate and Ester

| | | Cyclohexylcarboxylate | | | Ester | | |
|---|---|---|---|---|---|---|---|
| | | $C_nH_{2n+1}$-⟨H⟩- COO -⟨O⟩- R | | | $C_nH_{2n+1}$-⟨O⟩- COO -⟨O⟩- R | | |
| n | R | C — N — I , ΔT | | | C — N — I , ΔT | | |
| 3 | $C_4H_9O$- | 42 | 72 | (30) | 72 | (59)* | — |
| 4 | $CH_3O$ - | 42 | 59 | (17) | 61 | (25)* | — |
| 4 | $C_4H_9O$- | 38 | 69 | (31) | 48 | (46)* | — |
| 4 | $C_6H_{13}O$- | 25 | 69 | (44) | 29 | 50 | (21) |
| 5 | $C_4H_9O$- | 49 | 81 | (32) | 49 | 58 | (9) |

* : Monotropic

three rings (phenyl- or cyclohexyl-). The dopant reforms the temperature dependence of $V_{th}$ of the system with little increase of viscosity of the system. Examples of dopants and their effects on temperature dependence of $V_{th}$ of the system are shown in Fig. 14.

## CHARACTERISTICS OF ECH AS $N_n$ FOR BINARY $N_n$+$N_p$ SYSTEM

Brief reference will be made to the characteristics of ECH from the viewpoint of display performance as our investigation on ECH is still in progress.

### Temperature Dependence of the Optical Threshold Voltage

Temperature dependence of the optical threshold voltage $V_{th}$ is a dominant factor in the figure of merit (M). It is worth analyzing the relations between $\Delta V^T$, or $M_T$ and LC material properties, or molecular structures. Concerning physical constants, viz., dielectric anisotropy and elastic constant, both of which govern the threshold voltage of LC, we made some investigation of temperature dependence of $V_{th}$ for each LC in relation to that of ECH:

$$\left(\frac{1}{V_{th}}\right)\left(\frac{dV_{th}}{dT}\right) \propto \left(\frac{1}{V^*_{th}}\right)\left(\frac{dV^*_{th}}{dT}\right) = \left(\frac{1}{2}\right)\left(\frac{1}{K}\right)\left(\frac{\partial T}{\partial T}\right) - \left(\frac{1}{2}\right)\left(\frac{1}{\Delta\varepsilon}\right)\left(\frac{\partial \Delta\varepsilon}{\partial T}\right) \quad (8)$$

where

$V_{th}$ = optical threshold voltage

$V^*_{th}$ = electrical threshold voltage

$K = K_1 + (1/4)(K_3-2K_2)$.

We measured each term of Eq. (8), using a method developed by one of the authors. A small temperature dependence of $V_{th}$ could be realized if an effective cancellation between the first and second terms on the right side of Eq. (8) takes place. Good cancellation is realized in HN-1 ($N_n$(ECH)+$N^w_p$(PCH)+$N^s_p$) and in $N_n$(azoxy)+$N^s_p$, while in biphenyl and in cyanophenylcyclohexane (ZLI-1132), the cancellation is poor as shown in Table 4. In azoxy, absolute value of both first and second terms are small and well cancelled, while in HN-1, both absolute values are larger but fairly well cancelled.

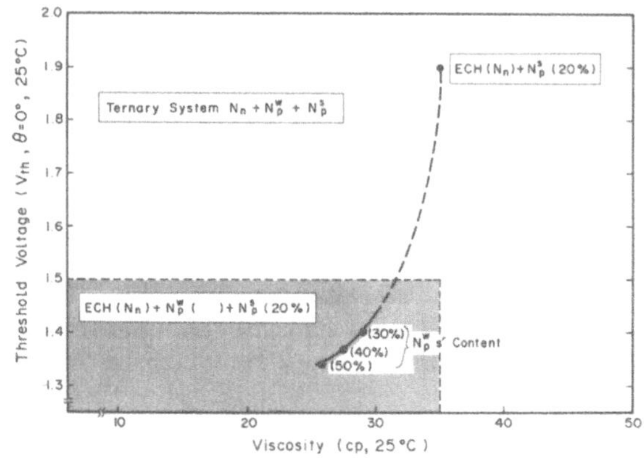


Fig. 12. Correlation between threshold voltage ($V_{th}$, 25°C, θ=0°) and viscosity in $N_n$+$N_p^W$+$N_p^S$.

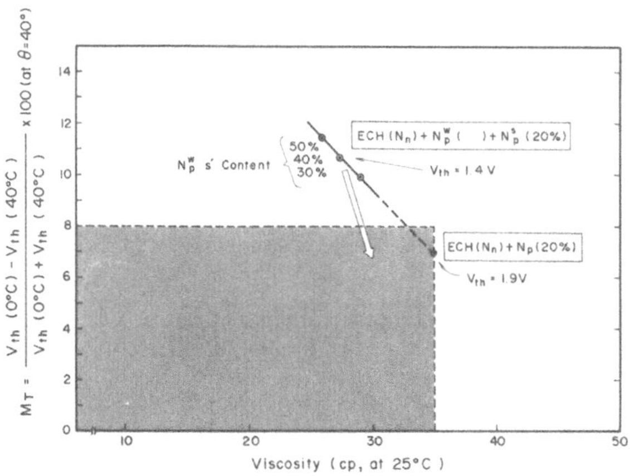


Fig. 13. Correlation between temperature dependence of $V_{th}$ and viscosity in $N_n$+$N_p^W$+$N_p^S$.

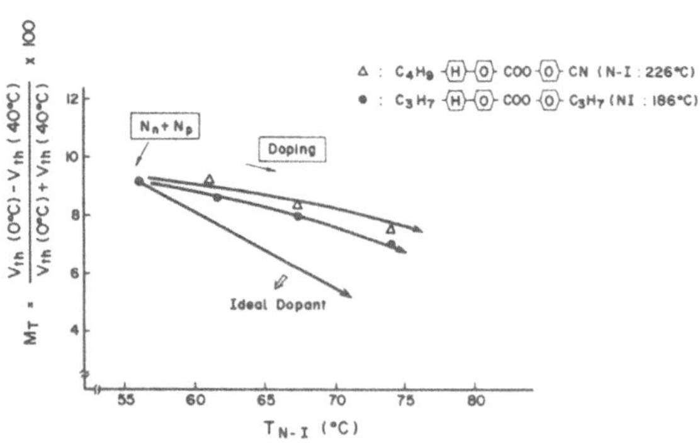


Fig. 14. Reforming of temperature dependence of threshold voltage by doping.

## Threshold Sharpness and Viewing Angle Dependence of $V_{th}$

In the system of $N_n$+$N_p$, a decreasing tendency of figure of merit (M) corresponding to the reduction of $V_{th}$ (in other words, to the increase of the $N_p^s$ component in the system) of LC materials is found.[4]

This tendency is common to all system of LCs, azoxy, Schiff's base, ester, and ECH.

Our explanation of this phenomenon is as follows: large dielectric anisotropies of LCs which are necessary for reducing threshold voltage are likely to have an adverse effect on the transmission curve, making it less steep. This may be a consequence of the relative decrease of electric field in the center of the LC layer, where the effective dielectric constant is larger. The experimental results suggest that the reduction of $V_{th}$ of LC has a strong correlation not with threshold sharpness ($\gamma$) but with both $\gamma$ and $\Delta V^{\theta}$, or the factor $\gamma/\Delta V^{\theta}$, as shown in Fig. 15. The mechanism of the correlation between $V_{th}(\Delta\varepsilon)$ and the factor $\gamma/\Delta V^{\theta}$ is not understood yet.

Anyhow, the decreasing tendency of figure of merit (M) corresponding to the lowering of the $V_{th}$ of LC is an obstacle to the design of LC materials.

As we have already stated, we know by experience that LCs consisting of dielectrically positive ($N_p$) type molecules have poor $\Delta V^{\theta}$ values, while those consisting of mixtures of dielectrically negative ($N_n$) and dielectrically positive material ($N_p$) give good $\Delta V^{\theta}$ values as shown in Fig. 8.

And also it is very important to note that among $N_n$ materials, those with small negative $\Delta\varepsilon$ values ($N_n^-$) and materials with positive but very small $\Delta\varepsilon$ are better than those with larger minus $\Delta\varepsilon$ for $N_n$+$N_p$ systems, because less doping of the system with $N_p^s$ is required for an approach to lower $V_{th}$ value of a mixed system. The ranking of azoxy ($N_n^+$)>ECH($N_n^{\pm}$)>Schiff's base ($N_n^+$) is verified by the corresponding results shown in Fig. 15.

## Optical Properties and Contrast Ratio

Another advantage of ECH over other $N_n$ such as azoxy, Schiff's base, and ester is its small birefringence ($\Delta n$). This small $\Delta n$ ensures a wide viewing angle of display as illustrated in Fig. 16, where contrast ratio's profiles for HN-1($N_n$(ECH)+$N_p^w$(PCH)+$N_p^s$) and azoxy($N_n$(azoxy)+$N_p^s$) are compared.

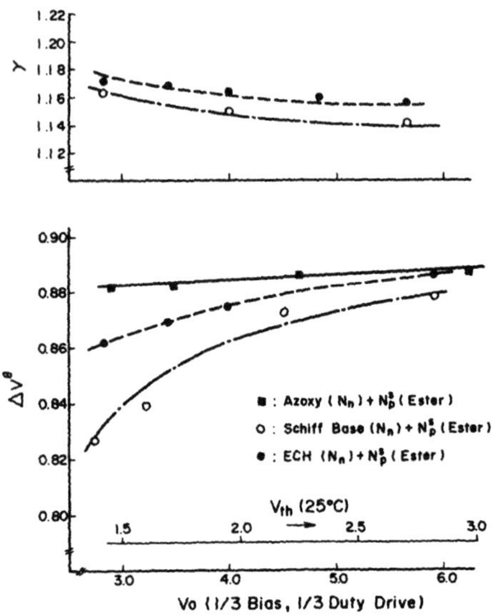


Fig. 15. Relation between threshold voltage and display performance parameter, $\Delta V^{\theta}$ (viewing angle dependence of $V_{th}$), $\gamma$ (threshold sharpness).

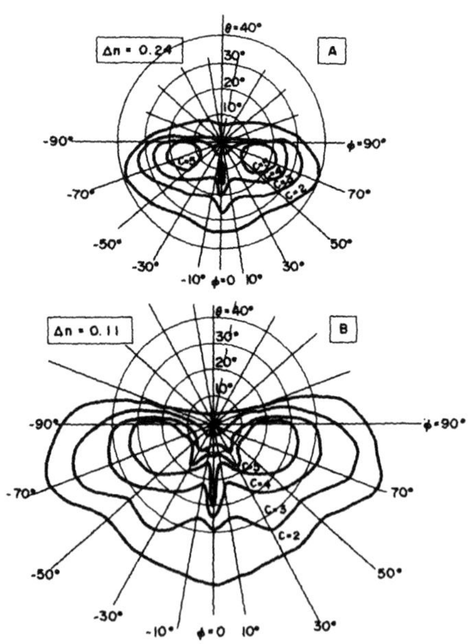


Fig. 16. Contrast ratio profile for typical LCs
A: $\Delta n$=0.24, B: $\Delta n$=0.11.

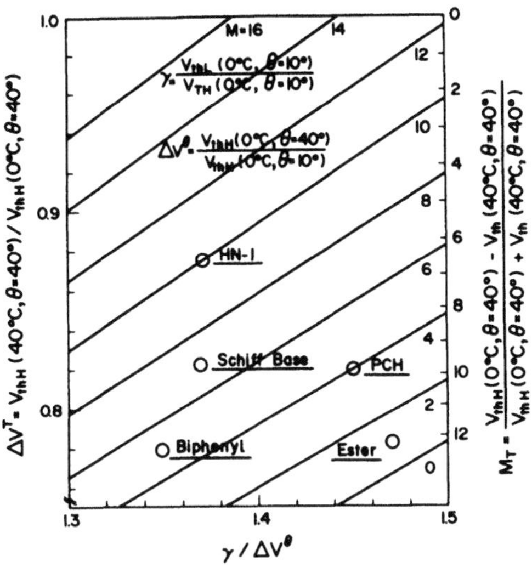


Fig. 17. Relation between figure of merit and factor $\gamma/\Delta V^{\theta}$.

Table 4. Temperature Performance Parameter $M_T$ of LCS Practically Used and Comparison with Relative Temperature Variations of Materials

| | $M_T$(%) | $-\frac{1}{V_{th}}\frac{\partial V_{th}}{\partial T}$ | (1) $\frac{1}{2}\frac{1}{K_1}\frac{\partial K_1}{\partial T}$ | (2) $-\frac{1}{2}\frac{1}{\Delta\varepsilon}\frac{\partial \Delta\varepsilon}{\partial T}$ | (1)+(2) |
|---|---|---|---|---|---|
| TN-200(SB) | 9.0 | 0.0048 | -0.0068 | 0.0036 | -0.0032 |
| Azoxy | 6.7 | 0.0034 | -0.0044 | 0.0038 | -0.0006 |
| E-7 | 16.1 | 0.0081 | -0.0067 | 0.0026 | -0.0041 |
| PCH1132 | 9.7 | 0.0050 | -0.0038 | 0.0021 | -0.0017 |
| HN-1 ($N_n+N_p^W+N_p^S$) | 8.8 | 0.0045 | -0.0052 | 0.0038 | -0.0014 |

Thus, we have selected a ternary mixture system (HN-1) of $N_n$(ECH), $N_p^w$(PCH), and $N_p^s$ for multiplexed TN-LCD. Figure 17 shows characteristics of display performance of HN-1 in comparison with other LCs.

## CONCLUSION

In developing LCs for multiplexed TN-LCD, we clarified the importance of binary $N_n$+$N_p$ systems and ternary $N_n$+$N_p^w$+$N_p^s$ systems. It was also verified that the component $N_n$ of those binary and ternary systems must be carefully examined. As an $N_n$ component, it was shown that cyclohexanecarboxylate (ECH) has good potentiality for display performance in multiplexed TN-LCD. Using ECH as $N_n$, we designed ternary system $N_n$(ECH)+$N_p^w$(PCH)+$N_p^s$ applicable to 3 volt, 1/3 bias, 1/3 duty TN-LCD which has made feasible the production of TN-LCDs with high quality and high reliability.

## ACKNOWLEDGMENT

The authors wish to thank Mr. M. Sato and Mr. H. Morishita for their helpful suggestions and advice. They also acknowledge the considerable support of Mr. H. Sato and Mr. U. Fujita (Dai-Nippon Toryo) for supplying the nematic liquid crystal materials.

## REFERENCES

1. (a) D. Demus, "Chemical Composition and Display Performance," _in_: Nonemissive Electrooptic Display, A. R. Kmetz and F. K. von Willisen, Eds. New York, Plenum, (1976).
   (b) F. J. Kahn and R. A. Burmeister, Jr., "Temperature Dependence of Multiplexed Twisted Nematic Liquid Crystal Displays," ibid.
   (c) W. H. de Jeu and J. van der Veen, Philips Tech. Rev. 37:131 (1977).
   (d) M. Schadt and F. Müller, IEEE Trans. Electron Devices, ED-25:125 (1978).
2. K. Suzuki, K. Toriyama and A. Fukuhara, Appl. Phys. Lett. 33(7):561 (1978).
3. F. Nakano, T. Umeda, H. Morishita and M. Sato, presented at the 7th International Liquid Crystal Conference, Bordeaux, (1978).
4. K. Toriyama, K. Suzuki, T. Nakagomi, T. Ishibashi and K. Odawara, "A Design of Liquid Crystal Material for Multiplexed Liquid Crystal Display," Journal de Physique, in press.

5. Chan S. Oh, Mol. Cryst. Liq. Cryst. 42:1 (1977).
6. G. Heppke and E. J. Richter, Z. Naturforsch, Teil A33(2):185 (1978).
7. H. J. Deutscher, B. Lasser, G. W. Döllin and H. Schubert, J. Prakt. Chem. 320(2):191 (1978).

# GUEST-HOST DISPLAYS

T. J. Scheffer and J. Nehring

Brown Boveri Research Center
CH-5405 Baden-Dättwil
Switzerland

## INTRODUCTION AND BACKGROUND

Guest-host displays were first described by Heilmeier and Zanoni[1] in 1968 and later in more detail by Heilmeier, Castellano and Zanoni.[2] The basic electrooptic effect consists of reorienting pleochroic dye molecules (guest) that are dissolved in a nematic liquid crystal (host) by means of an applied electric field acting on the host. Early prototype displays employing this effect were unsatisfactory because, requiring a polarizer and uniform parallel surface orientation, they were just as complex to produce as the then emerging twisted nematic displays,[3] in addition to having a lower contrast ratio because of the unavailability of suitable pleochroic dyes.

An important advance occurred in 1974 when White and Taylor[4] added a chiral compound to the guest-host mixture which made it possible to eliminate the external polarizer and the need for uniform parallel surface orientation in the cell. This modification provides the advantages of bright, shadow-free, wide-angle viewing combined with a simpler cell technology and lower production costs than the conventional twisted nematic display. In this paper we will concentrate exclusively on the White-and-Taylor-type guest-host display, reviewing the recent advances in the physics, materials and technology of these devices.

The basic features of a state-of-the-art guest-host display are illustrated in Fig. 1. The display cell consists of front and rear glass plates that are partially coated with transparent electrodes and spaced 8-10 μm apart. This space between the plates

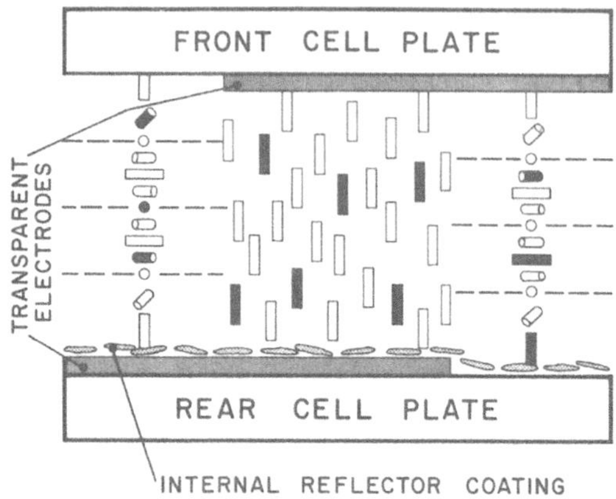


Fig. 1. State-of-the art guest-host display. The black cylinders represent pleochroic dye molecules dissolved in the nematic host (white cylinders).

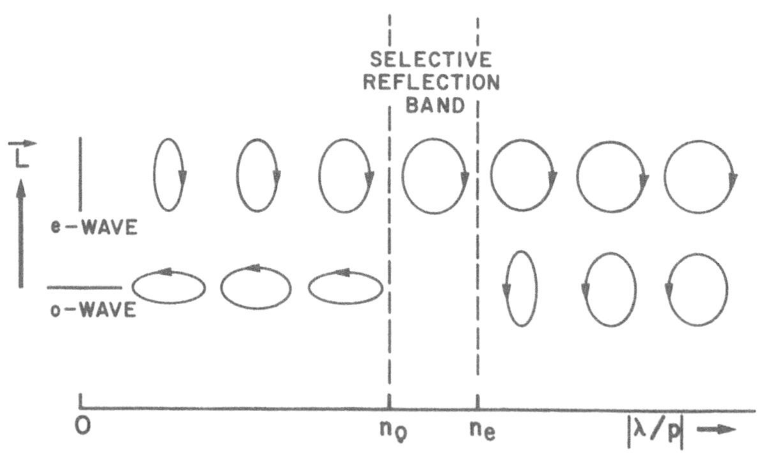


Fig. 2. Qualitative description of the ellipticity of the two eigenwaves propagating parallel to the helix axis in a cholesteric liquid crystal.

is filled with the guest-host mixture, which has a positive dielectric anisotropy, and is represented in Fig. 1 by the black and white cylinders (dye and nematic molecules). The pleochroic dye molecules, being anisotropic in shape, orient with their long axes parallel to the local optic axis in the display cell. The dye/host systems used in this type of display have positive dichroism, i.e., maximum absorbance occurs when the electric field vector of the light vibrates parallel to the local optic axis. The chiral component in the guest-host mixture gives the layer a spontaneously twisted structure in the regions of the cell where there is no applied electric field. These twisted regions appear dark because, as is shown below, such a structure is a very efficient absorber of unpolarized light. Applying a large enough electric field in the cell completely unwinds the twisted structure and reorients the optic axis so that it is uniformly parallel to the applied electric field. This is the optically bright state. The perpendicular boundary orientation and the internal reflector coating shown in Fig. 1 will be discussed later on.

## OPTICS IN TWISTED ANISOTROPIC ABSORBING MEDIA

To understand how displays employing the White and Taylor guest-host scheme can have high contrast ratios without polarizers we need to investigate the optical problem of light propagating in a twisted anisotropic absorbing medium. For uniformly twisted nematic (cholesteric) layers without absorption, Mauguin[5] and de Vries[6] have shown that light propagating parallel to the helical axis may be described by a superposition of two elliptically polarized eigenwaves of opposite sense. In a local Cartesian coordinate system (x,y,z) with the x-axis parallel to the local optic axis $\underset{\sim}{L}$ and the z-axis parallel to the direction of propagation, the electric field vectors in the (x,y)-plane, $\underset{\sim}{E}^+$ and $\underset{\sim}{E}^-$, of the two eigenwaves may be written in complex notation as[6]

$$\underset{\sim}{E}^{\pm} = \begin{bmatrix} A_x^{\pm} \\ iA_y^{\pm} \end{bmatrix} \cdot e^{i\left(\omega t - \frac{2\pi}{\lambda} m_{\pm} z\right)} \tag{1}$$

The amplitudes $A_x^+$, $A_y^+$ and $A_x^-$, $A_y^-$ are related to each other through

$$\frac{A_x^{\pm}}{A_y^{\pm}} = \frac{2m_{\pm}\left(\frac{\lambda}{p}\right)}{n_e^2 - m_{\pm}^2 - \left(\frac{\lambda}{p}\right)^2}, \tag{2}$$

with the "characteristic exponents" $m_+$ and $m_-$ being given by

$$m_\pm^2 = \left(\frac{\lambda}{p}\right)^2 + \frac{1}{2}\left(n_e^2+n_o^2\right) \pm \left[4\left(\frac{\lambda}{p}\right)^2 \frac{n_e^2+n_o^2}{2} + \left(\frac{n_e^2-n_o^2}{2}\right)^2\right]^{1/2} . \tag{3}$$

In Eqs. (1) through (3), $n_e$ and $n_o$ are the extraordinary and ordinary refractive indices of the locally uniaxial system, $\lambda$ is the wavelength of light in vacuum and p is the pitch of the twisted structure defined as the distance measured along the helical axis for the local optic axis to twist around a full 360°. It follows from Eqs. (1)-(3) that the major and minor axes of vibration of both elliptical eigenwaves are either parallel or perpendicular to the local optic axis $\underset{\sim}{L}$. The ellipticities of the eigenwaves, $e_+$ and $e_-$, are defined by

$$e_\pm = \left|\frac{A_x^\pm}{A_y^\pm}\right| . \tag{4}$$

Figure 2 qualitatively illustrates the effect of the $|\lambda/p|$ ratio upon the ellipticity of the two eigenwaves. For the case of an infinitely long pitch, i.e., nontwisted structure, the two eigenwaves are linearly polarized waves having electric field vectors parallel (e-wave) and perpendicular (o-wave) to the local optic axis. As $|\lambda/p|$ increases, the eigenwaves become elliptical and in the limit of large $|\lambda/p|$, circular. The ellipticity of the e-wave decreases monotonically with increasing $|\lambda/p|$, whereas the ellipticity of the o-wave reaches a relative maximum value in the range $0 \le |\lambda/p| \le n_o$. Figure 3 shows $e_+$ and $e_-$ as functions of $|\lambda/p|$ computed from Eqs. (2)-(4) with typical values $n_e=1.7$ and $n_o=1.5$.

The treatment of the optical problem when anisotropic absorption is present can be carried out in the same manner as in the nonabsorbing case, except that now complex refractive indices

$$n^*_{e,o} = n_{e,o} - i\,\frac{\lambda}{4\pi}\,\alpha_{e,o} \tag{5}$$

have to be used to characterize the nematic state. In Eq. (5), $n_e$, $n_o$, $\alpha_e$ and $\alpha_o$ are the principal refractive indices and absorption constants of the untwisted guest-host mixture. It can be shown that the effect of the absorption on the real part of the refractive index is so small that it may be completely neglected, even for the highest practical dye concentrations. The $n_e$ and $n_o$ of Eq. (5) may therefore be taken as the extraordinary and ordinary refractive indices of the guest-host mixture without the added dye. $\alpha_e$ and $\alpha_o$ are related to the transmittances $T_e$ and $T_o$ for the extraordinary and ordinary waves

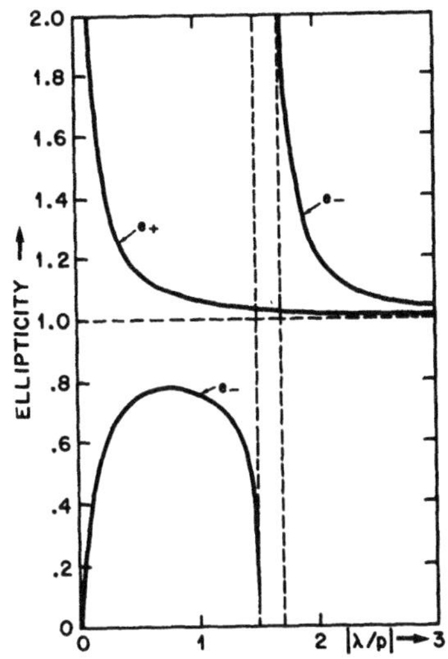


Fig. 3. $\lambda/p$ dependence of the ellipticity of the two eigenwaves propagating parallel to the helical axis in a twisted non-absorbing medium with $n_e$=1.7 and $n_o$=1.5. Region between dashed lines indicates selective reflection band.

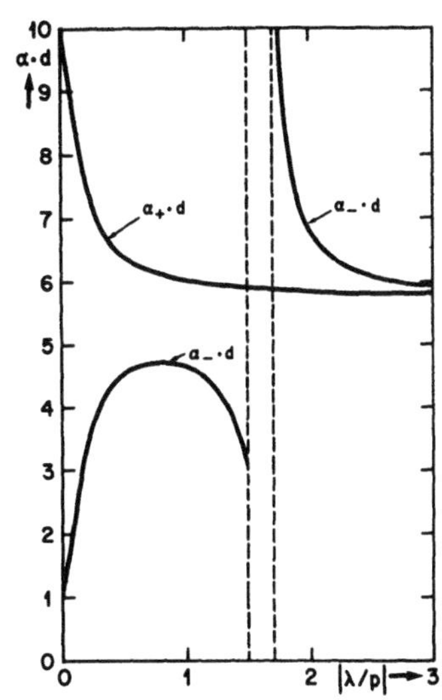


Fig. 4. $\lambda/p$ dependence of the eigenwave absorbances $\alpha_+ \cdot d$ and $\alpha_- \cdot d$ for the typical case where $n_e$=1.7, $n_o$=1.5, $\alpha_e \cdot d$=10.0, $\alpha_o \cdot d$=1.0, $\lambda$=500 nm and d=10 μm. Region between dashed lines indicates selective reflection band.

passing through a layer of thickness d of the untwisted guest-host mixture by

$$T_{e,o} = e^{-\alpha_{e,o} d} . \tag{6}$$

As in the nonabsorbing case, two plane elliptical eigenwaves are found which are described by Eqs. (1) and (2) after $m_\pm$ and $n_e$ have been replaced by $m_\pm^*$ and $n_e^*$, where $m_\pm^*$ is given by[7]

$$m_\pm^{*2} = \frac{\lambda}{p} + \frac{1}{2} n_e^{*2} + n_o^{*2} \pm \left[4\left(\frac{\lambda}{p}\right)^2 \frac{n_e^{*2}+n_o^{*2}}{2} + \left(\frac{n_e^{*2}-n_o^{*2}}{2}\right)^2\right]^{1/2} . \tag{7}$$

In analogy to Eq. (5), $m_\pm^*$ may be expressed as

$$m_\pm^* = m_\pm' - i \frac{\lambda}{4\pi} \alpha_\pm , \tag{8}$$

where $m_\pm'$ and $\alpha_\pm$ are real quantities, with $\alpha_\pm \geq 0$. Referring to the analog of Eq. (1) for the absorbing case, it is seen that $\alpha_+$ and $\alpha_-$ can be interpreted as the absorption constants of the two elliptical eigenwaves. Numerical values for $\alpha_+$ and $\alpha_-$ are easily obtained from Eqs. (7) and (8) by computer using standard complex algebra. For unpolarized light, the transmittance of a twisted absorbing layer of thickness d is approximately given by

$$T = \frac{1}{2}\left(e^{-\alpha_+ d} + e^{-\alpha_- d}\right) . \tag{9}$$

For a reflective system with a metallic reflector placed behind the layer, the reflectance I can then be shown to be

$$I = e^{-(\alpha_+ + \alpha_-)d} + \frac{1}{2} \frac{(x_+ + x_-)(\bar{x}_+ + \bar{x}_-)}{(x_+ - x_-)(\bar{x}_+ - \bar{x}_-)} \left(e^{-\alpha_+ d} - e^{-\alpha_- d}\right)^2 , \tag{10}$$

where $x_+$ and $x_-$ are defined by the complex quantity

$$x_\pm = \frac{2 m_\pm^* \left(\frac{\lambda}{p}\right)}{n_e^{*2} - m_\pm^{*2} - \left(\frac{\lambda}{p}\right)^2} ,$$

and $\bar{x}_\pm$ denotes the complex conjugate of $x_\pm$. Equation (10) takes into account the fact that the handedness of the ingoing elliptical waves is reversed upon reflection.

Our expressions (7) and (8) reduce to the approximate expressions for $\alpha_+$ and $\alpha_-$ derived by Saupe[7] and found in Ref. 8, which in our notation may be written

$$\alpha_{\pm} \simeq \left[ \frac{n_e\alpha_e+n_o\alpha_o}{2m_{\pm}} \cdot \left\{ 1 \pm \frac{2\left(\frac{\lambda}{p}\right)^2 + \frac{n_e^2-n_o^2}{2} \cdot \frac{n_e\alpha_e-n_o\alpha_o}{n_e\alpha_e+n_o\alpha_o}}{\left[4\left(\frac{\lambda}{p}\right)^2 \frac{n_e^2+n_o^2}{2} + \left(\frac{n_e^2-n_o^2}{2}\right)^2\right]^{1/2}} \right\} \right] \tag{11}$$

with $m_{\pm}$ given by Eq. (3). Equation (11) holds for $|\lambda/p|$-values falling outside the selective reflective band under the condition that in Eqs. (5) and (8) the imaginary parts are small compared with the real parts. For $(n_{\parallel}-n_{\perp})<<1$, Eq. (11) may be further simplified to give

$$\alpha_{\pm} \simeq \frac{n_e\alpha_e e_{\pm}^2+n_o\alpha_o}{n_e e_{\pm}^2+n_o} \tag{12}$$

with $e_{\pm}$ taken from Eqs. (4) and (2), i.e., the nonabsorbing case. Equation (12) is in agreement with a corrected version of the perturbation expression given by White and Taylor,[4] where the dielectric constants are replaced by the respective refractive indices.

Figure 4 shows the $|\lambda/p|$ dependence of the eigenwave absorption constants $\alpha_+$ and $\alpha_-$ computed from Eqs. (7) and (8). Note the similarity of the $\alpha_{\pm}$ curves with the $e_{\pm}$ curves of Fig. 3. The relationship between absorption and ellipticity is also seen from the approximate expression (12). The similarity between these curves is reasonable because the greater the ellipticity, the greater is the component of the E-vector along the major absorption axis. Note that the layer absorbs the most unpolarized light when $|\lambda/p|\sim n_e$. For applications in displays, however, values of $|\lambda/p|$ greater than $n_o$ are not of practical interest because the operating voltage would be prohibitively high (see Eq. 18).

## APPLICATIONS OF THE OPTICAL THEORY FOR PRACTICAL GUEST-HOST DISPLAYS

The theory developed in the last section can be applied to guest-host displays to obtain results which are of immediate practical interest. In this case the photometric brightness and contrast ratios will be determined because this is what the eye actually perceives. The perceived brightness of the display is given by the integral[9]

$$B = \frac{\int H(\lambda)\bar{y}(\lambda)I(\lambda)d\lambda}{\int H(\lambda)\bar{y}(\lambda)d\lambda}, \tag{13}$$

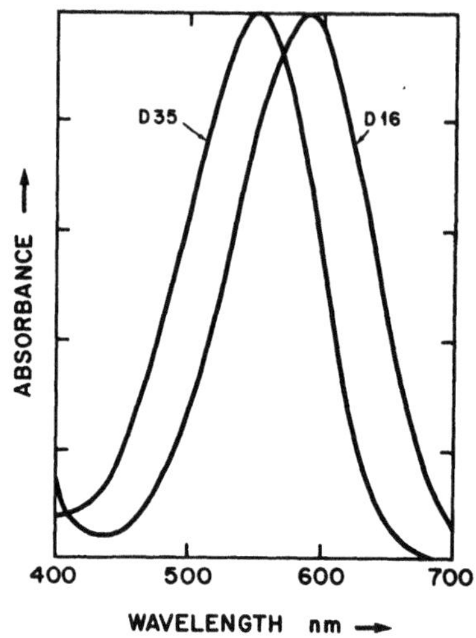


Fig. 5. Normalized absorbance spectra for two anthraquinone dyes D35 (rose) and D16 (blue) |10| dissolved in the nematic host RO-TN-403 |11|.

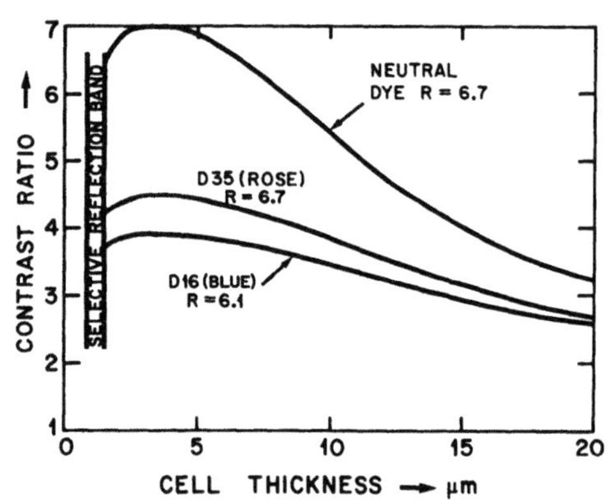


Fig. 6. Computed dependence of perceived contrast ratio on cell thickness for a reflective display with three different guest-host dyes having the indicated dichroic ratios R. Dye concentration adjusted to maintain an on-state brightness of 50%. Concentration of chiral dopant adjusted to maintain 3 turns of the cholesteric helix in the cell.

where $H(\lambda)$ is the intensity distribution of the illumination (CIE standard source C), $\bar{y}(\lambda)$ is the standard CIE spectral sensitivity function of the eye and $I(\lambda)$ is the reflection spectrum of the display in its field-off or field-on state. For these computations we assume a metallic reflector with 100% reflectivity. For the field-off state, $I(\lambda)$ is obtained from Eq. (10). $I(\lambda)$ for the field-on state is given by

$$I(\lambda) = e^{-2\alpha_o(\lambda)d} , \tag{14}$$

since the optic axis in the field-on state is uniformly perpendicular to the plane of the layer and the electric vector of the unpolarized light interacts only with the ordinary absorption constant $\alpha_o$.

We compute the display brightness in the field-off and field-on states, B(off) and B(on), by evaluating the integral (13) with the appropriate $I(\lambda)$ function. The perceived contrast ratio C of the display is defined by

$$C = B(on)/B(off) . \tag{15}$$

In order to obtain $I(\lambda)$ from Eqs. (10) or (14), it is necessary to know the wavelength dependence of the principal absorption constants of the untwisted guest-host mixture, $\alpha_e(\lambda)$ and $\alpha_o(\lambda)$. We have determined these data for two of the commercially available dyes, D35 (rose) and D16 (blue)[10] (see Section V), whose absorption spectra were measured in the nematic host mixture RO-TN-403.[11] The normalized absorbance curves for D35 and D16 are given in Fig. 5. Only one curve is shown for each dye because the dichroic ratio $R=\alpha_e/\alpha_o$ was found to be independent of wavelength. For 1% concentrations by weight of D35 and D16 in RO-TN-403, R was measured to be 6.7 and 6.1 at room temperature. In the computations to follow we have assumed that the normalized absorption curves and dichroic ratios measured for these dyes in RO-TN-403 are also valid in other host materials. We also carried out the computations with a hypothetical neutral dye whose absorption constants are independent of wavelength, assuming a value R=6.7 for the dichroic ratio.

Figure 6 shows the computed dependence of the display contrast ratio on the cell thickness for reflective guest-host systems employing the three dyes mentioned above. The host system is assumed to have a birefringence of 0.26 (with $n_o$=1.50), corresponding to a phenylpyrimidine-type material.[11] For these curves the cell thickness-to-pitch ratio (d/p), corresponding to the number of helical turns in the layer, is kept fixed at 3.0, which gives a constant critical voltage of about 5.4 volts when RO-TN-103 is employed as a host (see Fig. 11). The dye concentration was adjusted to maintain the display brightness in the field-on state, B(on), at the constant value of 50%. The contrast curves of Fig. 6 have maximum values for cells that are about 3.5 μm thick. This behavior is related to the maximum in the $\alpha_{-} \cdot d$ curve seen on the left-hand side of Fig. 4.

In practice the liquid crystal layer in guest-host display cells using commercial quality glass plates cannot be made much thinner than about 7 μm, since the large fractional variations in the thickness of the guest-host layer begin to introduce a cloudy appearance to the display background. If optically polished cell plates are used, the minimum layer thickness would be determined by the maximum solubility of the dye in the liquid crystal host. For anthraquinone-type dyes having a 5% solubility, this minimum layer thickness would be about 2 μm.

According to Fig. 6, a 7 μm-thick cell containing the neutral dye has a contrast ratio of 6.4, which is more than the 5:1 contrast ratio measured for an ordinary newspaper.[12] Higher contrast ratios can generally be obtained at the expense of display brightness by increasing the dye concentration. Figure 7 illustrates this point for the above three dyes and the typical values of a 10 μm-thick layer with d/p=3.0, $n_o$=1.5 and Δn=0.26. All three curves start on the upper left at the point B=100% and C=1.0, corresponding to zero dye concentration and no absorption. Increasing the dye concentration decreases the display brightness and increases the contrast ratio. For the neutral dye this process continues indefinitely, whereas the contrast ratios of the two colored dyes, D16 and D35, reach maximum values. This surprising behavior is a general phenomenon for any dye which has practically no absorption in a wavelength region where the eye is still sensitive. A very high concentration of such a dye would give a contrast ratio near 1.0.

Figure 8 illustrates the effect on the display contrast ratio of varying Δn and the total number of turns of the cholesteric helix while maintaining a constant 10.0 μm layer thickness. The three curves shown correspond to the neutral dye dissolved in hosts having $n_o$=1.5 and birefringences of 0.06, 0.14 and 0.26, corresponding to the substance classes cyclohexylcyclohexane (CCH),[13] phenylcyclohexane (PCH),[14] and phenylpyrimidine (PPYR).[11] The dye concentration is chosen to make B(on)=50%, and a dichroic ratio of 6.7 is assumed for all three guest-host mixtures. Figure 8 shows that increasing the number of helical turns in the 10.0 μm layer increases the display contrast ratio. The number of turns also increases the critical voltage of the display (Fig. 11) and therefore a compromise must be reached between display contrast ratio and operating voltage. Note that a host with small Δn requires fewer helical turns to obtain the same display contrast ratio.

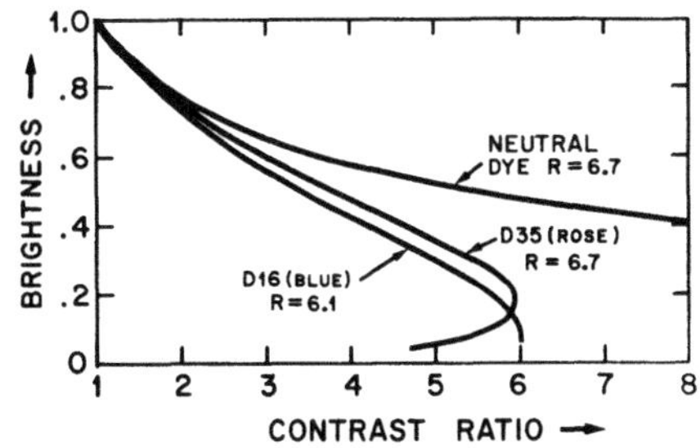


Fig. 7. Reflected display brightness versus perceived contrast ratio for three guest-host dyes having the indicated dichroic ratios. Increasing the concentration of the dye generally decreases the display brightness and increases the contrast ratio. Values used for this computation: $n_e$=1.76, $n_o$=1.50, d/p=3.0 and d=10.0 μm.

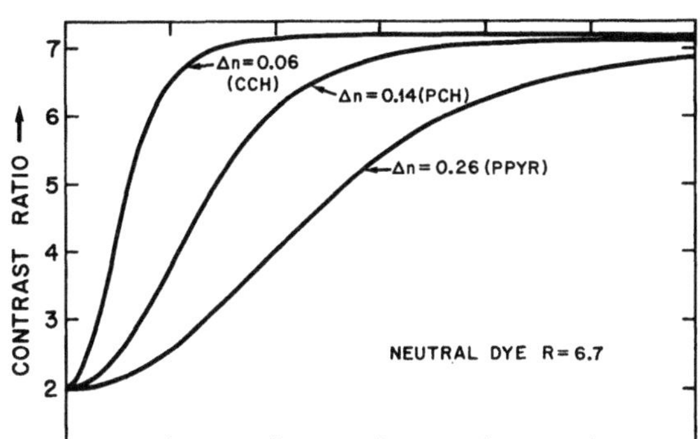


Fig. 8. Dependence of contrast ratio upon the number of turns of the cholesteric helix in the layer for different host birefringences Δn. The three curves are computed for a hypothetical neutral dye with R=6.7 dissolved in different liquid crystal hosts. Indicated birefringence values are typical for the cyclohexylcyclohexane (CCH), phenylcyclohexane (PCH) and phenylpyrimidine (PPYR) liquid crystal substance classes. Values used for the computation: $n_o$=1.50, d=10.0 μm and on-state brightness=50%.

The effect of the host birefringence and the number of helical turns upon the contrast ratio of the display is directly seen in Fig. 9. The curves of constant contrast were computed for the case of a 10 μm-thick guest-host layer with $n_o$=1.5 containing enough of the neutral dye so that B(on)=50%. The fact that these curves are nearly straight lines passing through the origin indicates that for a given cell thickness and small enough Δn, the contrast ratio is only a function of the product Δn•p. Without changing the contrast, the operating voltage of a guest-host display could therefore be significantly reduced by increasing the pitch and employing a host material with a smaller Δn. In practice, however, all known nematic liquid crystals having small Δn values also have small dielectric anisotropies Δε, and so an increase in the pitch with a decrease in the number of turns in the cell does not necessarily mean a decrease in the operating voltage of the display.

A useful index f for selecting a host material that will give the lowest operating voltage for a given contrast ratio can be obtained by combining the linear relationship observed in Fig. 9 with de Gennes' expression for the critical unwinding voltage (see Eq. 18) to give

$$f = \frac{1}{\Delta n}\left[\frac{\varepsilon_o \Delta\varepsilon}{k_{22}}\right]^{1/2}, \qquad (16)$$

where $k_{22}$ is the twist elastic constant of the nematic host, and $\varepsilon_o$ is the permittivity of free space=$8.85\times10^{-12}$ MKS.

Values of f are summarized in Table 1 for a number of commercially available nematic mixtures. The substance with the highest f value has the lowest operating voltage for a given contrast ratio. The index does not include the effect of the host material on the dichroic ratio, which can be large.[15,16] Other important criteria such as stability and viscosity are also not considered in f.

## EXPERIMENTAL

The importance of surface orientation on the performance of guest-host displays was recognized by White and Taylor,[4] who investigated displays having both parallel and perpendicular boundary orientations. In this section we will look more closely at the textures and electro-optical properties of guest-host layers having these two types of boundary orientation.

The cholesteric texture occurring under the conditions of parallel boundary orientation is a featureless, twisted planar

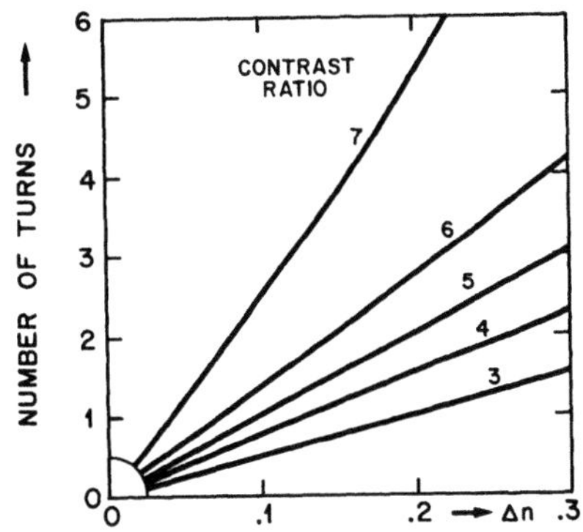


Fig. 9. Curves of equal perceived contrast ratio in dependence of the number of helical turns and the host birefringence Δn. The contrast curves are computed for a hypothetical neutral dye with R=6.7. Values used for this computation: $n_o$=1.50, d=10.0 μm and on-state brightness=50%.

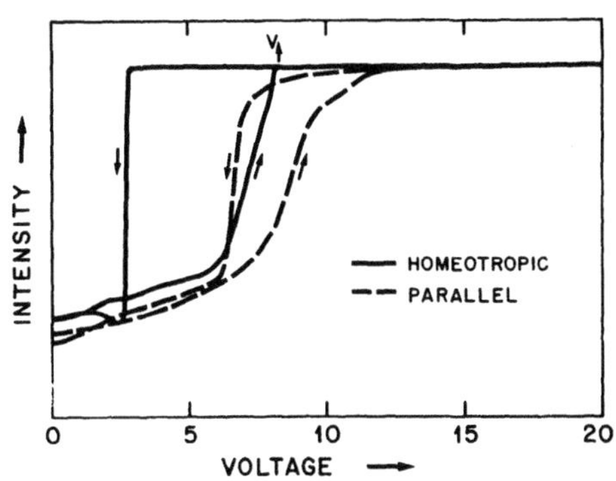


Fig. 10. Intensity versus voltage curves for a guest-host mixture under conditions of homotropic and parallel boundary orientation. Arrows in the region of hysteresis indicate the direction that the curves were traced. Curves are for a guest-host mixture containing 2.0% D16 and 3.7% CB 15 in nematic host RO-TN-403. Cell thickness is 15.8 μm. Voltage units is volts.

structure where the helical axis is perpendicular to the plane of the layer. This texture, known as the Grandjean texture, was assumed for the optical theory given above. Parallel boundary orientation is unsuitable for display applications because the Grandjean texture in deactivated regions of the display is full of metastable disclination lines. These disclination lines strongly scatter the incident light and give the display an objectionable "after image" that can persist for several minutes after a display element has been turned off.

The cholesteric texture occurring with perpendicular boundary orientation, on the other hand, is not a uniform texture. Under the microscope the entire field of view is filled with right- and left-handed spirals. This cholesteric texture is known as the "scroll" texture[17] and appears very much like an end-on view of a bundle of rolled-up scrolls. Capacitance measurements show that the cholesteric helical axis in this texture is still predominantly perpendicular to the plane of the layer. An exact analysis of the structure of this texture, even for the case of equal elastic constants, seems difficult to achieve.[18] Important from the applications point of view is that the scroll texture is adopted without disclinations immediately after a display element is turned off and that this structure is essentially nonscattering. Homeotropic boundary orientation therefore solves the problem of display after-images.

The nonscattering scroll texture is quite different from the highly scattering cholesteric focal conic texture which occurs at intermediate voltages.[17] In the focal conic texture the helical axis is essentially parallel to the plane of the layer. This texture never occurs in a display under the condition that all the display elements are either fully turned on (untwisted state) or completely turned off (scroll texture). This condition is generally not fulfilled in multiplexed displays.

The type of boundary orientation employed in a guest-host display has a large effect on the shape of its brightness-voltage curve. Figure 10 shows the brightness-voltage curves measured in a single cell which was divided into two sections, one with parallel boundary orientation and one with perpendicular boundary orientation. For perpendicular boundary orientation the saturation brightness is approached quite sharply and at a much lower voltage than is the case for parallel boundary orientation, where the approach is very gradual. Well-known hysteresis effects are present for both types of orientation.[19,20] Several schemes have been published that exploit this hysteresis effect for multiplexing phase-change matrices,[21-23] but for a nonmultiplexed display, where either the full addressing voltage or no voltage at all is present, the hysteresis is of no consequence.

Figure 11 shows the dependence of the critical unwinding voltage $V_\uparrow$ on the cell thickness-to-pitch ratio for a cholesteric layer with perpendicular boundary orientation. For these measurements two reusable cells, having thicknesses of 7.8 and 10.9 μm, were filled with various mixtures of the nematic liquid crystal RO-TN-103[11] and the chiral substance CB 15.[24] We used the Cano wedge technique[25] to determine the pitch of the mixtures and found the simple inverse relationship

$$p(\mu m) = 13.5/c(wt.\%)$$

to hold even at the highest CB 15 concentrations c of 8-10%. According to Fig. 11 the critical voltage $V_\uparrow$ depends only upon the ratio of cell thickness to pitch, d/p, which is the number of turns of the cholesteric helix in the layer. Furthermore, for $d/p \gtrsim 3$. the critical voltage $V_\uparrow$ obeys the simple linear relationship

$$V_\uparrow = A\left(\frac{d}{p}\right) , \qquad (17)$$

where A=1.7 Volts. The electrical analog of de Gennes' critical unwinding field[26] derived for an infinitely thick cholesteric layer where there are no torques exerted by the boundaries is given by

$$V_\uparrow = \pi^2\left(\frac{d}{p}\right)\left[\frac{k_{22}}{\varepsilon_o\Delta\varepsilon}\right]^{1/2} . \qquad (18)$$

Substituting into Eq. (18) the material parameters for RO-TN-103[11] (twist elastic constant $k_{22}=7.30\times10^{-12}$N, dielectric anisotropy $\Delta\varepsilon=26.20$) gives A=1.75V, in good agreement with the experimental value of 1.77V. The departure of the experimental curve in Fig. 11 from the straight line determined from Eq. (18) therefore gives an indication of the increasing influence that the boundaries have on the bulk as the number of helical turns in the layer becomes small. Note that for values of d/p below a certain critical value, the effect of the boundaries is so strong that the critical voltage is zero and a homeotropic nematic structure is adopted. Fischer's analysis[27] shows that the homeotropic structure is stable as long as the inequality $d/p \le k_{33}/(2k_{22})$ holds. From the elastic constant data available for RO-TN-103[11] $k_{33}/(2k_{22})=1.27$, which is of the same order as the critical value of 1.05 obtained for d/p from Fig. 11.

Table 1. Index f for Various Host Substances

| Substance[11] | $\Delta\varepsilon$[11] | $\Delta n$[11] | $k_{22}$[11] $[10^{-12}N]$ | f |
|---|---|---|---|---|
| RO-TN-200 | 18.4 | 0.2755 | 8.35 | 16 |
| RO-TN-103 | 26.2 | 0.2220 | 7.30 | 25 |
| RO-TN-132 | 22.2 | 0.1930 | 7.80 | 26 |
| RO-TN-403 | 19.2 | 0.2575 | 10.80 | 15 |
| RO-TN-430 | 17.5 | 0.2335 | 9.10 | 18 |
| E7 | 13.8 | 0.2240 | 10.00 | 16 |
| 1132 | 10.4 | 0.1420 | 9.00 | 23 |

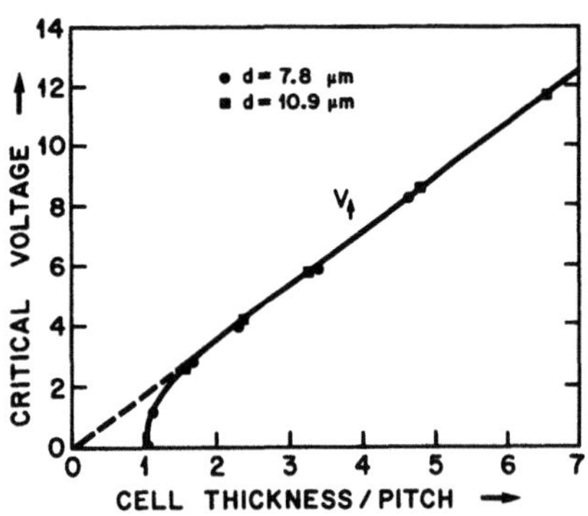


Fig. 11. Measured dependence of the critical unwinding voltage V↑ in volts upon the cell thickness-to-pitch ratio. Homeotropic boundary orientation. Dashed line is theoretical curve for infinite medium without boundaries. Host substance is RO-TN-103.

## MATERIALS AND TECHNOLOGY

### Guest-Host Dyes

Until recently, a major obstacle preventing the serious application of guest-host displays has been the poor photochemical stability of the dyes. Research in the past has primarily focussed on the azo dyes because their simple linear structures assured a high dichroic ratio. The three azo dyes, D1, D2 and D3 in Fig. 12, for example, are reported to have dichroic ratios in the biphenyl mixture E8 at 20°C of 8.0, 10.0 and 12.3.[28] The photochemical stability of azo dyes, however, is insufficient for use in guest-host displays. We measured practical lifetimes of 10, 220 and 15 hours for solutions containing the dyes D1, D2 and D3 that were exposed to illumination approximating the intensity and spectral distribution of noon sunlight.[29] Blue azo dyes, of greatest practical interest, seem to be the least stable. Attempts to stabilize the azo dyes by adding side groups to the benzene rings have generally only led to a decrease in solubility and dichroic ratio.

A completely new class of guest-host dyes based on the anthraquinone structures has recently been reported.[10] The chemical structures of these dyes are of the two types illustrated in Fig. 13 where R denotes a general substituent. R=$-OC_9H_{19}$(I) for the dye D16 (blue) and $-C_2H_5$(II) for the dye D35 (rose). 1% concentrations of these two dyes in RO-TN-403 at 20°C have dichroic ratios of 6.1 and 6.7, which is surprisingly high, considering the comparatively broad structures of these molecules. While these values are not as high as those of the linear azo dyes, they are still adequate for good display performance. The advantage of the anthraquinone dyes is their excellent photochemical stability, which is at least 100 times higher than the stability of the blue azo dyes. Liquid crystal solutions of the anthraquinone dyes D16 and D35 are able to withstand several thousand hours of continuous noon sunlight illumination, which is more than adequate for practical applications. Because the molar extinction coefficients of anthraquinone dyes are typically three to four times smaller than those of azo dyes, their concentration must be correspondingly increased in the liquid crystal host in order to obtain adequate absorption. A typical guest-host mixture for a 10 µm-thick display requires a 1.5-2.0% concentration of the anthraquinone dye. This presents no problem since the solubility of many of these dyes is more than 5% in biphenyl liquid crystal hosts.[10]

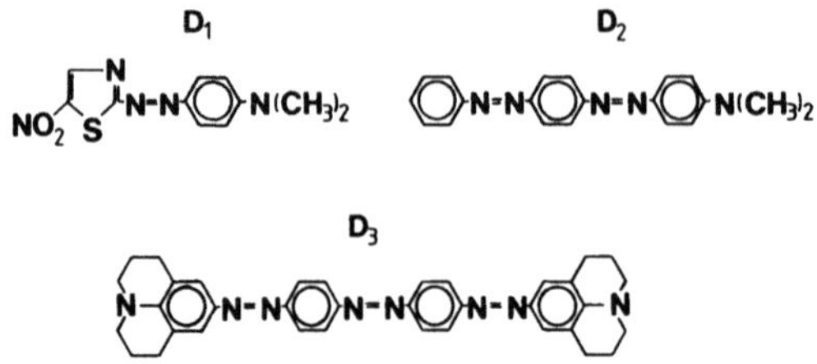


Fig. 12. Examples of 3 guest-host azo dyes. The linear structure of these molecules assures a high dichroic ratio.

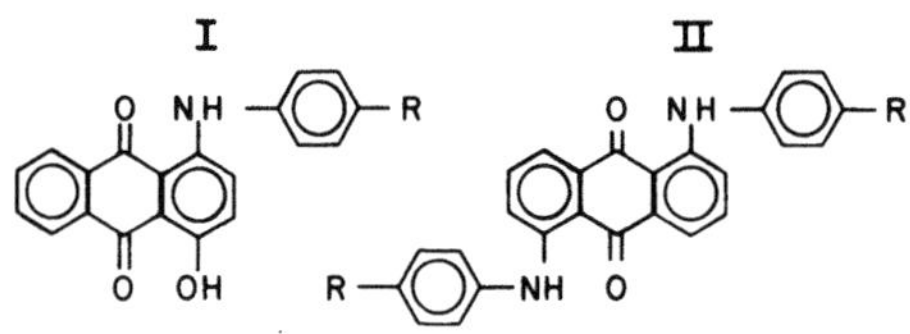


Fig. 13. Two classes of guest-host dyes that are based on the anthraquinone structure. R denotes substituents such as $-C_4H_9$, $-OC_4H_9$, $-C_2H_5$, etc. The photochemical stability of these dyes is adequate for display applications.

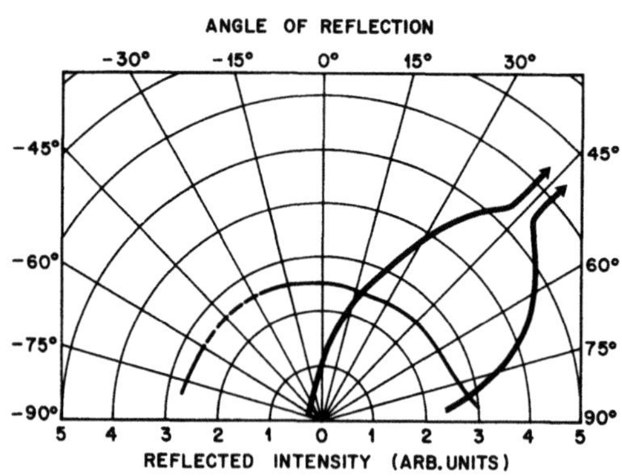


Fig. 14. Angular dependence of the reflected light intensity in the plane of incidence from a MgO reflectance standard (thin line) and an internal reflector in a guest-host cell (heavy line). Collimated white light is incident to both samples at -45°. Arrows denote specular reflection from front surface of display cell.

Internal Reflector

Unlike the twisted nematic displays which require an external reflector because of the external polarizers, the reflector for a guest-host display can be located inside the cell in direct contact with the liquid crystal layer. The internal reflector completely eliminates the parallax and shadow effects that are encountered in displays with external reflectors, and allows full use to be made of the intrinsically wider field of view available to the guest-host effect. An internal reflector can also improve the uniformity of the layer thickness because thicker cell plates can be employed without introducing parallax effects.

One obvious material to use for an internal reflector would be a thin layer of $TiO_2$ powder. In practice, however, this gives the display an unacceptably dark appearance because the reflectivity of $TiO_2$ is greatly reduced when it is in contact with the liquid crystal. Furthermore, the Lambertian nature of the $TiO_2$ reflectance means that too much of the incident light is scattered into low angles where it is readily absorbed in the guest-host layer and never reaches the viewer.

One way to obtain a non-Lambertian reflector having a high reflectivity, even in contact with the liquid crystal, is to use a metallic layer. This, however, presents some technical difficulties since the reflector must also be an electrical insulator if the conduction paths leading to the electrode segments on the front cell plate are to remain invisible when the display is in operation. We have developed an internal reflector that consists of a monolayer of commercially available aluminum pigment flakes.[30] This gives the desired metallic optical properties and at the same time is electrically insulating because there is a thin oxide coating on each pigment flake. We deposit the reflector onto the rear cell plate by a silk-screen printing process using a mixture of the aluminum flakes in a special binder. A subsequent heat treatment burns off the binder and leaves a monolayer of the aluminum flakes lying flat on the rear cell plate. The aluminum flakes themselves are about 10 μm in diameter and 0.5 μm thick.

These reflectors have a total reflectance of about 65% and a transmittance of about 6%, which makes it possible to operate the display also in the transmissive mode. These values can be varied by adjusting the concentration of aluminum pigment in the binder, but an upper limit for the reflectivity seems to be about 70%. The fact that this value is less than the reflectance of 91% for freshly evaporated aluminum is due to the strongly oxidized state of the aluminum flakes.

Figure 14 shows the angular dependence of the reflectance of an aluminum pigment internal reflector in a guest-host display cell filled with a liquid crystal, but without the added dye. The angular dependence of a white MgO reflectance standard is illustrated for comparison. For these measurements the angle of incidence of a narrow beam of collimated white light was fixed at -45° and the reflected intensity was measured as a function of angle in the plane of incidence. The MgO reflectance curve is nearly a semicircle, indicating Lambertian behavior. The internal reflector, on the other hand, has a maximum at the specular angle with a gain of about 2.2. The arrows in Fig. 14 indicate the specular reflection from the front surface of the display cell.

## Positive Contrast Displays

It is generally recognized that positive contrast passive displays (dark characters on a light background) are more readable than negative contrast displays (light characters on a dark background), especially under conditions of low ambient illumination. Twisted nematic displays can be either of the positive or negative contrast types, depending upon whether the polarizers are crossed or parallel. The conventional guest-host effect, on the other hand, is only a negative-contrast-type display since the characters are generally defined by the regions where the front and rear electrodes overlap and these regions are always bright in the field-on state.

## Materials with $\Delta\varepsilon<0$

One possible scheme for a positive contrast guest-host display is illustrated in Fig. 15.[31] The display cell has perpendicular boundary orientation, as with the conventional scheme, but here the nematic mixture has a negative dielectric anisotropy and the concentration of the chiral compound is reduced below the critical amount required to sustain a twisted structure in the field-off state. The optically bright homeotropic nematic structure is therefore present in the field-free regions of the display (background). In the display regions where an electric field is applied (characters), the optic axis tips away from the electric field direction and forms a twisted structure due to the presence of the chiral compound. Only one turn of the helix is shown in Fig. 15 because this is the largest amount of twisting that is possible with presently known materials. This limit follows directly from Fischer's relationship $d/p<k_{33}/(2k_{22})$,[27] which must hold in order for the homeotropic nematic structure to be stable in the field-off state, and the fact that for typical nematic materials $k_{33}/(2k_{22})\simeq 1$. As Fig. 8 shows, however, even one turn of the cholesteric helix can give good display performance if

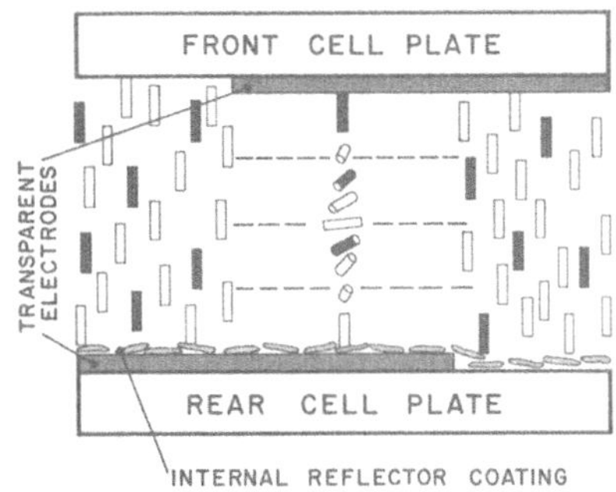


Fig. 15. Scheme for a positive contrast guest-host display (i.e., dark characters on a light background). The nematic host has a negative dielectric anisotropy. The concentration of the chiral additive is reduced below the critical amount required to sustain a twisted structure in the field-off state.

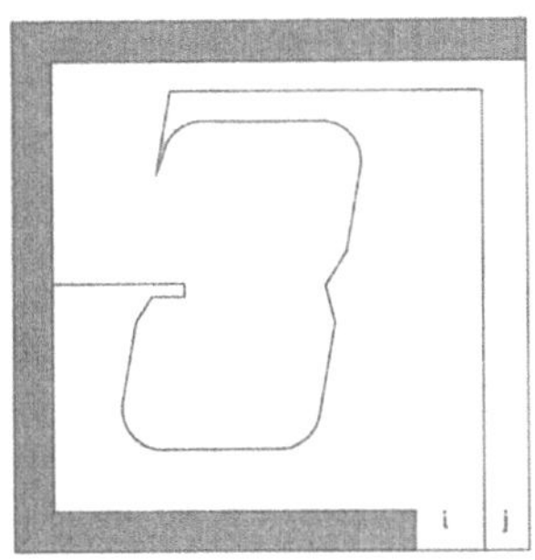


Fig. 16. Rear electrode pattern for positive contrast, 7-segment numeric display using inverse addressing. A 30 µm wide etched line separates rear cell plate into two electrodes i and j.

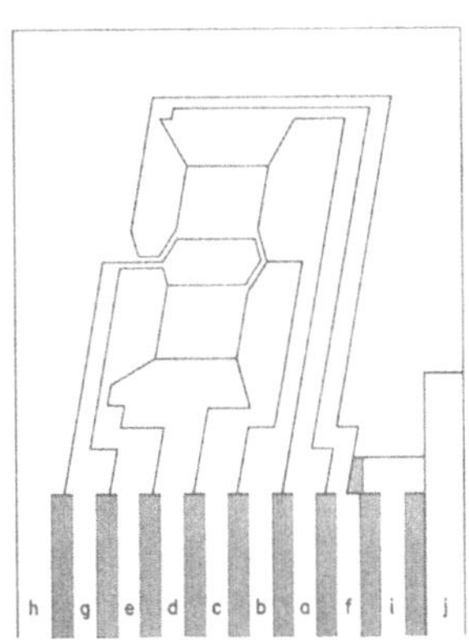


Fig. 17. Front electrode pattern corresponding to Fig. 16 with 8 electrodes a-h for positive contrast, 7-segment numeric display using inverse addressing.

the host birefringence is small. In order for this scheme to become practical, completely new host materials will have to be developed which have large negative $\Delta\varepsilon$, small $\Delta n$ and large $k_{33}/k_{22}$.

## Inverse Drive Scheme

Another positive contrast guest-host scheme uses a conventional guest-host mixture with positive dielectric anisotropy, but the display is inversely addressed, i.e., the optically selected elements are turned off (dark) and the remainder of the display is turned on (light).[32] This arrangement requires an electrode layout to make every region of the display addressable, which means that both the front and rear cell plates must be completely convered with electrodes.

Figure 16 shows the electrode pattern on the rear cell plate for one digit of a 7-segment numeric display. The white areas correspond to conducting electrode areas and the black areas and lines correspond to insulating regions where the electrode has been etched away. The rear cell plate is divided into two electrodes by the 30 μm wide separation line shown. Two electrodes on the rear cell plate are required in order to prevent the conduction paths leading to the activated segment electrodes on the front cell plate from becoming visible when the display is in operation. The separation line is placed so that electrode i (the background electrode) overlaps with no display segment, and electrode j overlaps with all display segments. Electrodes i and j are brought out to the lower edge of the cell plate so that they can be through-contacted to the connector raster on the front cell plate.

The electrodes of the front cell plate are shown in Fig. 17 with the black and white areas having the same meaning as before. The 30 μm wide separation lines are placed so that each of the electrodes a-g overlaps with a corresponding display segment and that none of the conduction paths overlaps with the rear electrode j. An additional front electrode h is provided which overlaps with both rear electrodes i and j.

One possible addressing scheme for this display is illustrated in Fig. 18. For clarity the display electrodes here have been represented by a topologically equivalent 8×2 matrix, the eight rows representing the front electrodes a-h and the two columns representing the rear electrodes i and j. The points where the rows and columns intersect on the matrix represent the regions where the front and rear electrodes overlap in the real display. The overlapping of the front electrodes a-g with the rear electrode j, for example, defines the region of the seven display

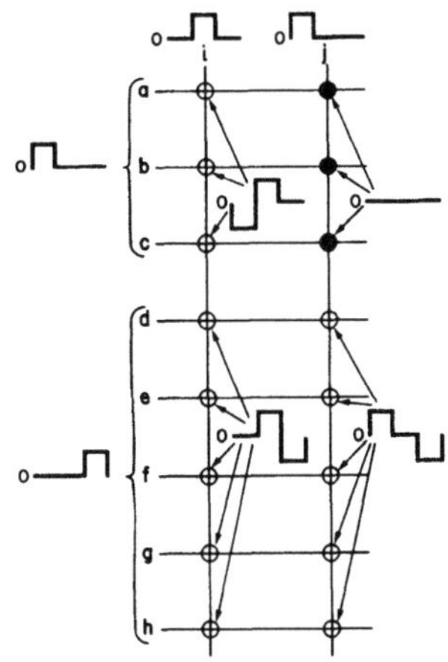


Fig. 18. Addressing waveform for inversely driven guest-host display with the electrodes of Figs. 16 and 17. Front and rear electrodes corresponding to an optically selected segment receive the same waveform which gives a zero potential difference between them and these regions of the display are turned off to the dark state. Overlapping electrodes corresponding to unselected segments and background regions of the display receive different waveforms and these regions are turned on to the light state. Letters a-j correspond to electrodes of Figs. 16 and 17. For this example segments a, b and c are turned off corresponding to a dark number 7 on a light background.

segments which is shown in Fig. 19a. This region of overlap can be expressed as $j \cap (a,b,c,d,e,f,g)$. All other regions of electrode overlap, i.e., $i \cap (a,b,c,d,e,f,g)$ (Fig. 19b), $i \cap h$ (Fig. 19c) and $j \cap h$ (Fig. 19d), occur in the display background.

For the driving scheme shown in Fig. 18, the two rear electrodes i and j are continuously driven with a 120°C and 0° phase-shifted train of amplitude V, 1/3 duty cycle pulses. The front electrodes corresponding to optically selected display segments (a,b,c in Fig. 18) are driven with this same 0° waveform so that there is no net potential difference in regions $j \cap (a,b,c)$, and they appear dark. The potential difference in the display background region defined by $i \cap (a,b,c)$ is given by the difference waveform shown in Fig. 18, which has an rms value of $\sqrt{2/3}$ V. The supply potential V must be chosen so that $\sqrt{2/3}$ V exceeds the critical unwinding voltage $V_{\uparrow}$ (Fig. 10) in order for this background region of the display to be fully turned on. The optically unselected segment electrodes (d,e,f,g in Fig. 18) are

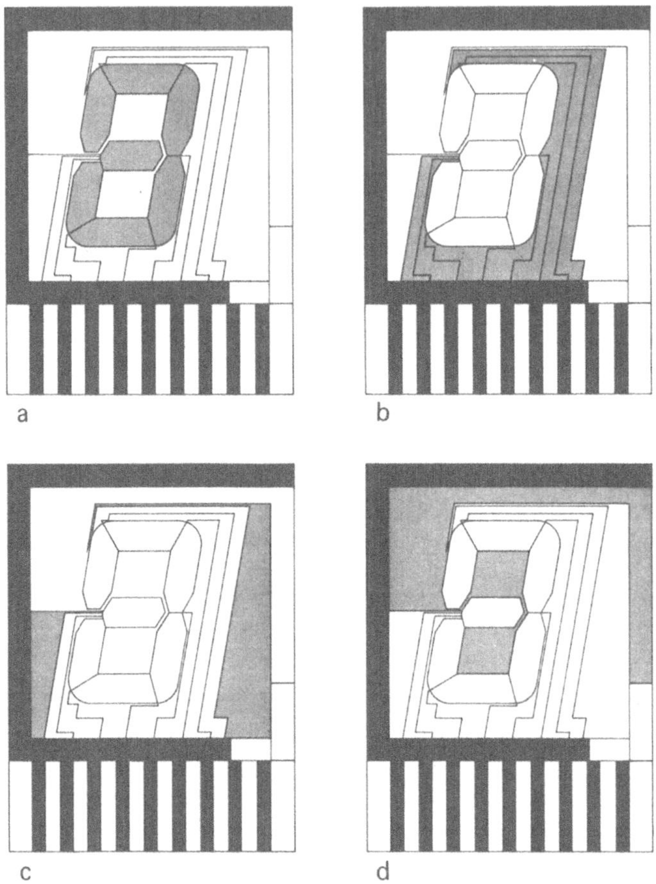


Fig. 19. x∩y denotes the region of overlap between rear electrode x and front electrode y. The shaded regions in the diagrams indicate the overlap given by (a) j∩(a,b,c,d,e,f,g); (b) i∩(a,b,c,d,e,f,g); (c) i∩h; (d) j∩h; where the letters a-j correspond to the electrodes of Figs. 16 and 17.

driven with a third pulsed waveform having a 240° phase shift. Here the rms potential difference is also $\sqrt{2/3}$ V in both segment and background regions. The h-electrode is treated like a segment electrode which is always turned on. For the example shown in Fig. 18, the dark numeral 7 is displayed against a light background. Other numbers can be displayed by applying the 0° and 240° waveforms to the appropriate segment electrodes a-g. The display background therefore always remains bright and the conduction paths leading to the activated segments are never seen.

This inverse drive scheme has the advantage of using present materials and cell technology to obtain the same brightness and contrast ratios that are available with the conventional negative contrast guest-host displays. One disadvantage, however, is the increased power consumption of the display which is in direct proportion to the increase in the active display area. This scheme may therefore not be suitable for wristwatches were the trend is toward smaller batteries with longer intervals between replacement.

## REFERENCES


1. G. H. Heilmeier and L. A. Zanoni, Appl. Phys. Lett. 13:91 (1968).
2. G. H. Heilmeier, J. A. Castellano and L. A. Zanoni, Mol. Cryst. Liq. Cryst. 8:293 (1969).
3. M. Schadt and W. Helfrich, Appl. Phys. Lett. 18:127 (1971).
4. D. L. White and G. N. Taylor, J. Appl. Phys. 45:4718 (1974).
5. Ch. Mauguin, Bull. Soc. Franc. Mineral. 34:71 (1911).
6. Hl. de Vries, Acta Crystallogr. 4:219 (1951).
7. A. Saupe, private communication.
8. H. S. Cole and S. Aftergut, J. Chem. Phys. 68:896 (1978).
9. See, for example. D. B. Judd and G. Wyszecki, "Color in Business, Science and Industry," 2nd ed., Wiley, New York, (1963).
10. J. Constant, M. G. Pellat and I. H. C. Roe, 7th International Liquid Crystal Conference, Bordeaux, France (1978); Many of these dyes are available from British Drug House Ltd., Poole, Dorset, England.
11. M. Schadt and F. Müller, IEEE Transactions on Electron Devices, Ed-25:1125 (1978).
12. T. J. Scheffer and J. Nehring, IEEE Transactions on Electron Devices, ED-24:816 (1977).
13. L. Pohl, R. Eidenschink, J. Krause and G. Weber, Phys. Lett. 65A:169 (1978).
14. L. Pohl, R. Eidenschink, J. Krause and D. Erdmann, Phys. Lett. 60A:421 (1977).

15. E. P. Raynes, Conference Record 1978, Biennal Display Research Conference, Cherry Hill, New Jersey (1978).
16. A. Bloom, P. L. K. Hung and D. Meyerhofer, Mol. Cryst. Liq. Cryst. 41:1 (1977).
17. M. Kawachi, K. Kato and O. Kogure, Japan. J. Appl. Phys. 17:1245 (1978).
18. M. J. Press and A. S. Arrott, J. Physique 37:387 (1976).
19. W. Greubel, Appl. Phys. Lett. 25:5 (1974).
20. M. Kawachi and O. Kogure, Japan. J. Appl. Phys. 16:1673 (1977).
21. C. J. Alder and I. A. Shanks, ELectron. Lett. 12:326 (1976).
22. T. Ohtsuka, M. Tsukamoto and M. Tsuchiya, Japan. J. Appl. Phys. 12:371 (1973).
23. M. Tsukamoto and T. Ohtsuka, Japan. J. Appl. Phys. 13:1665 (1974).
24. G. W. Gray and D. G. McDonnell, Electron. Lett. 11:556 (1975); CB 15 is compound III with n-1 in this reference and is available from British Drug House Ltd., Poole, Dorset, England.
25. R. Cano, Bull. Soc. Franc. Mineral. Crystallogr. 91:20 (1968).
26. P. G. de Gennes, Solid State Commun. 6:163 (1968).
27. F. Fischer, Z. Naturforsch. 31a:41 (1976).
28. J. Constant, E. P. Raynes, I. A. Shanks, D. Coates, G. W. Gray and D. G. McDonnell, J. Phys. D. 11:479 (1978); D1, D2 and D3 are available from British Drug House Ltd., Poole, Dorset, England.
29. T. J. Scheffer, in Solid State Devices 1977, Inst. Phys. Conf. Series No. 40, Techno House, Bristol (1978), pp. 65-86.
30. German Patent Disclosures DT-OS 26 29 765, D-OS 26 39 675 and DT-OS 27 16 609.
31. German Patent Disclosure DT-OS 28 35 863.
32. German Patent Disclosure DT-OS 27 38 679.

# ANALOG DISPLAYS - THE OTHER APPROACH TO LCD's

P. Andrew Penz

Texas Instruments Incorporated
Dallas, Texas

Before the recent development of digital displays, electronic information was indicated by the position of a hand or length of a bar relative to some scale. Such a display is referred to as an analog display and has certain advantages over the digital display. Analog displays are easier to read when rapid acquisition of approximate information is needed. One can interpret the distance from the current reading to another point without arithmetic. Another advantage is the natural manner in which rate of change of information comes from the velocity of the hand. These advantages are directly related to the long learning process which is necessary to understand an analog display and the ease with which the position of a hand can be recognized versus the presence of several digital segments.

Liquid crystal displays (LCD's) are best known in the digital format. It will be assumed throughout this paper that the reader is familar with LCD digital technology. The subject can be reviewed in articles by Sobel[1] and Heilmeier.[2] It will be assumed that the LCD display mode is the twisted nematic, but the discussion will not necessarily be limited to this mode. Several schemes, however, have been disclosed for producing analog type displays. The various electronics/display design schemes proposed for accomplishing the solution of this problem will now be reviewed.

## SOREF DESIGN

In 1970, Soref[3] was first to demonstrate an analog LCD designed around the fact that current flow in a resistor produces a voltage gradient along the resistor. The standard LCD has low resistivity transparent conductors which define equipotential surfaces on either side of the liquid crystal layer. Soref's idea, as shown in Fig. 1a, was to make one of the surfaces out of high resistivity material. When leads are applied to either end of the bar and a voltage applied ($V_1$), a potential gradient is set up along the bar as shown in Fig. 1b. The high resistivity is necessary to minimize power loss in the electrodes. It is assumed that the impedance of the liquid crystal layer is such that it does not draw enough current to spoil the voltage profile. The other electrode has low resistivity and is biased at a voltage $V_o$ with respect to one end of the high resistivity layer.

Potential as a function of position (X) along the bar is shown in Fig. 1b. The zero of the potential scale is chosen as the low resistivity plate which is an equipotential surface. The left hand side of the high resistivity electrode is held at $V_1+V_o$ while the right hand side is held at $V_o$. One particular voltage is of special interest for all LCD's: the threshold voltage $V_c$. Figure 2 shows that the optical properties of an LCD exhibit a threshold behavior at $V_c$. The threshold voltage is plotted on Fig. 1b. The position on the bar graph which corresponds to the threshold voltage is shown as $X_c$. The area to the left of $X_c$ will be on and the area to the right of $X_c$ will be off.

How does $X_c$ depend on $V_1$ and $V_o$? Ideally one wants $X_c$ to depend linearly on an input voltage so that the length of the bar accurately represents the magnitude. The equation for potential across the liquid crystal, $\Delta V$, as a function of position, X, is:

$$\Delta V = V_o + V_1(1-X/L) ,$$

where L is the length of the high resistivity bar. There will be a transition from "on" to "off" at $X-X_c$ for $\Delta V=V_c$:

$$V_c = V_o + V_1(1-X_c/L) .$$

The trick is to pick $V_o$ or $V_1$ so that $X_c$ varies linearly with one or the other. Clearly the choice is to vary $V_o$ with the signal the bar graph is to display. Inspection will show that a choice of $V_1=V_c$ will place the transition of the LCD at X=0 for $V_o=0$ and X=L for $V_o=V_c$

$$\frac{X_c}{L} = \frac{V_o}{V_c} .$$

Thus by properly scaling the signal ($V_o$) to be $V_c$ at maximum, a linear analog display can be achieved. Pictures of Soref's device in operation are shown in Fig. 3. The scheme described here is

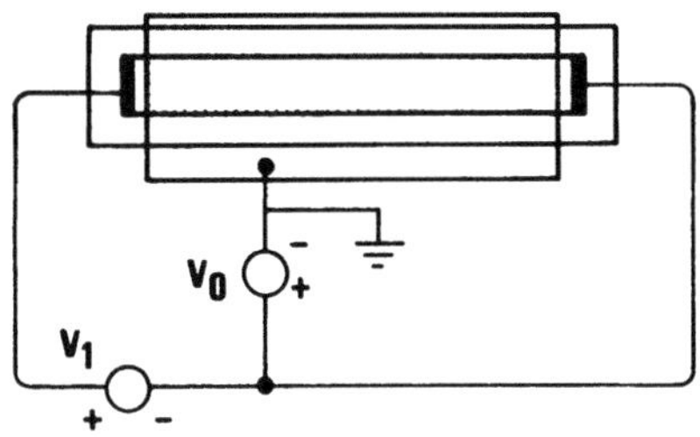


Fig. 1a. The physical design of Soref's analog bar graph is shown. The long substrate has a high resistivity transparent conductor so that a potential drop $V_1$ is produced down the substrate from X=0 to X=L. The narrow substrate has a low resistivity transparent conductor so that it is an equipotential plane at zero potential.

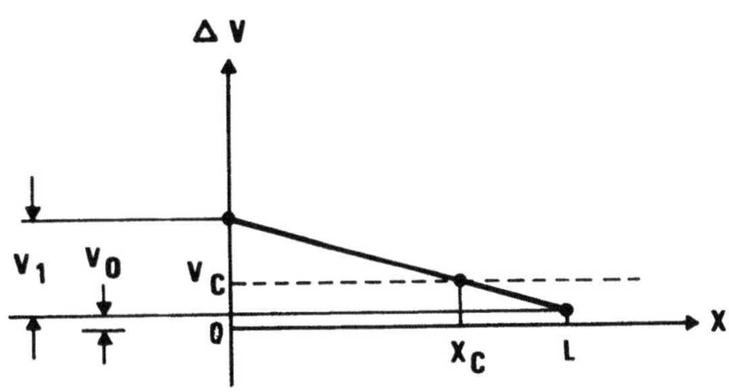


Fig. 1b. The potential as a function of position on the display is presented. The critical voltage $V_c$ and the position $X_c$ at which the voltage across the liquid crystal is equal to the critical voltage are shown. The display is on for $X<X_c$.

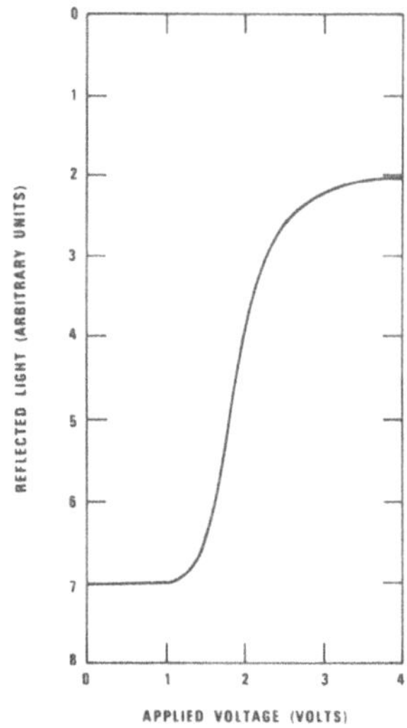


Fig. 2. The magnitude of light reflected from a typical nematic LCD is plotted as a function of voltage across the nematic layer. At a critical voltage of approximately 1V, the reflected light drops to a level that saturates at approximately 3V.

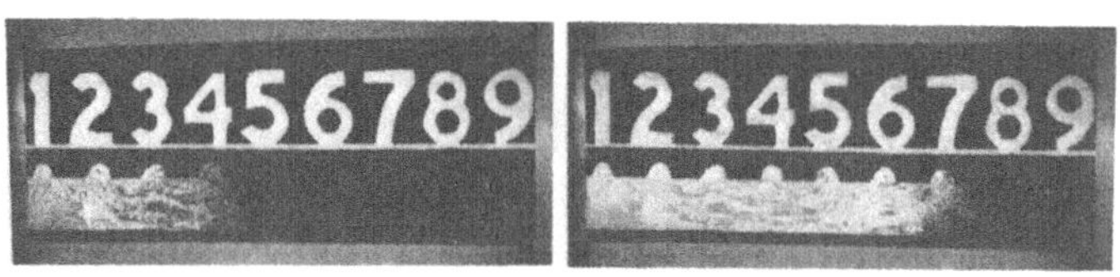


Fig. 3. A dynamic scattering bar graph LCD is shown where the end of the active region is a 4V at 7V in the two pictures.

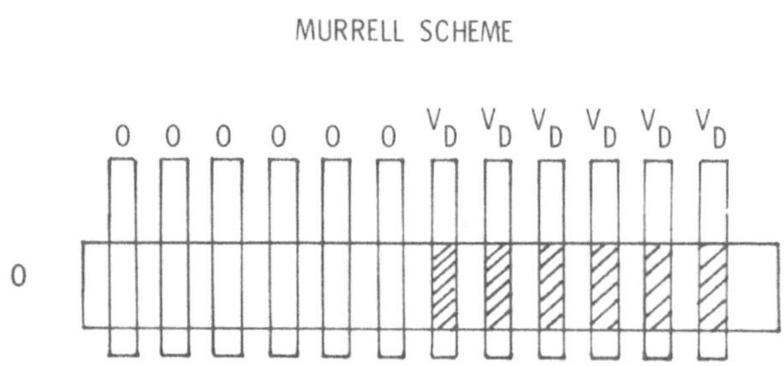


Fig. 4. A discrete bar graph design is shown where each of the elements can be addressed separately. A drive voltage $V_d > V_c$ is applied to the elements which are to be turned on.

a slight variation of Soref's original design, the change being made for pedagogical reasons and to insure device linearity.

There are two disadvantages to the true analog design. The first is that the transition from off to on in LCD's is gradual, as shown in Fig. 3. This leads to problems in definition of the bar edge. The second is that electrical instruments are moving increasingly to digital data processing, thus providing a digital output. It is often more convenient to use this digital format for input to an analog display rather than convert the digital data stream to analog form.

## DISCRETE ANALOG DISPLAYS

It has been shown that a true analog display is not terribly desirable for fabrication as well as system reasons. The alternative is to make the bar graph out of discrete elements and to address these elements digitally. One possible design for such a display is shown in Fig. 4. For L resolution elements, L+1 leads are required for addressing the display. One percent resolution would require 101 interconnections. This is prohibitive for several practical reasons, especially when a seven segment digital format will give 0-100 readings with $16(2\times7+1+1)$ interconnections. In this scheme, any number of points can be addressed simultaneously so that it is inherently random access. Murrell[4] used this principle in designing a discrete analog watch. Two discrete electrode patterns were placed side by side in a radial format. The sixty discrete radial electrodes were covered by two azimuthal electrodes which defined an inner and outer annules each with sixty resolution patterns for 62 leads. By half duty cycle multiplexing between the inner and outer annulus, a conventional watch appearance could be achieved.

There is a basic flaw in the Murrell scheme since it is only intended to display information in a bar graph mode. Each element in the display can be addressed independently of the other elements. An analog display, however, only presents information in two positions on the display at one time: the two ends of the bar. Thus, random access capability represents overkill with respect to the system requirements. As will be shown below, considerable savings in interconnections can be achieved by reformating the $1\times L$ design into an $N\times M$ matrix where $N\times M\geq L$ and $N+M<L+1$.

The discrete analog display can be run in two modes which need to be explicitly distinguished. The first is the bar graph mode where a bar extends from the zero point all the way to the position where information is specified by the end of the bar.

A second mode is where the information is specified by the position of a single element - a pointer. Both the bar and the pointer modes require the same amount of information: the place where the "bar" begins and ends. The use in a given situation will depend on human factors considerations. The traditional watch display uses the pointer modes.

## SHERR DESIGN

If one wants to address only one element in a display during any given interval of time, the number of leads can be reduced using a design proposed by Sherr.[5] The key to the idea is to break the display up into regions which contain several resolution elements. Electrodes are arranged so that any one of the given regions can be "energized" and then electrodes within the broad area are selected with increasing resolution. Since high resolution data is displayed in only one broad area, the high resolution electrode patterns within the various broad areas can be connected in parallel and thus avoid a large number of leads. The "energization" scheme for the broad areas must be such that the off broad areas do not display the data carried by the high resolution electrodes. Sherr's display for accomplishing this is shown in Fig. 5a.[6] A special case of the Sherr design was proposed by Taylor.[7]

Sherr's design was originally proposed for area displays but he discussed the special case of a linear display in 1976.[8] The display design is shown in Fig. 5b. Two things should be noted in particular. The technique for connecting the high resolution electrodes in parallel is to fabricate meandering lines as shown in Fig. 5b. This scheme will work for linear displays but not for area displays.

The primary feature of Sherr's design is that he proposed using two LCD's in series to achieve the energizing feature on the broad electrodes. The first cell contains area electrodes matched to the meander lines. The second cell contains two sets of area electrodes and the two cells are registered so that the area electrodes line up. In both cells, the polarizers are aligned parallel so that the twisted nematic does not pass light. The high resolution cell has voltages applied to the area electrodes sufficient to turn on the cell everywhere except for the region where the detailed data is to be displayed. In this special area, the end of the bar is defined. The low resolution cell has area electrodes either on or off depending on which way the bar is extending. The example shown in Fig. 5 is for a solid bar, but a line indicator can be established in a similar manner. Note that the Sherr design does not depend on a threshold property for the display. The energization feature comes from the use of

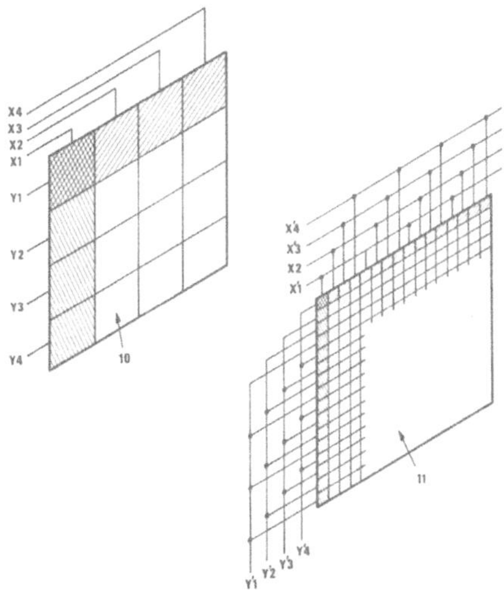


Fig. 5a. The Sherr design for addressing an area by applying voltages to different layers (planes 10 and 11) of a three dimensional sandwich. By time multiplexing plane 10 (a separate LCD from plane 11), the cross hatched area transmits light with 8 electrodes (X1 to X4, Y1 to Y4). Plane 11 is similarly addressed by electrodes X1' to X4' and Y1' to Y4' but light passes only through the cross hatched area of plane 11 because it is only in this area that light passed through plane 10. Light does not pass through the other equivalent positions (X1', Y4') in plane 11. Thus one addresses 256 positions with 16 leads at one fourth duty cycle.

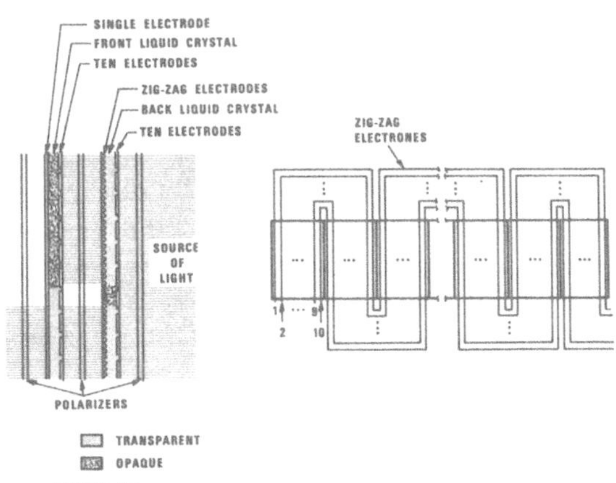


Fig. 5b. The area concept shown in Fig. 5a is shown as applied to a linear, discrete bar graph. The meander line concept is shown as the zig-zag electrode pattern.

the third dimension - stacking displays. Fabricating two separate displays is expensive and turns out to be unnecessary to achieve the energization property, as will be shown below.

## MEANDER LINES AS AN X-Y MATRIX

The number of drive lines to a 1×L array can be reduced by converting the array into an M×N matrix, as was discussed above. This means arranging the L elements into N groups and electrically connecting the M common elements in the N groups together. One particularly effective way of implementing an M×N matrix in a linear format is to employ the concept of a meander line. The fact that a meander line is equivalent to an x-y matrix is shown in Fig. 6. Note that the Sherr scheme shown in Fig. 5 also employs the meander line design. In Fig. 6, broad area electrodes RA and RC are also row electrodes CA, CB, CC, etc. The rest of the paper will be devoted to the nature of electrical signals which are applied to the electrodes to turn on one point in the meander line matrix. The ability to turn on two points will also be demonstrated. BA will be used for broad area and HR will be used for high resolution with respect to electrodes.

## THE CARL AND STEIN DESIGN

In 1976, Carl and Stein[9] disclosed a technique for energizing the BA electrodes using only one liquid crystal layer. This technique makes use of the fact that some liquid crystals have dispersion in their dielectric anisotropy. For low frequencies ($F_o$) the dielectric anisotropy is positive as shown in Fig. 7a. At higher frequencies (10 $F_o$) it is negative. High frequency voltages (R1) on the BA electrodes will prevent the LCD from turning on regardless of the signals on the HR electrodes provided these low frequency voltages are not too large. This is referred to as dielectric alignment (DA) and is indicated in the truth table in Fig. 7a. Other drive signals C1, C2 for the high resolution electrodes and R2, R3 for the area electrodes are also shown in Fig. 7. The net voltage across the liquid crystal is $V_o$ for C2, R2; $V_o$ for C2, R3; $V_o$ for C1, R3 and zero for C1, R3. Note that the numerical indices refer here to waveforms while the alphabetical indices in Fig. 6 refer to positions in the matrix.

Suppose an on segment is required at the second BA electrode and the third HR electrode, i.e., position RC, CC in the X, Y representation of the analog display shown in Fig. 6. Figure 7b shows how to use the Carl and Stein waveforms to achieve this result. Signal R2 is applied to the second BA electrode and signal C2 is applied to the third HR electrode to produce the

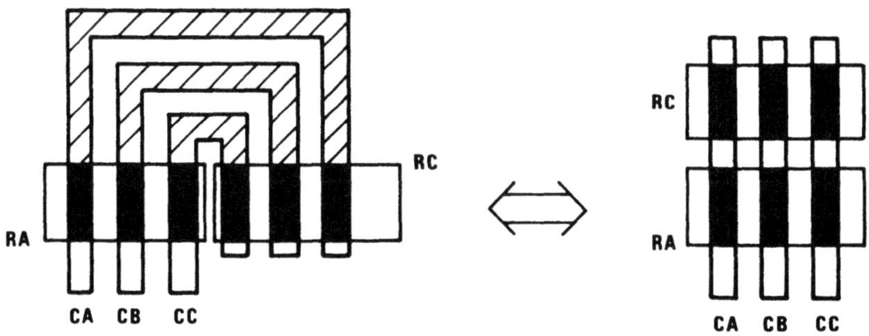


Fig. 6. The equivalence between a meander line electrode pattern in conjunction with broad area electrodes to an XY matrix is shown.

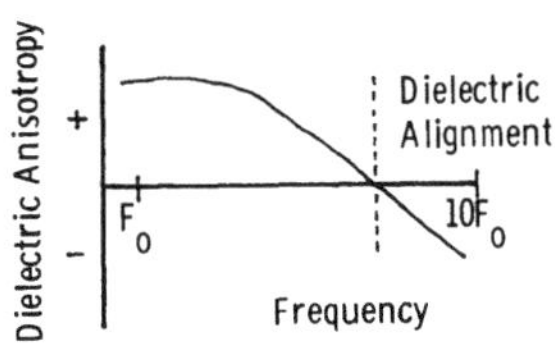


| | R1 | R2 | R3 |
|---|---|---|---|
| C1 | DA | 0 | $V_0$ |
| C2 | DA | $V_0$ | $V_0$ |

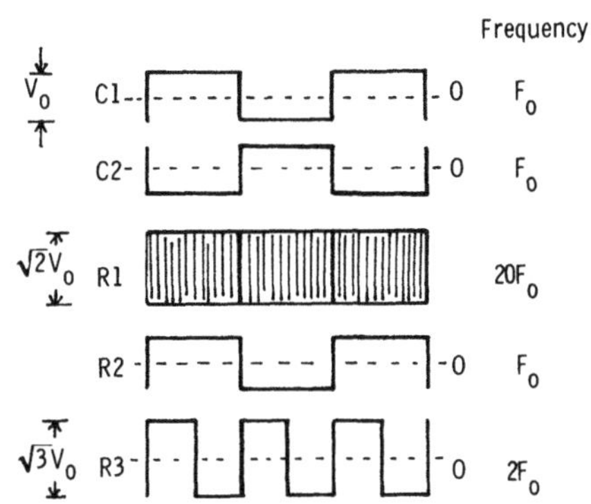


Fig. 7a. The Carl and Stein addressing scheme is indicated. The liquid crystal material has a dispersion in the dielectric anisotropy such that is is positive at $F_O$ and negative at 10 $F_O$. The waveforms used and the resulting truth table are shown.

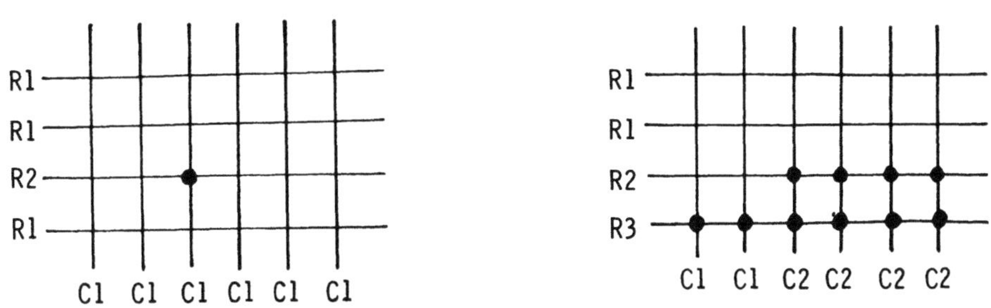


Fig. 7b. The Carl and Stein waveforms are used to place a pointer at the second row, third column position and to extend a bar graph to that point.

desired on segment in a pointer mode. Signal R1 insures that all the other rows are off by dielectric alignment. C1 signals on the other columns keep off other points on the second BA electrode. Note that it is not possible to turn on simultaneously this point and another arbitrary point in the array. The right hand side of Fig. 7b shows how a solid bar can be extended from the first row into the second row. The reader familar with traditional waveforms used to drive X, Y arrays will notice the lack of strobe (select) signals. All the waveforms are continuous square waves in Fig. 7a, as opposed to the pulse waveforms normally used. This is due to the fact that only one information point is being addressed in the analog display regardless of the number of points in the array. There are problems associated with the two frequency addressing scheme. First the higher frequency dissipates much more power, and second the dispersion crossover is temperature sensitive.

## FUKUMOTO DESIGN

The threshold characteristic of most LCD modes was used by Fukumoto[10] to achieve energization of the BA electrodes. The single level display and the single frequency improvements permitted the first commercial introduction of a bar graph LCD. The Fukumoto electrode design was basically the meander line scheme of Fig. 6. His electrical drive signals are shown in Fig. 8. Note that the BA electrode voltages (R1 and R2) and the HR electrode voltage (R1, R2 and R3) are all of the same frequency and amplitude, the difference lying in the phase of the voltages.

The truth table for the Fukumoto design is also shown in Fig. 8. Note that the net RMS voltages across the cell are never zero. Combinations (R1, C1), (R1, C2) and (R2, C1) all have $\sqrt{3}\times 0.25\ V_o$ RMS voltages. This means a ratio of "on" to "off" voltages of $\sqrt{3}$:1. The drive voltages are adjusted to $0.25\ V_o = V_c$, meaning that the off state is really at the threshold rather than V=0 condition. For most LCD's, $\sqrt{3}\ V_c$ is not very strongly "on" so that the display may be weak compared to standard digital LCD which are driven on by 2 to 3 times $V_c$.

This use of the Fukumoto voltage to achieve the same display elements shown for the Carl/Stein design (Fig. 6) is also shown in Fig. 8. One independent point can be addressed, but not two.

The Fukumoto design was used to make an analog display watch which is shown in Fig. 9. The outer display is a bar graph with 60 resolution elements bent around into a circle. The inner circle has twelve resolution positions. The two bar graphs are contained in one display package. It can be seen from Fig. 9 that the display does not faithfully reproduce the standard mechanical

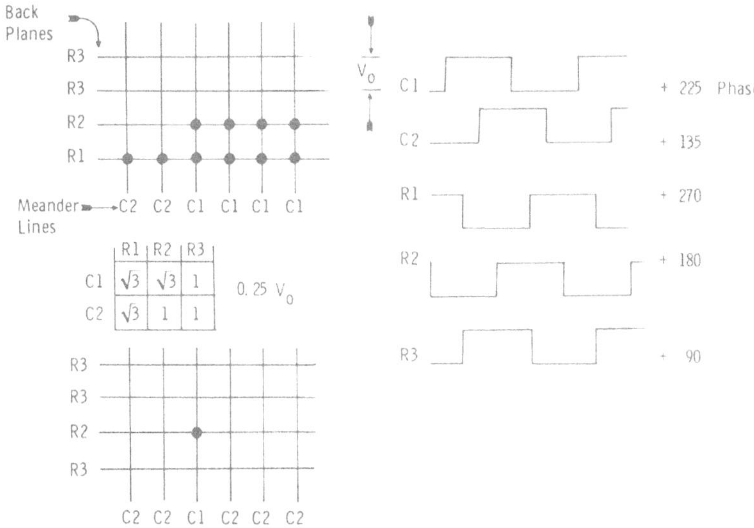


Fig. 8. The Fukumoto addressing scheme is shown, including waveforms and truth table. The bar graph and pointer modes are the same as in Fig. 7.

Fig. 9. The Fukumoto scheme is used to provide an analog watch display. The display format is non-traditional. The hour position is indicated by a short hand on the interior of the display. The minutes past the hour are indicated by a ring of short hands on the exterior part of the display. The time indicated is 1:21.

display. The traditional display has a long minute hand and a short hour hand. Now the significance of being unable to address two independent positions in the bar graph becomes apparent. The traditonal format requires two positions be on in the inner circle. Fukumoto's scheme does not permit this and so the observer must learn a new algorithm for telling time. It generally is difficult to give up habits developed early in life, and this may have been a factor in the limited success of the Fukumoto watch design.

## KMETZ DESIGN

The state-of-the-art of analog displays was advanced by Kmetz in 1977.[11] Kmetz used a matrix electrode design and employed the threshold characteristics of the LCD. Kmetz made two improvements. The first was a new voltage drive scheme shown in Fig. 10. Note that frequency of the BA and HR voltages are equal but the amplitudes are not all equal, as was the case in the Fukumoto scheme. This leads to an on/off RMS voltage ratio of 3:1 which is a significant improvement over the $\sqrt{3}$:1 result produced by the Fukumoto voltages. The Kmetz truth table is also shown in Fig. 10, as are the voltages required to produce the display results shown previously for the Carl/Stein and Fukumoto schemes. Again note that only one independent element can be addressed using the Kmetz truth table.

Kmetz's second contribution was to reintroduce the concept of time division multiplexing. This concept was inherent in Sherr's 1974 patent,[6] but had not been explicitly applied to bar graphs. Kmetz reasoned that two bar graphs arranged side by side could be addressed with the same HR electrode and two sets of BA electrode. The dual bar graph design is shown schematically in Fig. 11 along with the equivalent X-Y matrix. The Kmetz drive waveforms in Fig. 10 are used as follows. Half of the time the signals are applied to rows A and D and columns A, B and C. Rows B and C are held at zero potential, i.e., the rows are strobed. The other half of the time the signals are applied to RB and RC along with CA, DB and CC, RA and RD now being at zero potential. Any single position in either matrix can be chosen. There is a decrease in the ratio of on to off voltages from the 3:1 value due to the fact that the dual bar graph is being strobed at half duty cycle. Obviously the scheme can be extended to n graphs operated at 1/n duty cycle.

The details of time division multiplexing are beyond the scope of this article. They are given in another article by Kmetz.[12] The motive, as always, is to reduce the number of leads to the display. In the time division case, data is applied to the HR electrodes only half of the time for a given bar graph.

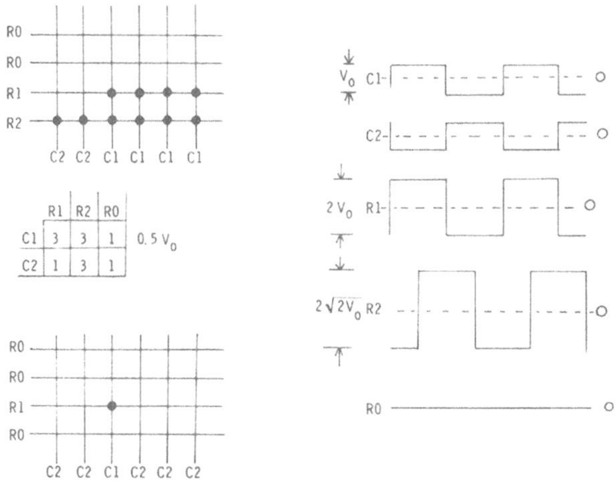


Fig. 10. The Kmetz addressing scheme is shown, including waveforms and truth table. Note that the selected voltage ratio in the truth table is 3 versus $\sqrt{3}$ for the Fukumoto design. This results in an improved contrast.

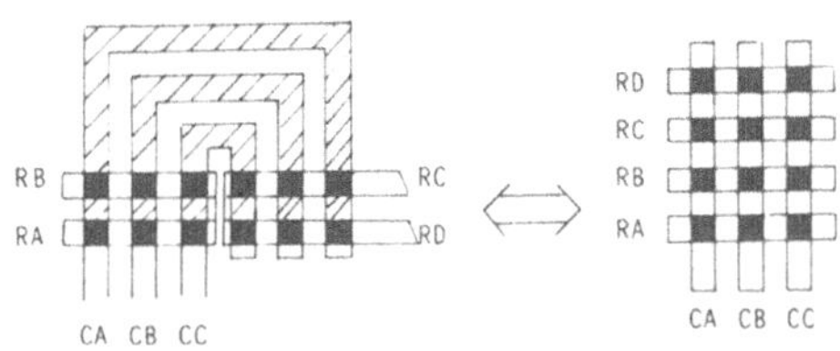


Fig. 11. The equivalence between a meander line array with two series of broad area electrodes and an X-Y area matrix is demonstrated. The scheme can be used to half duty cycle between the two broad area arrays to drive two bar graphs with the same meander line array.

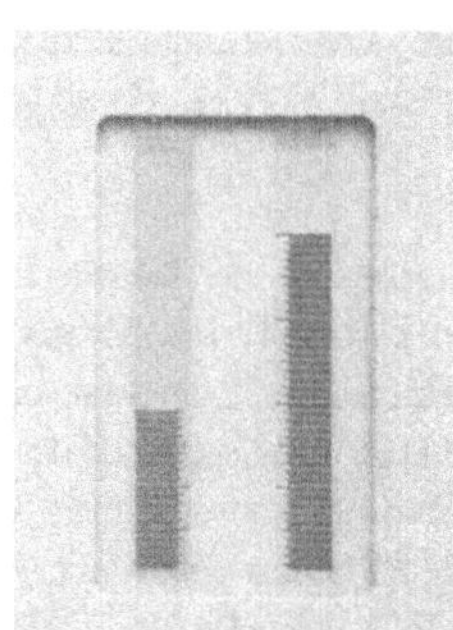

Fig. 12. Dual Bargraph. Twisted Nematic LCD color - selective polarizers, 2% resolution/bar, 20 leads (total) $<20$ μW, $<4$V.

Instead of requiring 2M+2N (where M and N are the number of HR and BA electrodes respectively) half duty cycle drive requires only M+2N interconnections. A dual bar graph which Kmetz made is shown in Fig. 12.

It should be obvious to the reader at this point that two types of multiplexing are involved in addressing a dual bar graph. First there is spatial lead sharing as provided by the meander lines. The second is time sharing provided by the strobe signals. Employing both concepts drastically reduces the number of leads, but the tolerances on the fabrication of the display are correspondingly decreased.

## TEXAS INSTRUMENTS DESIGN

Since the Suncrux analog LCD watch did not faithfully reproduce the mechanical watch appearance and since the utility of such a display is obvious, several of us at Texas Instruments invented a new design which would mimic the centuries old display.[13] Two faults in the Fukumoto watch display had to be corrected, from the point of view of the display. First, the hour hand had to have more than twelve resolution positions. In a mechanical watch, the hour hand moves continuously and this helps distinguish 12:01 from 12:59. Without the possibility of multiple resolution positions for the hour hand, the designer is forced to unusual formats as shown in Fig. 9. Second, the minute hand must extend into the center of the display. The electrode design for accomplishing these objectives is shown in cutaway view in Fig. 13. Basically two matrix addressed linear arrays of sixty resolution positions are arranged to be time division multiplexed. There are six meander lines on one plane and two sets of BA electrode of ten numbers each on the other. The display is addressed by 26 leads which compares favorably, with the 23 leads necessary for the standard 3 1/2 digit LCD for watch applications. A similar electrode design was proposed by Weigert and Vaucher.[14]

The Texas Instruments drive scheme is shown in Fig. 14 which has been kept in the same format as Figs. 7, 8 and 10 for ease of comparison. The truth table indicates a ratio of $\sqrt{5}$:1 for on/off voltages, a reasonable value considering the current state of the art in LCD's. The analog waveforms are continuous square waves with the BA signals (R0 to R2) at half the frequency of the HR electrode signals (C1 to C4). Although the traditional strobe format is missing, the scheme produces results similar to a half duty cycle strobe design. Again, the difference between analog and strobe waveforms has to do with the limited information to be displayed in the analog equivalent matrix. Note that the TI truth table has a row of ones as well as a column of ones.

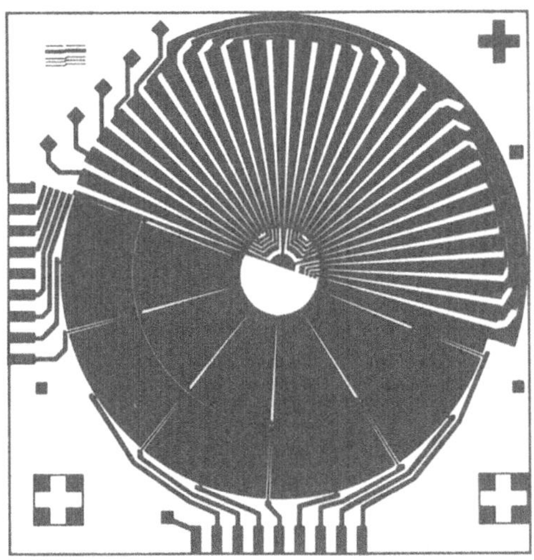

Fig. 13. The Texas Instruments electrode design for an analog watch display. Six meander lines intersect two arrays of broad area electrodes each having ten leads. One hundred twenty positions can be accessed by 26 leads. The drawing is a cutaway view, looking at half of one plane of electrodes and half of the other plane.

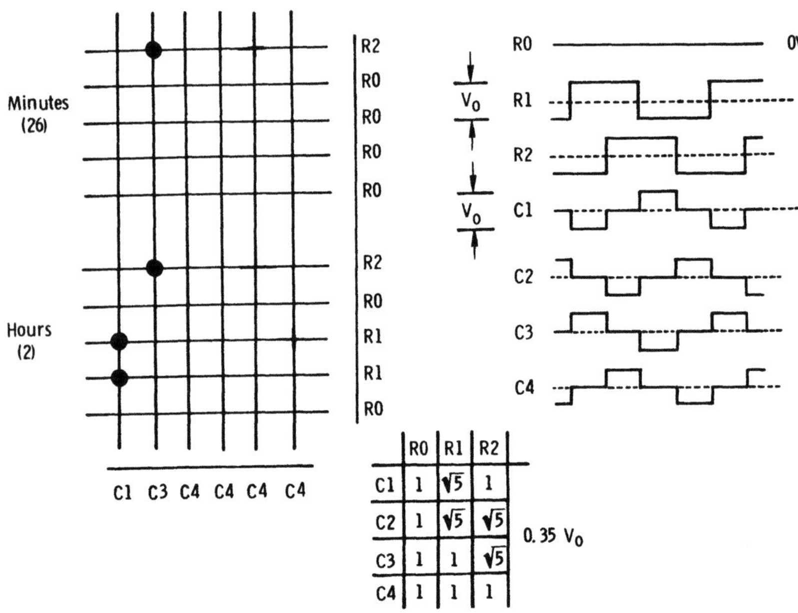


Fig. 14. The Texas Instruments addressing scheme is shown including waveforms and truth table. The waveforms are equivalent to half duty cycle drive ($\sqrt{5}$ ratio on selected to non-selected elements). The inner and outer rings of broad area electrodes are shown from the 0 to 30 min. position. The inner ring is represented by the lower X-Y array and the outer ring by the upper X-Y array. The time of 2:26 is shown. Note that two separate pieces of information are placed on the inner ring: the minute hand and a "fat" hour hand. The display has a hybrid pointer and bar graph presentation.

This means that two positions can be simultaneously addressed in one analog display as well as either or both of the same positions in a second display. As has been noted above, this capability is essential for reproducing the appearance of the mechanical watch display.

The use of the TI drive scheme to address the TI electrode scheme to produce 2:26 is shown in Fig. 14. The lower 6×5 array indicates half of the inner circle, from 0 to 30 minutes. The top 6×5 array indicates the outer half circle. The full 6×10 arrays are not reproduced in interest of saving space. Starting at the lower left hand corner of both arrays, one counts 26 positions, remembering the fact that the meander lines double back over the BA electrodes, to reach the two positions of the minute segments at 26 minutes. The hour hand has five resolution positions per hour and so can change positions at 0, 12, 24, 36 and 48 minutes past the hour. The hour hand is placed in the lower array only at the 12 and 13 positions. The hour hand is made fat, i.e., has two segments, to make it easier to read. The application of the drive voltages to achieve this clock setting is indicated in Fig. 14. Texas Instruments is marketing this design currently, and one of the products is shown in Fig. 15. Note the difference in format between Figs. 9 and 15. In addition to the traditional watch format, the TI scheme can be used to show day of week and date (using graphics on top and bottom of the display), a fan type sweep second mode and a kaleidoscope mode termed "starburst."

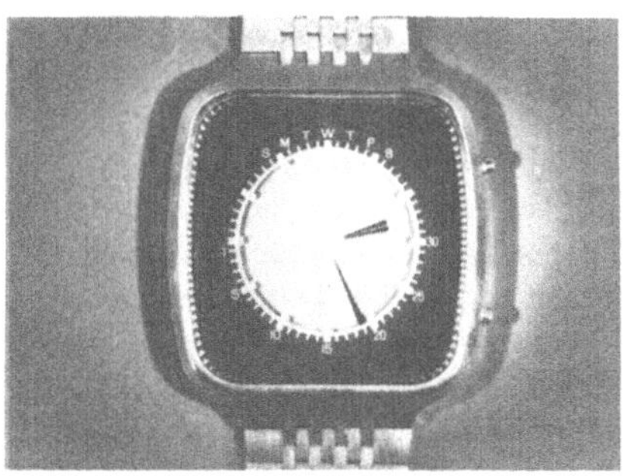

Fig. 15. Texas Instruments discrete analog display is shown in the watch application. The long minute hand and the short hour hand are in the traditional format as opposed to the display shown in Fig. 9.

## CONCLUSION

The concept of the analog or bar graph LCD has been reviewed. Soref demonstrated that voltage variations produced by current flow in appropriately constructed area electrodes would give a true analog display with analog input. Sherr fabricated a "discrete" analog display addressed digitally, by converting the linear array to an area array by stacking two LCD's in series to "energize" the BA electrodes. Carl and Stein demonstrated that only one LCD was necessary if two frequency multiplexing was employed. Fukumoto further simplified the drive scheme by taking advantage of the threshold property of many LCD's. Kmetz improved the on/off ratio for the threshold design and produced dual bar graphs using half duty cycle addressing. Gruebal et al. developed an electrode design and a half duty cycle drive scheme which permitted the accurate simulation of the traditional mechanical watch display.

LCD's have become the dominant display for portable applications mostly in the digital format. The technology is now at hand to produce displays which are not limited to the digital format and more bar graph type applications can be expected in the 1980's.

## ACKNOWLEDGMENTS

The author would like to thank A. R. Kmetz for several helpful discussions. Dr. Kmetz's familiarity with the literature regarding analog displays and his willingness to share this knowledge has been an invaluable asset during the composition of this paper.

## NOTE ADDED IN PROOF

Recently, a new analog scheme has been invented.[15,16] It uses correlated drive voltages to produce a single pointer position which has a zero voltage whereas all other resolution positions have finite voltage. Thus, this scheme does not require a display with a threshold and is especially useful for dye LCD's. The "pointer only" scheme cannot produce a bar format and cannot display two pointer positions in the same array. Watch type formats can be achieved at the expense of segment width or resolution. This scheme will find use in wide temperature range applications where the temperature dependence of threshold voltages make the Fukumoto/Kmetz/TI schemes difficult to implement.[17]

REFERENCES

1. A. Sobel, Scientific American, Vol. 22, No. 6, 65 (1973).
2. G. Heilmeier, Scientific American, Vol. 222, No. 4, 100 (1970).
3. R. A. Soref, Appl. Optics 9:1323 (1970).
4. N. J. Murrell, U.S. Patent 3,959,963 (1976).
5. S. Sherr, IEEE Conf. on Display Devices 52 (1972).
6. S. Sherr, U.S. Patent 3,848,247 (1974).
7. G. W. Taylor, Proc. IEEE 58:1812 (1970).
8. S. Sherr, SID Symp. Dig. 42 (1976).
9. W. L. Carl and C. R. Stein, SID Symp. Dig. 40 (1976).
10. S. Fukumoto, U.S. Patent 3,969,887 (1976).
11. A. R. Kmetz, SID Symp. Dig. 58 (1977).
12. A. R. Kmetz, IEEE Trans. Electron Devices ED-20:954 (1972).
13. R. L. Gruebel, B. W. Marks, R. T. Noble, P. A. Penz and K. H. Surtani, Conf. Record 1978 Biennial Display Res. Conf. 59 (1978).
14. H. Weigert, U.S. Patent 3,934,241 (1976); J. M. Vaucher, Swiss Patent 575,146 (1975).
15. A. R. Kmetz and J. Nehring, 1979 IBM Symposium on the Physics and Chemistry of Liquid Crystal Devices.

# PROJECTION STORAGE DISPLAYS USING LASER-ADDRESSED SMECTIC LIQUID CRYSTALS

A. G. Dewey

IBM Research Laboratory
San Jose, California 95193

## INTRODUCTION

A beam-addressed storage light valve (LV) can form the basis for a very high information-content display. In many cases the upper limit will be determined by the resolution limit of the optical system used to project the LV onto a screen, and this may be as high as $10^8$ picture elements (pels). In contrast, a refreshed display such as a CRT has a lower limit to the refresh rate in order to avoid flicker so that high information content requires a very high data rate. Practical extension of refreshed CRT displays beyond a few million pels only seems feasible if multiple beams are used. Even then, as with storage CRT's, an upper limit on the amount of information which can be resolved by the user is going to be determined by the physical size limitations of the tube.

In this paper a laser beam addressed liquid-crystal light valve is described. Its writing speed is limited by the available laser power and by the laser scanning system, but it does have selective erase capability. This feature, which allows local erase and update of information on the screen, is important in large storage displays and is not available with present storage CRT's.

## THERMAL WRITING ON LIQUID CRYSTALS

Early work on this technology was performed at Bell Labs and in 1972 a paper was presented[1] describing a LCLV system. A cholesteric/nematic mixture was used in a transmissive cell having cholesteric/nematic mixture was used in a transmissive cell having indium-tin oxide (ITO) coatings on both substrates. The LC mixture had negative dielectric anisotropy and the transparent conductor coatings served the dual purpose of electrodes for the erase field and absorbers for the writing laser beam. Erasure was achieved with an ac field which causes an in-plane alignment of the LC. The aligned cell is relatively transparent. Writing was done with a Nd:YAG laser beam which is partially absorbed by the ITO coatings. Following local heating of the LC into the isotropic state by the absorbed laser energy, it cools rapidly into the cholesteric state forming a focal conic texture. This texture is stable and hence written information is stored by the LC cell; however, it can be erased at any time. The focal-conic texture scatters light and hence when a cell is projected, even in a normal slide projector, written information appears dark on a light background. The optical system used by BTL is shown in Fig. 1.[2]

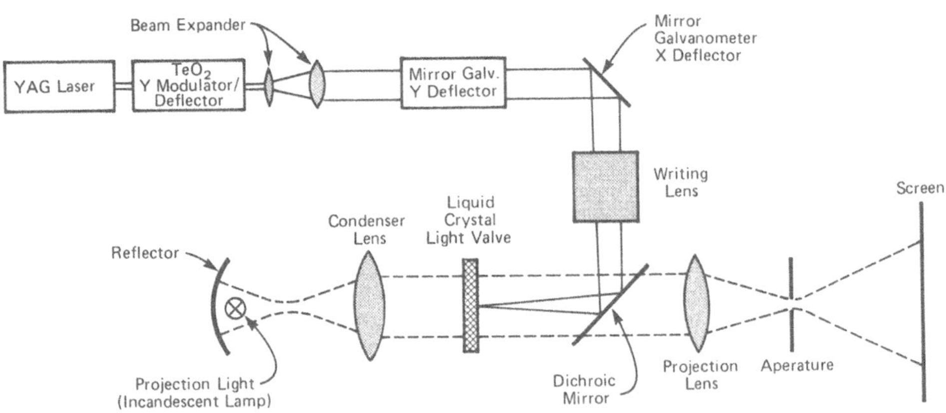


Fig. 1. Liquid crystal display system.

We subsequently performed some experiments with mixtures of MBBA plus 10 to 20% of cholesterol nonanoate or cholesterol-oleyl-carbonate. At room temperature the erase time, even with fields as high as $10^5$ V/cm, was on the order of 1 sec. There was some residual scattering in the aligned state which limited the achievable contrast ratio.

## SMECTIC LIQUID CRYSTALS

In 1973 the advantages of using smectic materials were recognized by the BTL researchers.[3] The material they used was the Schiff-base CBOA which was temperature biased in the smectic range (see Table 1) and heated by the laser beam through the nematic to the isotropic state. All of the smectic materials which have been investigated for use in laser addressed LV's have positive dielectric anisotropy. The aligned state is homeotropic and is completely transparent. In addition the scattering centers created by the thermal writing process are smaller and denser than those in the cholesteric cells, thus they scatter light more strongly. This combination leads to much higher CR in the projected image.

Table 1. Liquid Crystal Transition Temperatures (°C)

| | C-S | S-N | N-I (or S-I) |
|---|---|---|---|
| CBOA | see Ref. 3 | 61 | 72 |
| BTL ternary mixture[4] | 24 | 73.5 | 74 |

Cyano-Biphenyls:

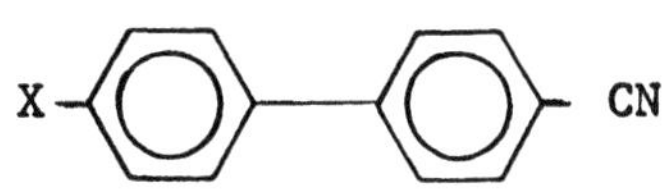


| X | | | |
|---|---|---|---|
| $C_8H_{17}$ | 21 | 32.5 | 40 |
| $C_9H_{19}$ | 40.5 | 44.5 | 47.5 |
| $C_{10}H_{21}$ | 43 | ---- | 50.6 |
| $C_{11}H_{23}$ | 52 | ---- | 56 |
| $C_8H_{17}$ | 54.5 | 67 | 80 |
| $C_9H_{19}O$ | 63 | 77.5 | 79 |

One of the problems with the CBOA is that it cannot be aligned in the smectic state with reasonable values of field strength. Erasure of the LV can only be achieved with a heating cycle up into the isotropic state followed by either slow cooling or else rapid cooling in the presence of a field. Even the latter took several seconds for a complete erase, which was undesirable. However, this problem was turned into an advantage by recognizing that it afforded a means of selective erasure of written information by using the laser beam for local heating in the presence of a field. This property, which will be discussed in a later section, is very desirable for a large storage display and is not available on current storage CRT displays.

Early in 1973 we became aware of the newly discovered cyano-biphenyl class of liquid crystals which includes a room-temperature smectic (the octyl - Table 1). We found that this material (OCBP) could be readily aligned at room temperature. For example, a 12 μm thick cell could be erased in about 40 msec with 70 $V_{rms}$ sine wave at 1 kHz. This effect was also studied extensively by a group at Thomson CSF.[5] Besides this possibility of fast erasure of the LV, OCBP had two additional advantages over CBOA. The first is that it did not crystallize at room temperature. The melting point is 21°C but it will supercool several degrees for long periods of time. Written information on the LC cell is lost if it crystallizes, but more important is the fact that the material contracts when it solidifies. If the cell is not tightly sealed, and most of our early cells were not sealed at all, it will draw in air when the LC crystallizes. On reheating the cell to the smectic state it will contain many air bubbles which are not easy to remove. The second advantage is that the biphenyls are considerably more stable than Schiff-base materials. To date we have discovered no degradation processes in our biphenyl LC mixtures.

## LIQUID CRYSTAL MIXTURES

Since the display operates with the LC in the smectic state, it is desirable to have the smectic range go from a point higher than that at which the cell is required to operate, to a point lower than that at which information is required to be stored on the cell. The high point could be 60°C or higher when the cell is being projected by a high-powered light source, or is operated in a warm environment. The low point could be 10°C or -10°C or even lower depending on the location. No single compound exists with such a smectic range and hence a mixture must be used. In comparison, mixtures of five or more components are used in twisted nematic displays for watches and calculators. These have a nematic range from about -10°C to 60°C.

A eutectic mixture of two smectic materials has a crystal to smectic transition which is lower than either of the components.

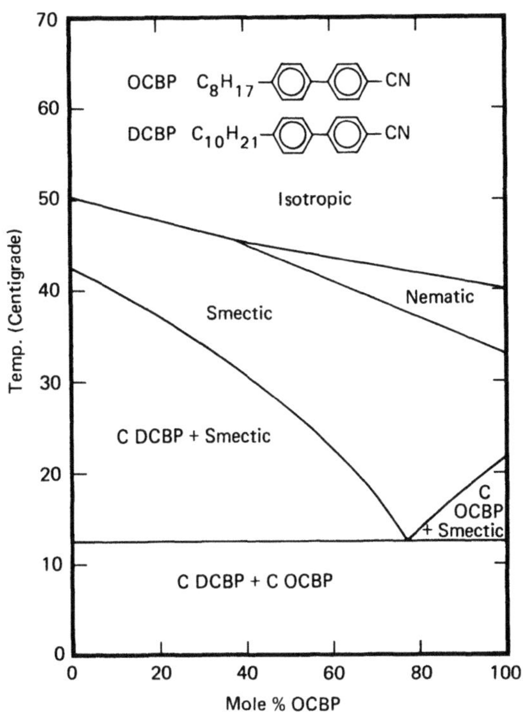


Fig. 2. Phase diagram for a mixture of two LC's.

In Fig. 2 is shown the phase diagram for OCBP (octyl) and DCBP (decyl). The eutectic mixture has a smectic range of 12°C to 37°C and an isotropic transition at 42°C. If the cell is temperature-biased at 35°C, the minimum temperature rise $\Delta T$ which must be created by the laser beam is 7°C. Since the writing energy is roughly proportional to $\Delta T$ we would clearly like to minimize it. This can be done in two ways: by biasing the cell at a point very close to the S-N transition or by making the nematic range very narrow or even nonexistent. As seen in Fig. 2 it is not easy to get a mixture which has both the widest smectic range and a narrow nematic range. Of course, it would be easy to make a mixture with no nematic range, however, it has been found experimentally[4] that a narrow nematic range is necessary in order that the cell be erasable with reasonable values of field at useful values of $\Delta T$. For example, a cell filled with DCBP could only be aligned when $\Delta T$ was <1°C.

The ternary mixture used by BTL has the desired narrow nematic range as shown in Table 1. However, it is still not possible to make ΔT very small without degrading the writing process as will be described in the section on contrast ratio. Much of our work has been performed with a mixture of biphenyls which has similar transition temperatures to the BTL mixture.

## LASER ABSORPTION

All of the work at BTL, the later work at Western Electric Labs[6] and at Thomson-CSF[7] employed custom-made ITO transparent conductors for the laser absorption, and the lasers used were Nd:YAG operating at 1.06 μm wavelength. As reported by Fraser,[8] ITO coatings could be made with good transmission in the visible range (<0.4 dB/μm @ λ=500 nm) and reasonable absorption of 1.7 dB/μm at Nd:YAG wavelength (λ=1.06 μm). However, even with a 1 μm coating on each substrate, the absorption is only 54%, and 1 μm thick ITO films are extremely difficult to make.

One of our major goals was to make this technology work with room-temperature GaAs lasers. These are small, efficient, can be current modulated, are potentially inexpensive and reliable but have limited power. At present, 10 to 20 mW CW is the most one can expect to achieve reliably. In comparison Nd:YAG lasers are bulky, expensive, inefficient, require water cooling and the pump-lamps have limited lifetime. In addition, an expensive modulator (e.g., acousto-optic) is needed for the LV system. Room temperature GaAlAs lasers operate at a wavelength of about 850 nm depending on the Al content of the active layer. This is so close to the visible range that it is virtually impossible to achieve good absorption of the laser energy with a visibly transparent coating. The absorption of the ITO films discussed earlier is only 0.6 dB/μm and because of the limited power available from a GaAs laser (compared with a Nd:YAG laser) it is important to achieve as high an absorption as possible.

Some attempts were made to create a transparent coating with strong absorption at λ=850 nm. A nickel-chelate dye was identified[9] with suitable properties and the idea of dissolving the dye in the LC was considered. This would create an ideal situation because the laser energy would heat the LC directly rather than heating the substrate. In comparison, heating the LC by conduction from a thin absorbing layer on one, or both sides of the cell has two problems. One is that lateral spreading of the heat causes a written spot that is larger than the beam. This creates a loss in resolution when one is trying to achieve the smallest spot size, but this property also allows variable spot size over a tremendous range (e.g., 10 to 1000 μm) simply by increasing the amount of energy delivered to the spot. This could be used to advantage in some applications.[6] The second problem when using glass substrates is that their thermal conductivity is about an order of magnitude greater than that of the LC. This implies inefficient use of the absorbed energy in heating the LC. Nevertheless, attempts at dissolving the absorbing dye in the LC were not successful. Small amounts of dye which did not give adequate absorption caused significant changes in the properties of the liquid crystal.

It was found that the dye could be dissolved in polyamide.[9,10] When a 1 μm layer was spin-coated onto the substrate and cured, the resulting film had good absorption at $\lambda$=850 nm and was light green in color. This was a possible solution to the absorption problem but it created several new problems. One is that spin-coating does not produce a surface with high optical quality, and alternative means of producing a 1 μm thick polymer film are not easy. The second problem concerns the transparent electrodes. It was difficult to deposit good-quality ITO films on the surface of the polyamide, but the polyamide could be spin-coated over the ITO. This created blocking capacitors in series with the LC and required a large increase in the erase voltage. Work on this approach was suspended when it was decided to concentrate on reflective cells.

## REFLECTIVE LC CELLS

With a reflective cell, one substrate is a mirror and the image is projected in reflection mode. As shown in Fig. 3 the cell is written from the back and since there are now few restrictions on the absorbing film, the laser absorption can approach 100% at <u>any</u> laser wavelength. This would allow the use of a visible laser, e.g., HeNe or Ar if so desired. For GaAs use, the goal is to produce on the substrate a multi-layer coating which has maximum reflectivity over the visible range from the top surface when it is covered with the aligned smectic LC; and maximum absorption in as thin a layer as possible of laser energy incident through the substrate. Several versions of this coating have been produced and have achieved >90% reflectivity of visible light and >95% absorption at 850 nm wavelength. In one design, a thin (~30 nm) Al reflector also serves as the absorber. The reflectivity of the Al on the substrate side is reduced to <10% by a dielectric A/R coating.

Besides the laser absorption problem, there was a second compelling reason for switching to reflective cells. With a Nd:YAG laser, the working distance of the writing lens can, as shown in Fig. 1, be quite long and still allow a small spot size. This is not the case with GaAs lasers. Since the beam divergence

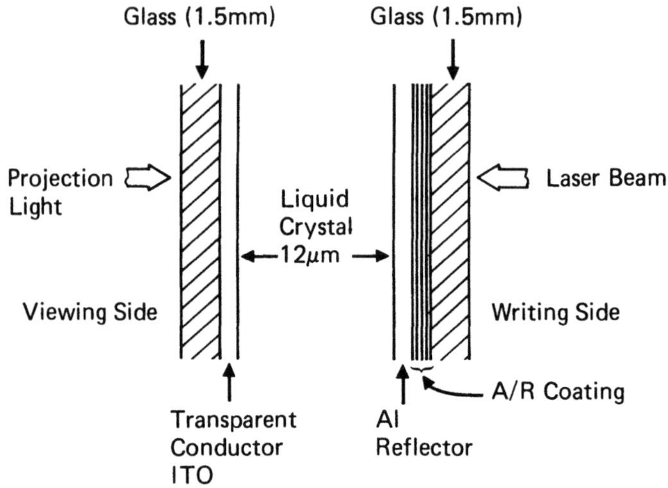


Fig. 3. Reflective cell structure.

in a plane normal to the junction is high (up to 60° full-width-half-power), a high numerical aperture lens must be used to collect the laser output efficiently. In order to achieve a small spot size, the writing lens must also have a high numerical aperture. In practice this implies a very short working distance, or at best one which is too short to include a 45° dichroic beam splitter in the optical system as is used in Fig. 1.

The use of a reflective cell has some disadvantages, however. One is that projection of the image is more difficult as will be discussed in a later section. The other is that the achievable contrast ratio is lower than that of a transmissive cell. The focal conic texture of a written spot scatters light strongly both in the forward and backward directions. As explained in Ref. 11, it would be difficult to achieve either efficient light collection or a uniform display if the projection system were operated in a white-on-black (dark-field) mode. In the black-on-white mode, most of the incident light is scattered out of the aperture of the projection system and very high CR is possible with a high f# system. In Fig. 4 the CR of some early cells is shown as a function of the f# and half-angle aperture $\alpha$ of the projection system. For comparison, the results from a sheet of ground glass and the theoretical curve for a perfect forward Lambertian scatterer ($CR=1/\sin^2\alpha$) are shown.

In a reflective cell the projection light passes through the LC layer twice, and in some cases this can increase the scattering

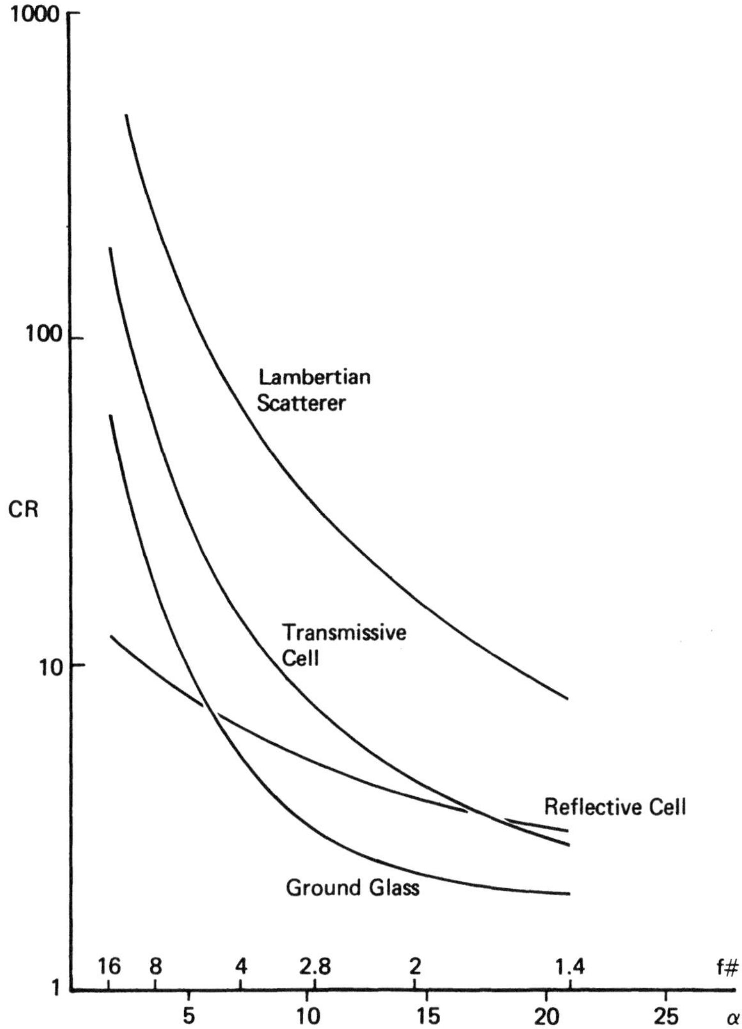


Fig. 4. Contrast ratio as a function of projection aperature.

and hence the CR for a given writing energy. However, if the information is fully written, a condition which will be discussed in the section on contrast ratio, there is no significant increase in scattering as the path length is increased beyond about 6 μm. Hence the double pass through a 10 μm thick cell does not necessarily increase the CR. Stray light and unwanted reflections from the optical system will reduce the CR of the image, and whereas these can be minimized in a transmissive system, this is not the case in one of the reflective systems. The major contributors to reduced CR with a reflective system are the reflections from the front surface and the interfaces on the front substrate. Whereas these can be minimized by careful design, their effect is not reduced by reducing the aperture of the projection system, and it is not possible to achieve the very high CR possible with a transmissive cell.

## OPTICAL PROPERTIES OF THE SUBSTRATES

The various coatings on the substrates are shown in Fig. 3. The laser-absorbing coating includes as its top layer a component designed to give high reflectivity of visible light. We have used only Al (covered by the LC aligning layer) and the reflectivity of a filled cell is typically >90%. This has been quite adequate for our purposes but in a high-magnification, high luminous output system, it may not be high enough. The problem is that energy absorbed from the projection lamp will heat the cell, but as long as the illumination is uniform and the temperature does not approach the operating level, the temperature control system will be able to maintain the required $\Delta T$. The reflectivity could be increased by using silver as the reflector and by enhancing the reflectivity with dielectric layers.

As mentioned earlier, the specular reflections from the front substrate are the major contributors to reduced CR. A broad band AR coating on the front surface can reduce this component from 4% down to less than 0.5%. However, if the thickness of the ITO transparent conductor is chosen incorrectly, a condition which existed in many of our early cells, the reflectivity $R_I$ of this layer, shown in Fig. 5, can be high. The expression for $R_I$ is:[12]

$$R_I = \frac{n_I^2(n_G-n_L)^2\cos^2\phi + (n_Gn_L-n_I)^2\sin^2\phi}{n_I^2(n_G+n_L)\ \cos^2\phi + (n_Gn_L+n_I)^2\sin^2\phi}$$

where $\phi$ is the one-pass phase shift in the ITO, and the n's are the respective indices of refraction. When $\varphi=\pi/2$ or $d_I=\lambda/4n_I$ (~700Å) $R_I$ is about 8%, but if $d_I$ is chosen to be $\lambda/2n_I$ or $\phi=\pi$, then since $n_L \approx n_G \approx 1.5$, $R_I \approx 0$.

By this means an optimized cell can have a total front surface reflection of <1% and a maximum CR (in a high f# projection system) of 100:1 or more. In comparison, a cell with no A/R coating and the wrong ITO thickness could have a front surface reflectivity of 12% and a maximum CR of only 8:1.

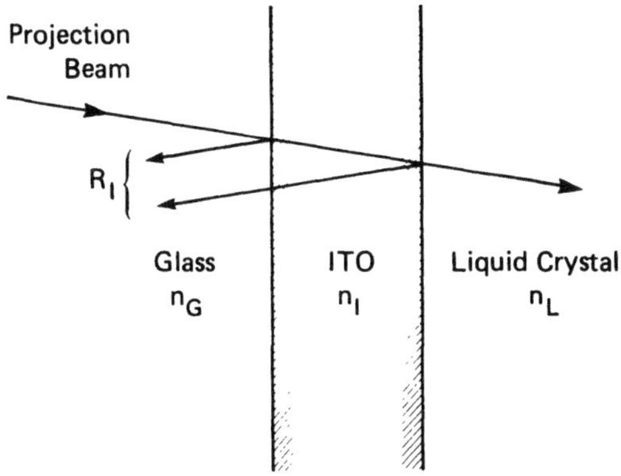


Fig. 5. Reflection from the ITO layer.

SURFACE TREATMENT

The optical properties of the two substrates are clearly important, but of equal importance in obtaining high CR, high sensitivity, long-term memory, good erase properties and a useful selective-erase window are the surface aligning layers. In previously reported work on this technology,[4,7] a silane surface treatment was given to the substrates in order to promote homeotropic alignment. Since the LC material has positive dielectric anisotropy, application of a field to a cell biased with a sufficiently small $\Delta T$ will create a perfect smectic structure throughout the cell. Thus the bulk erase properties of this cell are excellent. We have found, however, that at low values of $\Delta T$, the stability of a written spot in such a cell is poor and high writing energy is needed.

Our early cells employed substrates which had been cleaned in solvents and these produced very erratic results. Since that time we have employed a plasma etching (cleaning) and deposition system[12] for preparing all substrates. The coated glass substrates are first etched in an oxygen plasma which removes any hydrocarbon contaminants. This process also causes a dramatic increase in the resistivity of the ITO film but this can be reversed by exposing these substrates to a forming gas ($N_2/H_2$) plasma. Finally both sets of substrates are exposed to a silane plasma which creates a layer (~100Å) of $SiO_2$. This surface causes homogeneous alignment with all the cyano-biphenyl materials and mixtures we have used. Some experiments were performed with homeotropic surface alignment but, as mentioned earlier, the results were not encouraging. The surface treatment in this case involved exposure to a plasma containing $C_2F_4$ which produced a polyfluorocarbon film. It is possible that the homeotropic aligning forces produced by this surface are much stronger than those produced in the earlier work.[4,7]

The alignment of the LC molecules near the surfaces depends not only on the chemistry of the surfaces but on their topography. We have seen little variation in the Al surface (which is produced by evaporation) but there are several means of producing ITO and the roughness of the surface varies considerably. Examples of rough ITO produced by sputtering are shown in Ref. 14. Based on limited evidence, we suspect that smooth ITO produces the best overall results. However, further work is needed before we can optimize the surface properties of the substrates.

## CELL CONSTRUCTION

The glass substrates we have used are 1.5 mm thick and have a flatness of about 1 μm/cm. Cell sizes have been in the range of 1 to 3.5 cm square and the cell thickness, defined by a Mylar spacer, is typically 12 to 14 μm. Two holes are drilled in opposite corners of the ITO substrate prior to the coating steps. The finished substrates are assembled with the Mylar spacer, which is coated on both sides with a thermo-setting epoxy, clamped together and heated in an oven. The cell is filled at elevated temperature with the LC in the isotropic state. After cooling, the fill-holes are plugged to minimize leakage of the LC when the cell is subsequently heated.

When a new cell first cools it produces a Grandjean texture which is slightly scattering. This is because of the homogeneous surface alignment and is in contrast with cells which have homeotropic surface treatment and which cool to a clear homeotropic state. Cells with mixed alignment (homogeneous on one substrate and homeotropic on the other) cool to a scattering focal-conic texture.

Before writing on a cell it is heated to the bias temperature and the Grandjean texture is erased with a field (~5 kHz square wave, up to 200V, pk to pk). At low values of field, above a certain threshold, the Grandjean texture is transformed into a scattering focal conic texture. As the field is increased, this texture is erased and the cell becomes clear. This clear state is not completely homeotropic, however, because of the homeogeneous surface alignment. There must be a network of large focal conics in the cell as described in Ref. 15, but they are not visible in the projected image.

## CONTRAST RATIO AND WRITING ENERGY

The energy required to write on the cell depends strongly on $\Delta T$, the difference between the bias temperature and the isotropic point. Reducing $\Delta T$ will reduce the writing energy but since the scattering texture is created by rapid cooling of the LC from the isotropic state, reducing $\Delta T$ will lower the cooling rate, weaken the scattering texture and lower the contrast of the image. As in previous work,[4] however, we have found that there is no significant loss in CR unless $\Delta T$ is reduced below about 4°C. We have also noticed that if a cell is allowed to cool down to room temperature, the CR of written information will increase - sometimes as much as 40%. This implies that even with slow cooling, the scattering texture grows more efficient as the temperature is lowered.

The above results were obtained with the best LC cells we have produced. However, there are many variables involved and not all cells behaved this way. The variables include the roughness of the surfaces, the quality of the aligning layers and the purity and composition of the LC mixture. As mentioned earlier, on cells with strong homeotropic alignment, written spots would rapidly collapse and disappear. Stable spots could be written with about ten times the energy but these tended to be larger and of irregular shape. On some cells written spots faded slowly after they were written and the CR dropped to about 50% of the initial value. On other cells a written spot would bloom and slowly grow in size, and in severe cases focal-conic domains would spontaneously nucleate throughout the cell. All of these problems vary with the value of $\Delta T$ and depend strongly of the variables in the cell construction.

Much of the cell testing is performed with a HeNe laser which is scanned across the cell with a pair of galvanometer mirrors. A rectangular array of spots is written and the large-area CR of the projected image is measured with a photometer. The laser beam is focused to about 10×15 μm and is modulated with an A/O modulator. The array of spots is on 12.5 μm centers. Near threshold, the low CR is caused both by the low scattering power of a written spot and by the low packing density. At high energy levels the written spots overlap completely. Some results are shown in Fig. 6 which were taken with a $\Delta T$ of 4°C in an f/11 projection system. The cell absorption at $\lambda$=632.8 nm was 85% compared with 90% at $\lambda$=850 nm, so slightly better results would be expected with a GaAs laser. A CR of 10:1 is achieved with an energy per spot of about 0.2 μJ. This is considerably less than the figure of 8 μJ per spot given in Ref. 7.

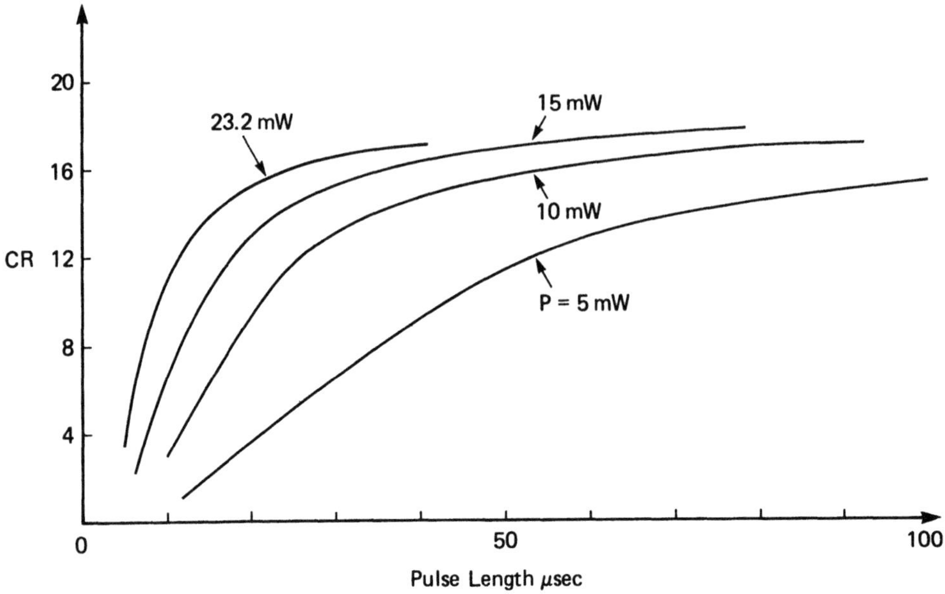


Fig. 6. Writing results using a HeNe laser.

As can be seen from the figure, grey scale writing is possible by varying the amount of energy per spot (power level or pulse length). A different method which uses constant energy and is related to the selective erase process, is described in the next section.

SELECTIVE ERASE

Selective erasure of information on the screen is highly desirable in a storage display. As mentioned earlier, this property of smectic LC cells was discovered with materials which could not be erased with a field alone, but which required simultaneous heating. With the biphenyl materials there is a threshold in the erase voltage below which no realignment of the focal conic texture occurs. This is shown in the upper curve of Fig. 7 where the threshold is about 120V pk-to-pk. This threshold which was studied in detail by Hareng et al.[5] decreases with increasing temperature. With voltages above threshold a partial erase occurs until at sufficiently high voltage the CR of previously written information is reduced to an imperceptible level. This behavior is different than that of cells with homeotropic surface alignment as were studied in Ref. 5. With these cells, erase was always complete and the time required was

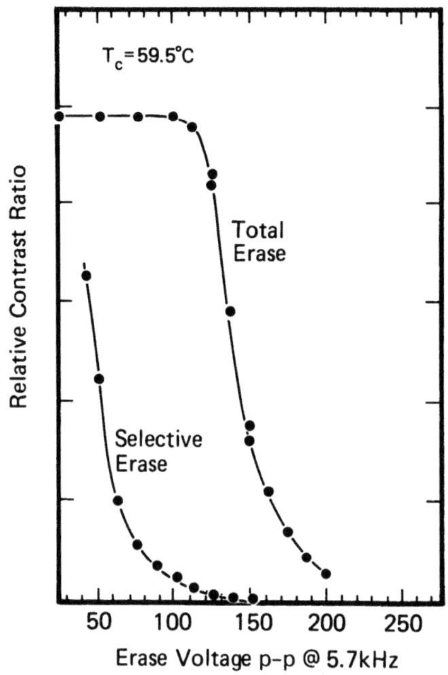


Fig. 7. Contrast ratio versus erase voltage.

inversely proportional to the square of the amount by which the voltage exceeded the threshold. Our present cells can only be completely erased by heating them into the isotropic state and cooling in the presence of a field, or by erasing the resultant Grandjean texture. This is not a practical limitation, however, as the residual scattering is not cumulative and remains imperceptible in the projected image after continued use of the cell. It is detectable, however, by directly viewing light which is scattered by the cell.

The selective erase process is shown in the lower curve of Fig. 7. Here the cell is subjected to a field and a previously written area is re-scanned by the laser beam at the same energy density. As the heated area cools the effect of the field is to discourage the creation of the focal-conic texture and to promote homeotropic alignment. At 120V a selective erase process exists for this cell because background information (upper curve) is virtually unaffected whereas the CR of the selected (re-scanned) area is reduced to an acceptably low level. The width of a practical selective erase "window" will depend on the acceptable values of CR and is affected by all of the cell variables mentioned in the previous section.

This selective erase behavior allows for a different method of grey-scale writing which was proposed by Hareng and Le Berre.[16] Referring to Fig. 7 it can be seen that by varying the erase field between zero and 120V during the writing of the cell, the CR will vary between maximum and minimum and moreover the field will not affect information which has already been written. The advantage of this technique for a Nd:YAG laser (and for other types) is that the laser power is constant and hence no modulator is needed. The disadvantage is that the writing speed is limited by the thermal time constants of the cell to a value of a few thousand points per second.

## PROJECTION OF THE CELL IMAGE

Projection of a reflective LV is more difficult than the corresponding transmissive one.[11] The two demonstration units which are described in following section both employ a reflex system in which the cell is illuminated through the lens as in Fig. 8. The lens is used in a telecentric manner and the aperture must be larger than the diameter of the cell. As the size of the cell is increased it becomes increasingly difficult to find a lens (for laboratory use) which will do the job. A custom-designed telecentric lens will also increase in cost rapidly as the cell size is increased. An alternative which does not have this problem is an off-axis reflective Type I system[11] in which the aperture of the lens may be much smaller than the cell. The problem with this system is that in order to have a short projection throw, which is necessary for a desk-top rear-projection screen, the lens must have a wide field of view (60° or more). This is not easy to achieve with a high resolution flat-field lens.

## DEMONSTRATION DISPLAY UNITS

These units were assembled in order to demonstrate the feasibility of the technology and have been described in detail elsewhere.[17,18] The A unit uses a single GaAs laser and was configured as a desk-top keyboard-entry display. It can write 2000 characters of stored text or from the keyboard at 20 char/sec. The laser is scanned by a pair of galvanometer mirrors and each of the 7×9 dot matrix characters is scanned by a mini-raster. The underline cursor is selectively erased and rewritten each time a character is entered. The laser, which has its heat sink maintained at 10°C by a thermoelectric cooler, emits about 15 mW CW. A pair of f/2 lenses collects and focuses about 3 mW into a 3×15 μm beam at the LC cell. Overlapping spots are written on 10 μm centers over a field of 7 mm diameter. The laser pulse time is 120 μsec which implies an energy of 0.36 μJ per spot. This is higher than the results quoted in an earlier section because of non-reciprocity at the low power and long pulse length needed.

One of the advantages of a black-on-white LV is that overlay information can be projected in series with the LV. As shown in Fig. 8 the LC is imaged into the plane of an overlay slide which can contain full-color, high resolution images of business forms, maps, etc. The combined image of the cell and the overlay is projected at 25x magnification onto a 10-inch square rear-projection screen. Dynamic information appears in the central 5-inch square as black characters, an example of which is shown in Fig. 9.

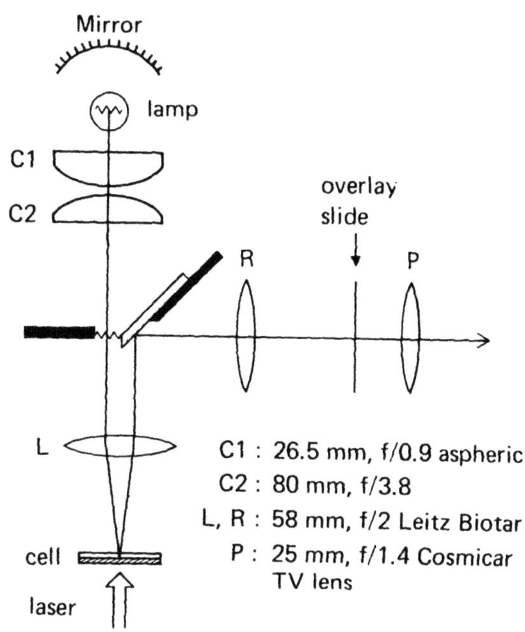


Fig. 8. Projection system for demonstration unit A.

Demonstration unit B was designed primarily to show the high information content capability not shown by the A unit. In order to achieve a higher writing speed than was possible with a single GaAs laser, we chose to use an Ar ion laser. The unit displays 1536×1536 pels in 3 sec or more, the speed being limited by the galvanometer mirror laser-beam deflectors. The laser beam is compressed 2:1, modulated by an A/O modulator, expanded 4:1 and focused by a 50 mm focal length lens into a 15 μm diameter beam.

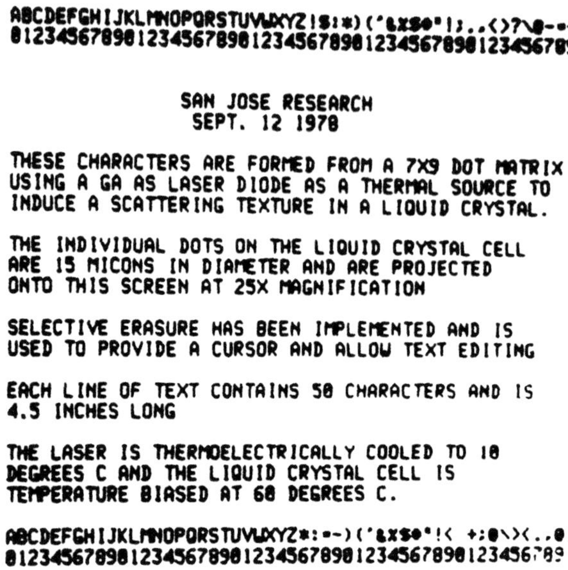


Fig. 9. Photograph of screen on demonstration unit A.

Up to 1W of modulated power is available at the LC cell but typically 200 mW is used with a pulse length of 2.4 μsec. The relatively high energy per spot is caused by the low absorption of the coatings on the cell (which were designed for GaAs).

An 18 mm square area of the LC cell is projected onto a 15-inch square screen. Actually the system is designed to project two LC cells for experimenting with a multi-color display. White light from a xenon arc lamp is separated by a dichroic beam-splitter into two spectral components. Each half of the system is similar to that of Fig. 8 and the two paths are recombined by a second splitter before the final projection lens. With no information on either cell, the screen appears white. Using a green/magenta splitter, information written on one cell appears as green, on the other cell as magenta, and on both as black. Other color combinations are possible by using a different splitter or by filtering one of the components.

Most of the displays demonstrated on unit B have been derived from documents which have been optically scanned. The resulting data set of 2.36 Mbits is stored on the disk of a S/7 computer and transferred to the display via a one-line buffer. An example of a display which was generated using a digital half-tone algorithm is shown in Fig. 10. Direct computer-generated alphanumeric images have also been implemented using 5×7 dot

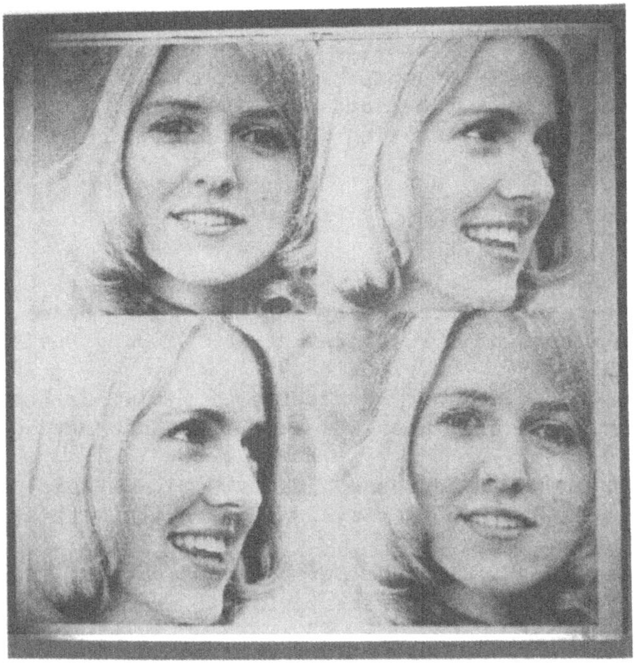

Fig. 10. Photograph of screen on demonstration unit B.

matrix characters (in a 7×10 box). With this format up to 33,000 legible characters have been displayed on the screen.

CONCLUSIONS

Although our goal of a high-information-content display using a GaAs laser has not been demonstrated, there seems little doubt that it would be possible to configure such a system using multiple lasers. Using an array of lasers, it should be feasible to achieve a writing speed which is adequate for a wide range of applications, especially when the selective erase and local update capability of the technology is used. With specially-designed laser scanning systems and writing optics it should be possible to write at least $10^4 \times 10^4$ spots on the LC cell. However, the upper limit on information content is going to be determined by the ability of the projection lens to resolve this large number of points, and ultimately on the cost of producing this lens.

ACKNOWLEDGMENTS

The author gratefully acknowledges the technical contributions to this project of J. T. Jacobs, A. Juliana, R. W. Koepcke, G. T. Sincerbox, G. J. Sprokel and especially C. E. Stratton for his work on cell fabrication, testing and evaluation.

REFERENCES

1. D. Maydan, H. Melchior and F. Kahn, Liquid Crystal Graphic Displays Thermally Addressed by Infrared Laser Beams, IEEE Conference Record of 1972 Conference on Display Devices, October 11-12, 1972, pp. 166-168.
2. D. Maydan, Infrared Laser Addressing of Media for Recording and Displaying of High-Resolution Graphic Information, Proc. IEEE 61:1007 (1973).
3. F. J. Kahn, IR-Laser-Addressed Thermo Optic Smectic Liquid-Crystal Storage Displays, Appl. Phys. Letters 22:111 (1973).
4. G. N. Taylor and F. J. Kahn, Materials Aspects of Thermally Addressed Smectic and Cholesteric Liquid Crystal Storage Displays, J. Appl. Phys. Vol. 45, No. 10, 1974, pp. 4330-4338.
5. M. Hareng, S. Le Berre and L. Thirant, Electric Field Effects on Biphenyl Smectic A Liquid Crystals, Appl. Phys. Letters 25:683 (1974).
6. R. A. Heinz and R. C. Oehrle, Rapid Generation of Complex Images with a Liquid Crystal, Western Electric Engineer, Vol. 21, No. 2, April 1977.
7. M. Hareng, S. Le Berre and R. Hehlen, Valve Optique a Crystaux Liquides Smectiques, Revue Technique Thomson-CSF, Vol. 9, No. 2, June 1977.
8. D. B. Fraser, Sputtered Films for Display Devices, Proc. IEEE, Vol. 61, No. 7, July 1973, pp. 1013-1018.
9. G. J. Sprokel, Beam Addressed Liquid Crystal Cells, U.S. Patent 3,999,838 (1976).
10. E. A. Cunningham, A. G. Dewey and G. J. Sprokel, Absorbing Films for Reflective Laser-Beam-Addressed, Thermally-Activated, Liquid Crystal Cells, IBM Tech. Discl. Bull. 18:3776 (1976).
11. A. G. Dewey, Projection Systems for Light Valves, IEEE Trans. Vol. ED-24, No. 7, 1977, pp. 918-929.
12. G. J. Sprokel, Molecular Order Induced by Cell Walls, Mol. Cryst. Liq. Cryst. 1977, Vol. 42, pp. 233-248.
13. E. Hecht and A. Zajac, "Optics," Addison Wesley, Massachusetts (1974), p. 314.
14. D. B. Fraser and H. D. Cook, Highly Conductive, Transparent Films of Sputtered $In_{2-x}Sn_xO_{3-y}$, J. Electrochem. Soc. Vol. 119, No. 10, October 1972, pp. 1368-1374.

15. M. Hareng, S. Le Berre and J. J. Metzger, Planar-to-Homeotropic Structure Transition under Electric Field in an A-phase Smectic, Appl. Phys. Lett. Vol. 27, No. 11, December 1, 1975, pp. 575-576.
16. M. Hareng and S. Le Berre, Formation of Synthetic Images on a Laser-Beam-Addressed Smectic Liquid-Crystal Display, Elec. Letters 11:73 (1975).
17. A. G. Dewey, et al., A 2000 Character Thermally-Addressed Liquid Crystal Projection Display, Proc. 1977 SID Intl. Symp., pp. 108-109.
18. A. G. Dewey, J. T. Jacobs and B. G. Huth, Laser-Addressed Liquid Crystal Projection Displays, Proc. SID, Vol. 19/1, 1978.

# THE ROLE OF SHADOWS IN THE VIEWABILITY OF TWISTED NEMATIC LIQUID CRYSTAL DISPLAYS

Arthur L. Berman and Chan S. Oh

Beckman Instruments, Inc.
2500 Harbor Boulevard
Fullerton, California 92634

## INTRODUCTION

Most of the data published on the viewability of twisted nematic liquid crystal displays (TNLCDs) has been measured in transmission.[1-3] This approach has been favored because transmissive data is easier to interpret from a fundamental point of view. On the other hand, characterizing displays in reflection is more realistic from the practical standpoint, since TNLCDs are primarily viewed in reflection. It was the purpose of this work to explore the relationship between reflective and transmissive viewability. To this end, data was collected on the angular variation of the contrast as a function of applied voltage. These measurements indicate that the viewability in transmission is much more limited than in reflection. The cause of this difference in viewability is shadows, which occur when the TNLCD is viewed in reflection. The "object" casting the shadow occurs in the plane of the liquid crystal. We call this object the real image. Ambient lighting projects an apparent shadow of the real image into the plane of the reflector. It should be noted that these results are characteristic of TNLCDs and, therefore, independent of the manner in which the TNLCD is manufactured, or the liquid crystal material used in the display.

## EXPERIMENTAL

We have built an optical device similar to that described by Barna.[4] In this device, the display is mounted on a section that is free to rotate about a vertical axis. Centered on this axis, and above the display, is a stationary fluorescent ring that

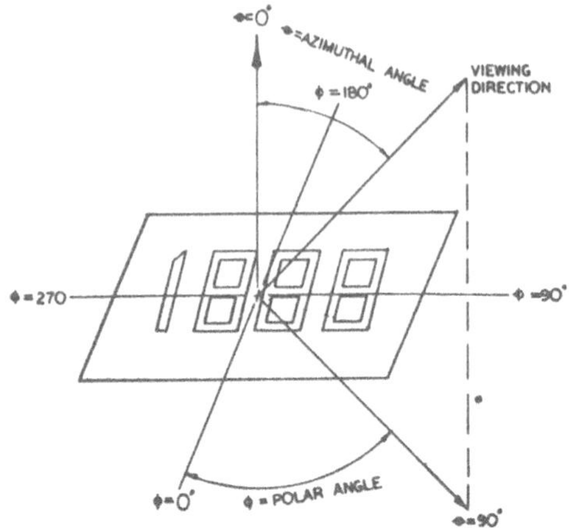


Fig. 1. Coordinate system.

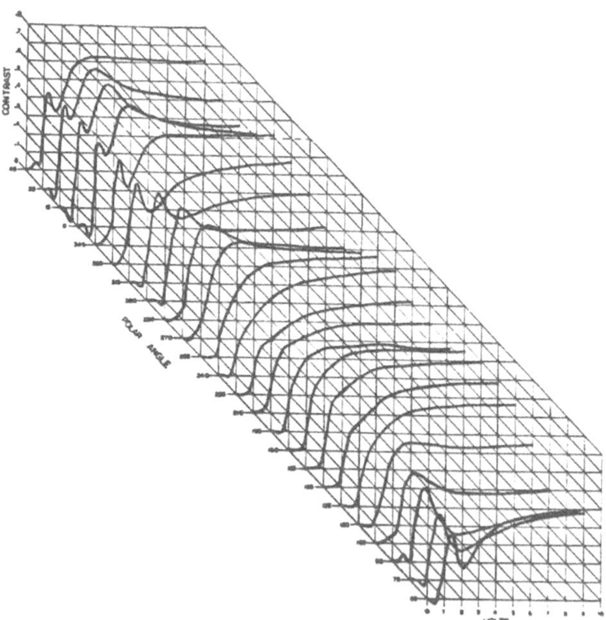


Fig. 2. C-V curves measured in the real image region of a reflective display.

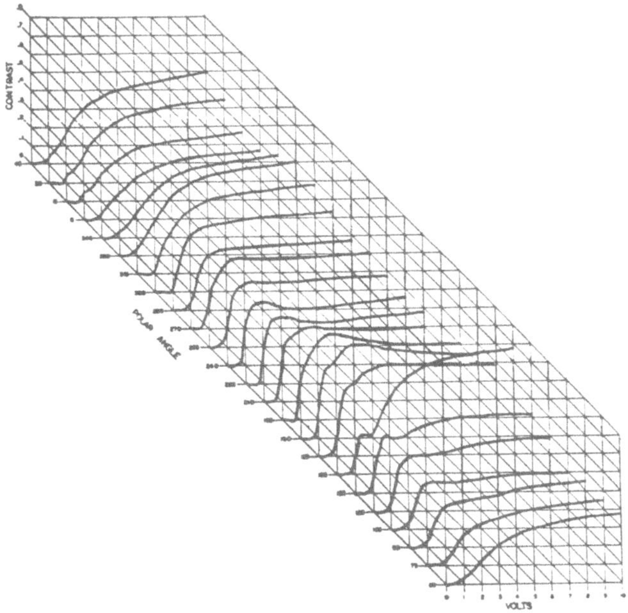


Fig. 3. C-V curves measured in the shadow region of a reflective display.

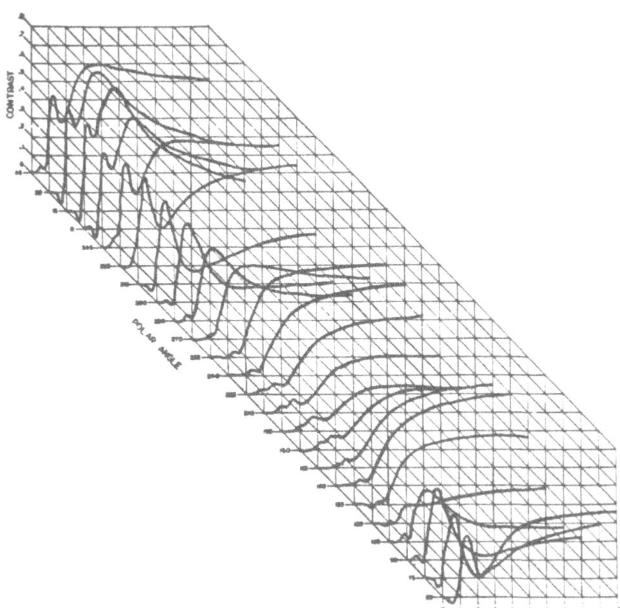


Fig. 4. C-V curves measured in the overlap region of a reflective display.

serves to illuminate the display for the reflective measurements. The display is observed through a microscope, mounted on an arm that can rotate about a horizontal axis. The combined motions of these two sections allow the microscope to remain focused on a single spot on the display and yet be oriented such that the display can be observed at any viewing angle within a cone having a half angle of 45°. The microscope has a fiber optics probe in the center of its eyepiece that collects the light from a spot on the display 150μ in diameter. The intensity of the light is measured by a digital photometer which drives the Y-axis of an X-Y recorder. The X-axis is drived by a DC signal proportional to the RMS voltage applied to energize the display. The applied voltage was a 32 Hz squarewave ramped from 0 to 10 volts peak at 150 millivolts per second.

The procedure for taking data was as follows: The fiber optics probe is centered on the real image of one of the segments. The zero reflection value ($R_o$) was obtained by closing the photometer shutter and setting the recorder pen on the top left side of a chart paper. With the shutter open the intensity of the light reflected from an unenergized segment ($R_{100}$) was set at a unit value less than $R_o$ on the Y-axis of the chart paper. The ramp voltage was then applied and the reflected light intensity ($R_V$) was plotted on the X-axis. In this manner we have directly graphed $C=1-R_{100}/R_V$, where C is the contrast and varies between 0 and 1. Proceeding as described above, data was taken at intervals of $\varphi=15°$ until all 360° were accomplished. The $R_o$ value of each successive graph was displaced downward and to the right (see Fig. 2). In this way, a series of C-V curves are plotted in perspective. Data presented in this manner allows trends to be easily identified and facilitates comparison between different data sets. Once the measurements in the real image were completed, the entire set of measurements was repeated with the fiber optics probe centered in the shadow region of the segment and then the overlap region between the real image and the shadow. Finally, transmissive measurements were made using an electroluminiscent backlight located directly underneath the display. The procedures for taking this data were the same as described above.

## RESULTS AND DISCUSSION

Figures 2-4 illustrate the C-V data, in reflection, of the real image, shadow and overlap regions respectively of a TNLCD. The polar plots in Figs. 5-8 are taken from these C-V data. In Fig. 5, the threshold voltage is plotted as a function of polar angle at the fixed azimuthal angle $\theta=45°$. The threshold voltage is defined as the voltage at which the contrast achieves a value 10% of its value when the maximum voltage is applied. Figures 6-8

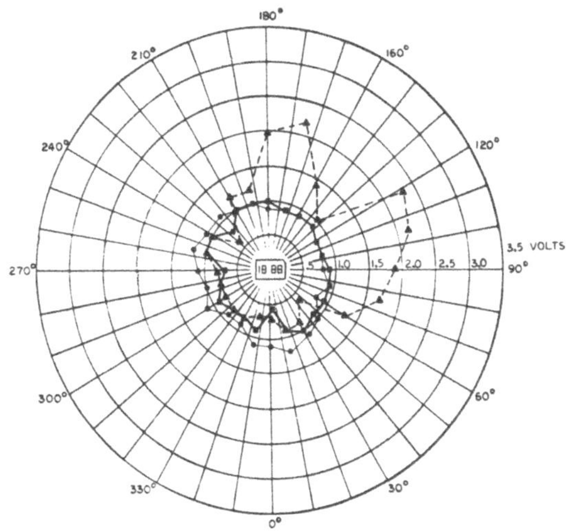


Fig. 5. Polar plots of the threshold voltage in reflection
▲ real image; ■ overlap; ● shadow.

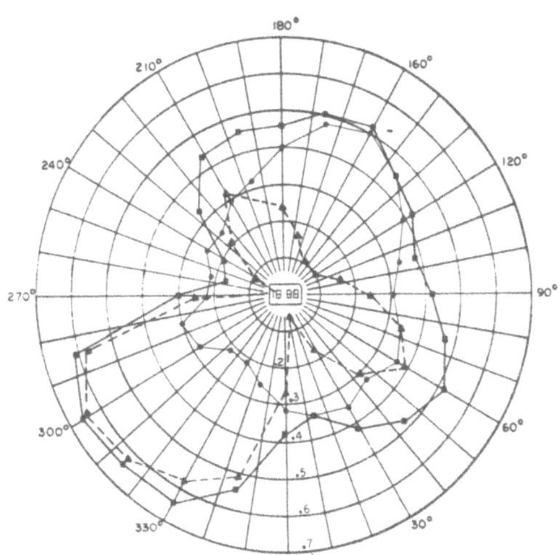


Fig. 6. Polar plots of the contrast at 3 volts in reflection
▲ real image; ■ overlap; ● shadow.

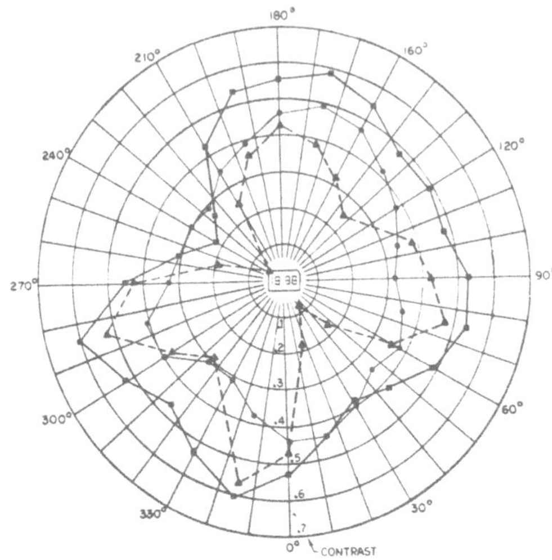


Fig. 7. Polar plots of the contrast at 6 volts in reflection. ▲ real image; ■ overlap; ● shadow.

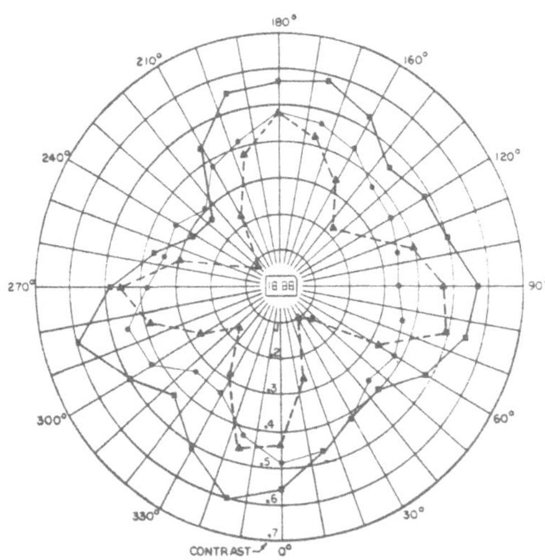


Fig. 8. Polar plots of the contrast at 10 volts in reflection ▲ real image; ■ overlap; ● shadow.

are polar plots of the contrast at the same azimuthal angle when 3, 6, and 10 volts, respectively, are applied.

Several general observations can be made regarding the behavior of each region. As seen in Fig. 5, the threshold voltage of the real image undergoes substantial variations with polar angle, varying between 0.6 and 2.3 volts. The variation of contrast with polar angle at 3 volts (Fig. 6) shows a strong 3 lobed pattern with the lobe centered around 310° being the largest. It should be noted that the large lobe is oriented in the same plane in which the mid-layer director of the TNLCD points upward. As the voltage was increased to 6 volts (Fig. 7) and then to 10 volts (Fig. 8), the strong variation of contrast with polar angle persisted with the 3 lobe pattern developing into a 4 lobe pattern. This is expected because at 10 volts the liquid crystal is almost homeotropic.[5] It is interesting to note that the additional lobe that develops by increasing the voltage from 3 to 10 volts is formed by the major lobe observed at 3 volts dividing down the center. The net effect of this is that the best viewing direction at 3 volts becomes a region of poor contrast when 10 volts is applied.

Next, consider the behavior of the shadow. As seen in Fig. 3, the threshold voltage of the shadow region varies relatively little taking on values between 0.8 and 1.2 volts. The shadow also exhibits relatively minor variations of contrast with polar angle when measured at 3, 6, and 10 volts as shown in Figs. 4, 5, and 6, respectively.

The general features of these results can be explained by considering the following simplified model. Let there be a collimated light source directly above the display. The observer can be located in an arbitrary off-axis viewing direction as illustrated in Fig. 9. As the light rays enter the display they are polarized at the top polarizer. The incoming rays that go through an unenergized area such as 1, 2, and 3 have their plane of polarization rotated a full 90° and will not suffer any attenuation at the rear polarizer. Those outgoing rays that travel through an unenergized area such as 6', 7', and 8' will be rotated almost a full 90° and will suffer very little attenuation at the front polarizer. The light rays that do travel through the energized area such as 4, 5, 6, 7, 2', 3', 4', and 5' will have their plane of polarization rotated to an extent which depends on the polar and azimuthal angle the ray makes with respect to the display. Consequently these rays will experience an attenuation that ranges between 0 and 100%. The rays that are attenuated on the way in, such as 4, 5, 6, and 7 form a shadow which occurs in the plane of the diffuse reflector. The rays that are attenuated on the way out, such as 2', 3', 4', and 5' form a real image that occurs in the plane of the liquid crystal.

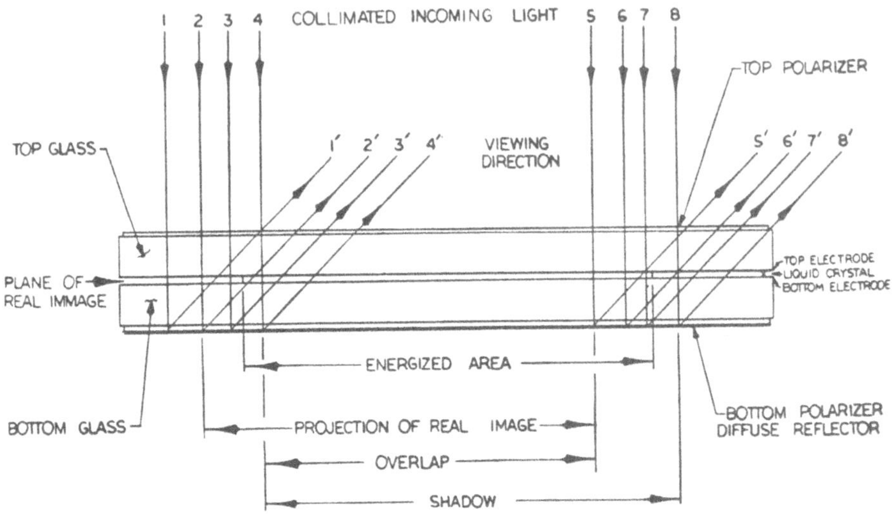


Fig. 9. Ray tracing in a TNLCD: a simple case.

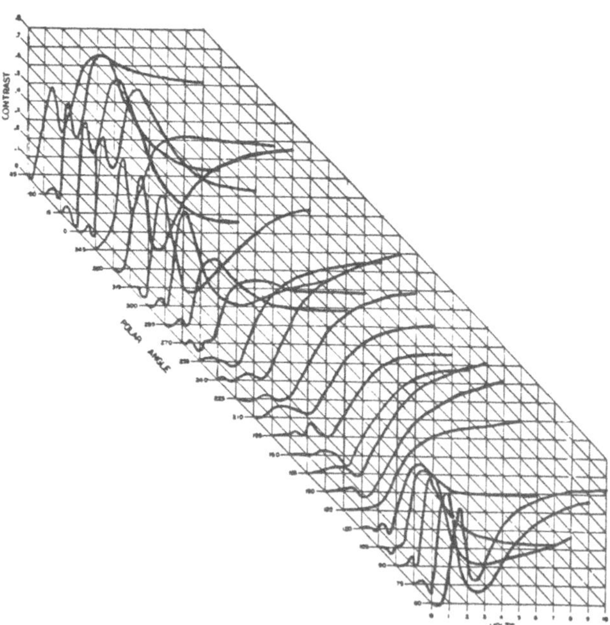


Fig. 10. C-V curves measured in transmission.

Let us consider the angular variation in the contrast of the real image and the shadow. (The contrast is, of course, a measure of the attenuation of the rays.) Furthermore, let us now consider the more realistic case in which the display is diffusely illuminated. Rays enter the energized area of the display from all directions and are attenuated to an extent that depends on their angle of incidence. These incoming rays form the shadow. The contrast of the shadow is the average attenuation of all these rays and is hence independent of viewing direction. Furthermore, the contrast will increase uniformly with voltage since the oscillations that do occur in some of the component rays are averaged out. The incoming rays now strike the back reflector and are diffusely reflected. These outgoing rays now form the real image. Since the extent to which an outgoing ray is attenuated depends on the direction it travels through the energized area, the observer will see a different real image contrast along each viewing direction.

As can be seen in Fig. 9, even though the shadow and the real image are vertically separated, it is possible for them to overlap in certain viewing directions. When an observer views a TNLCD the contrast that is perceived is primarily determined by the overlap region. If, as a rule of thumb, we assume that

$$\begin{pmatrix}\text{Attenuation}\\ \text{of overlap}\end{pmatrix} = \begin{pmatrix}\text{Attenuation}\\ \text{of shadow}\end{pmatrix}\begin{pmatrix}\text{Attenuation}\\ \text{of real image}\end{pmatrix}$$

there are several features that would be expected of the overlap region. First, the threshold of the overlap should be determined by the threshold of the shadow since, in general, the shadow has a lower threshold. As seen in Fig. 5, this is indeed the case. The threshold voltage of the overlap region is almost constant with polar angle as was the shadow and is of the same magnitude. Secondly, the contrast of the overlap region would be expected to follow the same variations with voltage as the real image since the turn-on of the shadow is so uniform. Comparing the general shapes of the C-V curves at equivalent polar angles in Figs. 2, 3, and 4 shows this to be the case. The role the shadow plays is to increase the contrast of the overlap at each voltage. For a similar reason, it is also true that at a fixed voltage the contrast of the overlap region will undergo polar variations that follow the real image.

Inspection of Fig. 6 shows that at 3 volts the overlap region does indeed exhibit a 3 lobe pattern similar to the real image. The contrast of the overlap is, however, uniformly increased by the presence of the shadow. It is clear that the shadow is playing a critical role in the viewability of this TNLCD. Along the viewing direction 20°, 130°, and 250° an observer would not

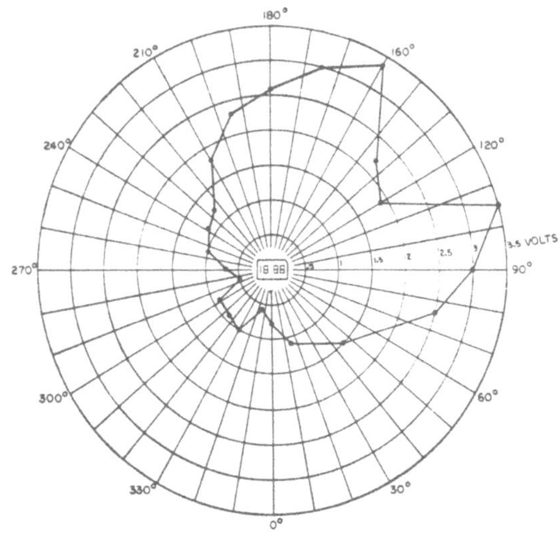


Fig. 11. Polar plots of the threshold voltage in transmission.

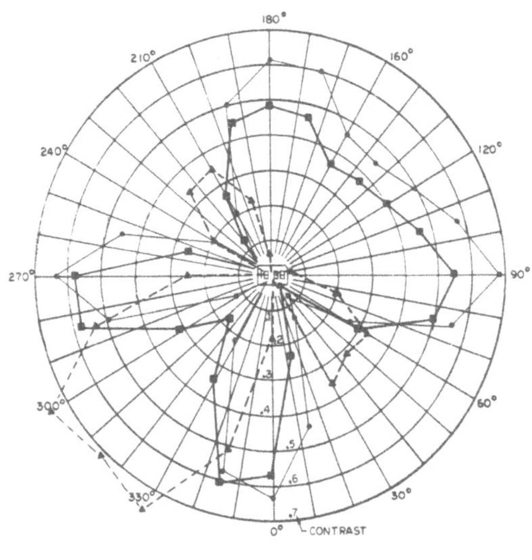


Fig. 12. Polar plots of the contrast in transmission ▲ 3V; ■ 6V; ● 10V.

be able to read the display without the presence of shadows. At 6 volts and 10 volts the same considerations are equally valid.

Let us now compare the reflective data with the data measured in transmission. In transmission there are no shadows and only the real image is present. Figure 10 presents the C-V data measured at $\theta=45°$ using the electroluminiscent backlight. The polar plots in Figs. 11 and 12 are taken from this data. As seen in Fig. 11, the threshold voltage in transmission undergoes substantial variations with polar angle varying between 0.5 and 3.5 volts. The variation of contrast at 3 volts, illustrated in Fig. 12, exhibits a 3 lobed pattern with the largest lobe centered at 310°. When the voltage was increased to 6 volts and then 10 volts (Fig. 12) the pattern developed into a 4 lobed pattern just as did the equivalent data for the real image. In fact, the transmissive data appears to be strikingly similar to the reflective real image data despite the very different type of illumination.

This last result is perhaps not so surprising. The light rays that form the real image in the reflective display are traveling along the observers viewing direction and are attenuated on the way out of the display. In the transmissive display, the rays that form its real image are traveling in the same direction. The attenuation is therefore substantially the same in both cases.

## CONCLUSION

The viewability of a TNLCD is much better in reflection than in transmission. The main reason for this is the presence of shadows in the reflective mode.

As voltage is applied to a reflective display the shadow turns on uniformly at a threshold voltage that is almost independent of polar angle. On the other hand, the threshold voltage of the real image undergoes considerable variation with polar angle and its contrast can exhibit oscillations with increasing voltage. Contributions from the real image and shadow determine the response of the overlap region which in turn dominates the viewability of the display. The threshold voltage of the overlap is primarily controlled by the shadow and is almost constant with polar angle. The contrast of the overlap region exhibits variations with polar angle similar to the real image, except at a consistently higher value of contrast. At 3 volts this pattern has 3 lobes of which 1 lobe is larger than the others. When 10 volts was applied, this pattern developed into 4 lobes; the largest lobe at 3 volts having divided at its centerline. The transmissive data was found to be almost identical to the reflective real image data.

A simple model that explains these results was presented. It was based on tracing light rays that form the real image and the shadow.

## REFERENCES

1. D. Meyerhofer, J.A.P. 48:1179-1185 (1976).
2. C. Z. Van Doorn and J. L. A. M. Heldens, Physics Letters 47A:135-136 (1974).
3. P. McIntyre, J. Opt. Soc. Am. 68:869-872 (1978).
4. G. Barna, Rev. Sci. Instrum. 47:1258-1260 (1976).
5. R. A. Soref and M. J. Rafuse, J.A.P. 43:2030-2037 (1972).

# DIRECTLY DRIVEN LIQUID CRYSTAL ANALOGUE METER AND HOROLOGICAL DISPLAYS

I. A. Shanks

RSRE
Malvern
Worcs, England

## INTRODUCTION

Digital liquid crystal displays are now in extensive commercial use. Their legibility in bright ambients and their minimal power consumption make them well suited for use in watches, calculators and a wide range of portable instrumentation. While digital displays may be excellent to display the absolute value of a measured quantity, it is necessary to perform subtractions to estimate changes in this quantity and they are often difficult to read when the value displayed is changing with time. This may be due to the measured quantity changing with time or to quantization noise in an analogue to digital converter which provides the digital value to be displayed. This makes it almost impossible to estimate the rate of change of the measured quantity with any degree of certainty.

These problems may largely be overcome by using a display which shows bars, hands or pointers to indicate the value of a measured quantity against an adjacent scale. Mechanical versions of these have been in use for centuries. Their mechanical nature, however, means that they exhibit inertial lag and dynamic overshoot when tracking a varying quantity. Their movements may also be expensive, subject to wear and they may be difficult to make both sensitive and robust.

There is therefore a demonstrable need for a cheap analogue display, with no moving parts, which can allow accurate estimation of the absolute change or the rate of change of a measured quantity in addition to its absolute value and permit quantization noise to be visually averaged without difficulty. It is also

Fig. 1a. Meander electrode geometry used in prototype meter display (shown 0.62 size).

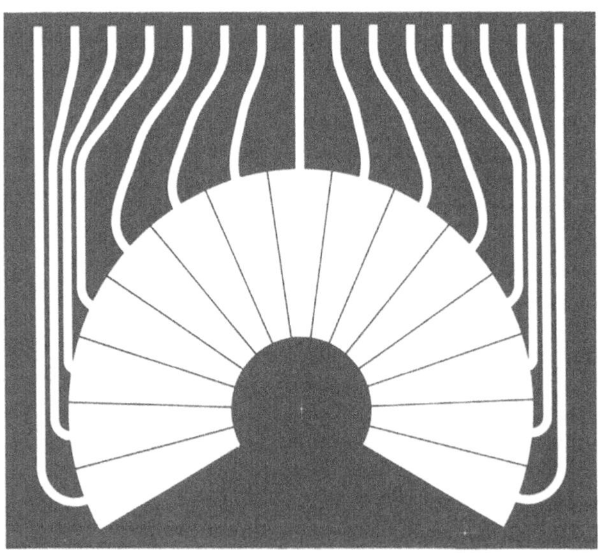

Fig. 1b. Segmented backplane geometry used in prototype meter display (shown 0.62 size).

highly desirable that this display should have very low power consumption, reasonably rapid dynamic response, legibility in all levels of ambient light and should be capable of being driven from digital logic signals.

Most of these requirements are met by analogue liquid crystal displays described by Kmetz[1] and by Greubel et al.[2] who use multiplexing techniques applied to a meander matrix format to overcome the large number of connections required when directly driving such a display. These techniques use the fact that liquid crystal displays respond to the RMS value of an applied voltage and they require the use of drive voltages having at least 5 voltage levels so that straightforward driving from two level logic circuitry is not possible. They also give finite values of the ratio of the RMS voltage on an "ON" element (the hands or pointer) to that on an "OFF" element (the background). This means that the display performance may be a compromise involving the operating temperature range and viewing angle and that the drive voltages may have to be compensated as a function of the display temperature.

The new methods reported here allow such an analogue display to be directly driven using two level waveforms while also drastically reducing the number of display connections. This allows the display to be connected directly to logic circuitry and to operate over the full temperature range of the liquid crystal composition used without the need for temperature compensation circuitry.

## ADDRESSING METHODS

The addressing methods used are those employed in a liquid crystal storage oscilloscope subsequently reported by the author.[3,4] They make use of the correlation properties of differently shaped drive waveforms and are applied in this instance to a meander matrix or modified meander matrix. This results in a display having a high RMS voltage on the background elements which are therefore "ON" and zero voltage at each pointer position so that the pointer is "OFF".

The method will be illustrated with reference to a single pointer display which has 120 pointer positions and may have only 23 connections. The electrode configurations used are shown 0.62 size in Figs. 1a and 1b but the display could alternatively be linear or full circular. Differently shaped binary sequence waveforms are applied simultaneously, one to each backplane segment. The meander electrode which will produce the pointer has the same binary sequence applied as the segment where the pointer is to lie. Thus there is zero voltage difference between

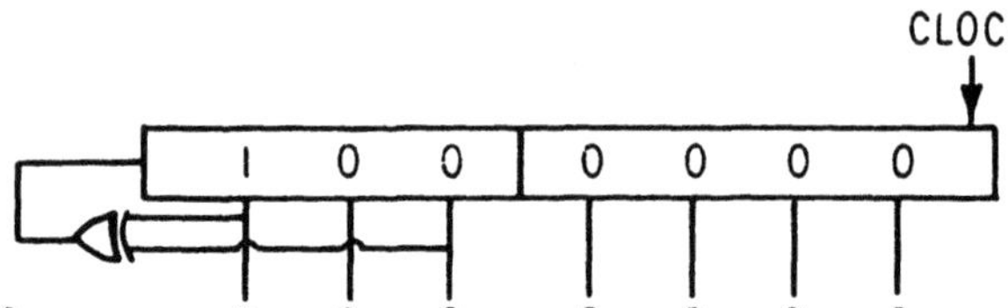


| CLOCK PULSE NO. | $Q_1$ | $Q_2$ | $Q_3$ | $Q_4$ | $Q_5$ | $Q_6$ | $Q_7$ | |
|---|---|---|---|---|---|---|---|---|
| 1 | 1 | 1 | 0 | 0 | 0 | 0 | 0 | |
| 2 | 1 | 1 | 1 | 0 | 0 | 0 | 0 | LOAD |
| 3 | 0 | 1 | 1 | 1 | 0 | 0 | 0 | PERIOD |
| 4 | 1 | 0 | 1 | 1 | 1 | 0 | 0 | |
| 5 | 0 | 1 | 0 | 1 | 1 | 1 | 0 | |
| 6 | 0 | 0 | 1 | 0 | 1 | 1 | 1 | |
| 7 | 1 | 0 | 0 | 1 | 0 | 1 | 1 | |
| 8 | 1 | 1 | 0 | 0 | 1 | 0 | 1 | |
| 9 | 1 | 1 | 1 | 0 | 0 | 1 | 0 | FIRST |
| 10 | 0 | 1 | 1 | 1 | 0 | 0 | 1 | OUTPUT |
| 11 | 1 | 0 | 1 | 1 | 1 | 0 | 0 | SEQUENCE |
| 12 | 0 | 1 | 0 | 1 | 1 | 1 | 0 | |
| 13 | 0 | 0 | 1 | 0 | 1 | 1 | 1 | |
| 14 | 1 | 0 | 0 | 1 | 0 | 1 | 1 | SECOND |
| 15 | 1 | 1 | 0 | 0 | 1 | 0 | 1 | OUTPUT |
| 16 | 1 | 1 | 1 | 0 | 0 | 1 | 0 | SEQUENCE |

ETC.

LOGIC "0" = 0 VOLTS
LOGIC "1" = V VOLTS

Fig. 2. Generation and properties of pseudorandom binary sequence waveforms.

Properties

1. Only two voltage levels are used throughout.
2. Easily generated using a shift register with exclusive OR feedback from outputs between 1 and N.
3. $2^N-1$ waveforms possible.
4. Voltage difference has high RMS value

   $(V_i-V_j)_{RMS} = V\sqrt{\frac{2^{(N-1)}}{2^N-1}}$ i.e., $V\sqrt{\frac{4}{7}}$ in case N=3 shown above.
5. Voltage difference is always A.C.

   e.g., from above
   0111010
   -0011101
   0100-11-1

this meander electrode and the segment at the required hand position and a large RMS voltage difference elsewhere along this meander electrode. The remaining meander electrodes all have a further differently shaped binary sequence applied to them so that, at all points along them, there is a large RMS voltage difference between them and all the backplane segments. Thus only the pointer is "OFF" and has zero applied voltage. It is desirable to have the RMS voltage difference applied to the background elements to be the same everywhere so that the background appears uniform at all viewing angles. This may be achieved using linearly independent isogonal or orthogonal functions, such as Walsh functions or pseudorandom binary sequences (prbs), as the drive waveforms. The latter may be easily generated using a shift register with exclusive OR feedback as illustrated in Fig. 2 and some of their properties are also summarized below this figure. Notice that they give an RMS background voltage of ~0.707 of the peak applied voltage.

The display used was a lecithin aligned dyed-phasechange display using E18 nematic liquid crystal containing 3.9% by weight of CB15 cholesteric liquid crystal and 1.2% by weight of D16, a stable blue anthraquinone dichroic dye (obtained from B. D. H. Chemicals Ltd., Poole, Dorset, England). The layer thickness was 10 μm and the threshold voltage was about 5 volts. The substrates were 4mm thick, glass plates coated with 30Ω/□ ITO which had been etched to give the required electrode patterns with the electrode areas separated by 250 μm gaps. Both ends of the meander electrodes were brought out of the cell to allow for possible electrode breaks. The reflector used was a white card tinted slightly yellow to give a neutral gray color in the display background in use. The hand position could be changed about 10 times/sec at 20°C. The circuitry used to drive the display consisted of 8 standard 15 volt CMOS integrated circuits which include an 800 Hz master clock. A circuit diagram is shown in Fig. 3 and two views of the complete display are shown in Figs. 4 and 5.

The addressing method can be simplified to use only three different waveforms at the expense of a slight increase in the complexity of the electronics. In this variant all the segments receive a voltage waveform $V_1$ except the segment containing the hand which receives a differently shaped waveform $V_2$. All the meander electrodes receive a further differently shaped waveform $V_3$ except the meander electrode required to form the pointer which receives $V_2$. Thus the background is again uniformly "ON" and the pointer is "OFF" with zero voltage applied. This method allows the use of a prbs having only 3 bits and thus the clock frequency can be reduced to 90 Hz with a consequential reduction in the already low power consumption. The RMS background voltage is also increased to 0.816 of the peak applied voltage so that

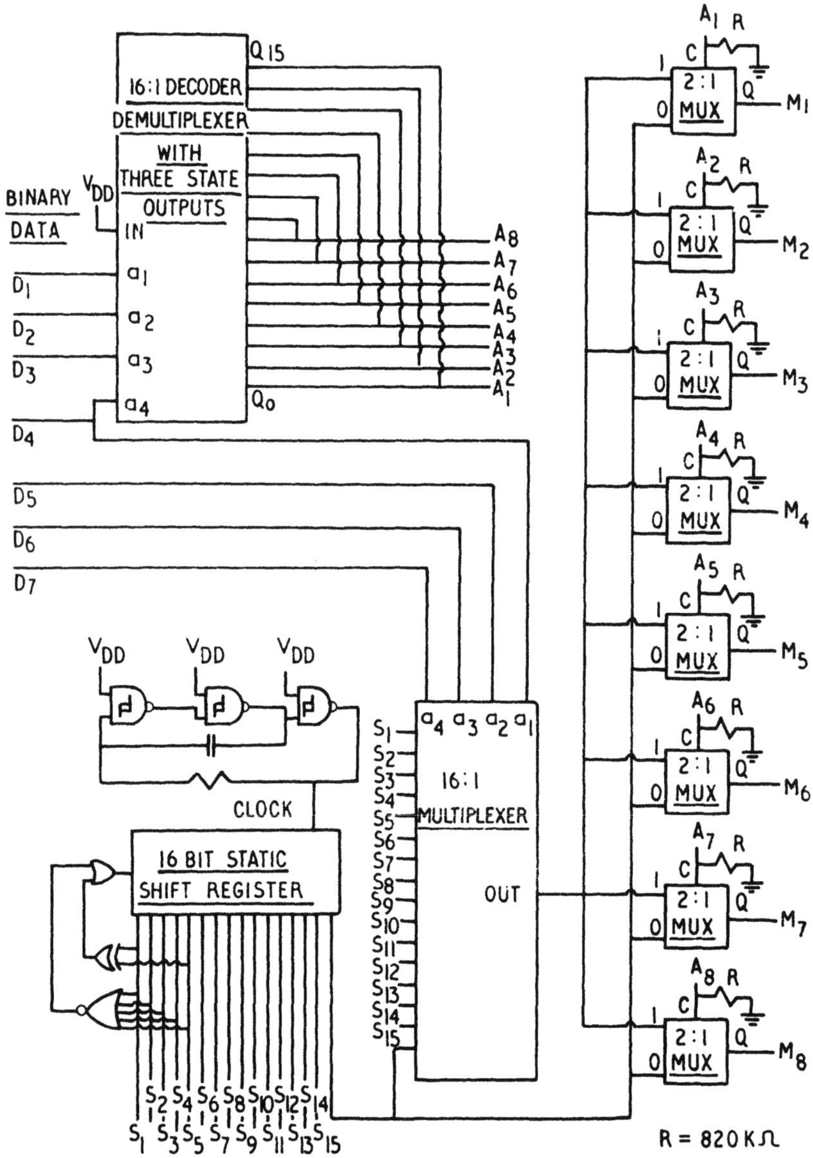


Fig. 3. Diagram of CMOS circuitry used to drive prototype meter displays.

the latter may now be reduced giving a further reduction in power consumption and a possible saving in battery count (if used).

The addressing methods can easily be extended to drive clock displays having two or three different length hands. Each meander electrode is dedicated to producing a given length hand in conjunction with a backplane consisting of patterned segments and a further electrode. This occupies areas where no hands are required and receives a further differently shaped drive waveform. In order that the fastest moving hand be equispaced around the display, to accurately resolve the displayed value, the meander is modified by intercalating single, commonly connected, radial electrodes in the folds of the meander. This also ensures that the different hands occur in the same order in each segment. Since each different length hand lies on a dedicated meander electrode it can be directly addressed independently of the other hands. This allows a two hand clock display which resolves individual minutes to have only 23 connections and a three hand display resolving individual seconds to have only 28 connections.

## DISCUSSION

The single hand display performed extremely well. It had, as expected, almost instantaneous dynamic response and no overshoot. It had good brightness and contrast, excellent viewing angle and was uniform in appearance. Even so, the brightness and contrast could both be further improved by reducing the electrode gaps to ~50 μm, to remove the dark radial lines in the display, and by using an internal reflector adjacent to the liquid crystal layer as described by Scheffer.[5] The use of an internal reflector would also eliminate the parallax which presently exists between the hand and its shadow on the reflector. Reductions in drive voltages and improvements in dynamic response may already be obtained using dyed phasechange displays incorporating recently developed low birefringence liquid crystal mixtures.

The drive methods can also in principle be used with twisted nematic displays. There are, however, some potential problems. In a large display there may be a significant voltage drop along the long meander electrodes involved. This can give a residual RMS voltage difference at the hand position due to incomplete cancellation of the meander and segment waveforms at that point. This would be worse in a thin cell and could cause the hand to come "ON" in an extreme case due to the lower threshold voltage of the twisted nematic effect. This need not be a problem with the dyed phasechange, however, since it can have a higher threshold voltage and, in any case, may have an internal reflector of roughened conducting metal[6,7] (e.g., silver) which has been etched to serve also as very high conductivity meander electrodes.

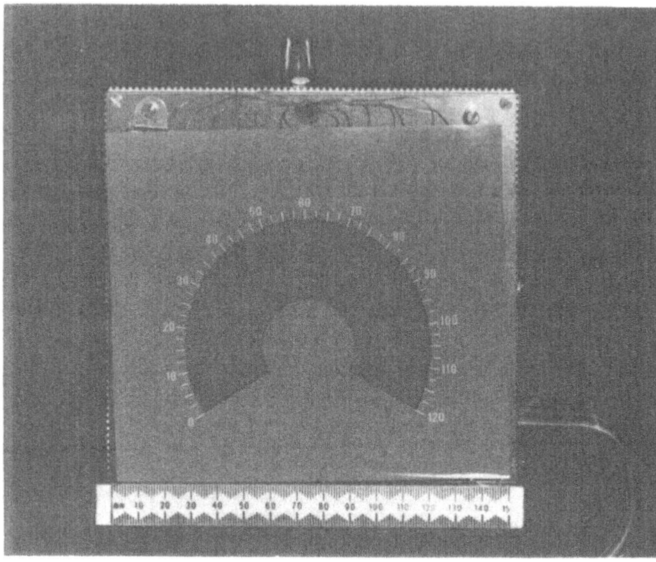


Fig. 4. Prototype meter display (front view).

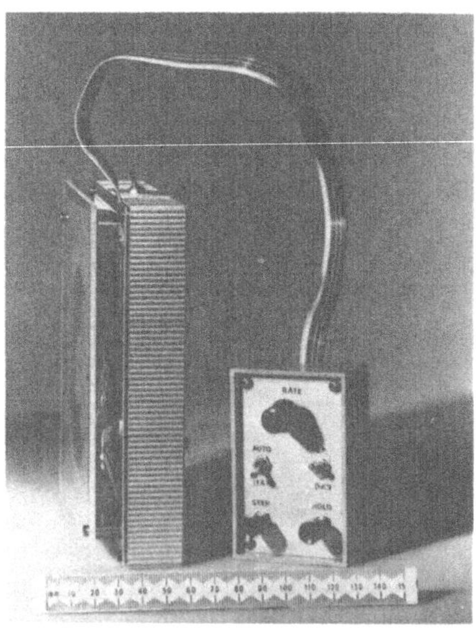

Fig. 5. Prototype meter display (side view).

Another problem with the twisted nematic effect is that, when used with parallel polarizers so that we have the preferred presentation of a black hand on a light background, the background exhibits birefringence effects which cause a reduction in the light transmitted at certain angles. If the display is driven at low voltages and is illuminated and/or viewed at these angles, the background may appear quite dark and although the display is still legible its appearance is aesthetically poor. This problem may be improved by using higher drive voltages and a thin (~6 μm) cell and this will also improve the sluggish dynamic response of the twisted nematic effect at low temperatures. These solutions may, however, exacerbate the voltage drop along the meander electrodes as mentioned above. Thus an individual compromise may have to be reached in each case.

## CONCLUSIONS

New, binary waveform, addressing methods, used initially in a liquid crystal storage oscilloscope, can be applied to drive analogue displays having one or more pointers. This allows the number of display connections to be significantly reduced while allowing the display to be directly driven and thus to operate over a wide temperature range. The feasibility of a single pointer, meter display has been successfully demonstrated using the dyed-phasechange effect and its performance has been evaluated.

## ACKNOWLEDGMENT

The author gratefully acknowledges constructional help from Mr. P. A. Holland and Mr. C. J. T. Smith of RSRE, Malvern.

## REFERENCES

1. A. R. Kmetz, "A Twisted Nematic Dual Bargraph System," SID Symposium Digest, (1977), pp. 58-59, 681:325-343 (1977).
2. R. L. Greubel, B. W. Marks, R. T. Noble, R. A. Penz and K. H. Surtani, "A Radial Format LCD/Semiconductor System for Analog Watch Applications," IEEE/SID Biennial Conference Record, October 24-26, 1978, pp. 59-60.
3. I. A. Shanks, P. A. Holland and A. J. Hughes, "Liquid Crystal Oscilloscope Displays," SID Digest (1978), pp. 98-99.
4. I. A. Shanks, P. A. Holland and C. J. T. Smith, "Non-multiplexed Addressing Methods for Liquid Crystal Oscilloscope Displays," Displays 1:1 (1979), pp. 33-41.

5. T. J. Scheffer and J. Nehring, "Progress in Guest-Host Displays," Symposium on the Physics and Chemistry of Liquid Crystal Devices (IBM Research Laboratories, San Jose, California, U.S.A., February (1979)), proceedings to be published.
6. Private communication from D. Castleberry of General Electric, Schennectady, New York.
7. Private communication from H. Kruger, A. G. Siemens, Munich, West Germany.

# LIQUID CRYSTAL MATERIALS FOR DISPLAY DEVICES

Joseph A. Castellano

Stanford Resources, Inc.
P.O. Box 20324
San Jose, California 95160

Kenneth J. Harrison

Kylex, Inc.
Mountain View, California

## INTRODUCTION

In 1963, Richard Williams observed the formation of very regular patterns or domains in a nematic liquid crystal when the material was subjected to an electric field. This report marked the beginning of a new era in research on the electro-optic properties of liquid crystals, a field which had laid dormant for nearly 30 years. During the remaining years of the 1960's and the early 70's, numerous studies of electro-optic effects in liquid crystals were performed, and at the same time, investigations into the synthetic and physical chemistry of these materials were conducted. As a result of these efforts, a whole new display industry evolved.

In this review, we will discuss the relationships between molecular structure and electro-optical behavior of various classes of liquid crystalline compounds.

Much of our discussion will deal with the nematic and cholesteric phases; some mention of smectic compounds which have been used in display devices will also be made.

I will present some of the early work on the development of nematic compounds and discuss briefly cholesteric materials and their application to displays. Dr. Harrison will follow with a discussion of recent developments in nematic and smectic preparations. He will also discuss the recent developments in dichroic dye-liquid crystal systems.

## PART I

### A. Azoxy Compounds

The aromatic azoxy compounds were among the first nematic liquid crystals to be studied and p-azoxyanisole (1) was the most widely used compound.

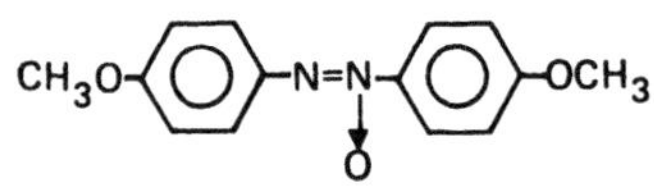


(1)

These compounds have strong negative dielectric anisotropy due to the dipolar nitrogen-oxygen group in the center of the molecule. Williams[1] discovered that thin layers of these materials between transparent conducting electrodes (NESA-coated glass or quartz) formed regular patterns or domains at low voltages. At higher voltages, these stationary patterns gave way to a turbulent condition. Similar observations were subsequently reported by Heilmeier[2] who named the effect, "dynamic scattering."

Most of the other azoxy derivaties prepared in the early days[3] were symmetrically substituted compounds prepared by oxidation of the amine or azo compound. Attempts to prepare azoxy compounds which exhibit nematic properties at ambient temperatures have been reported only very recently. Steinstrasser et al.[4] reported the synthesis of a series of p-alkyl-p'-alkoxy- and p-alkyl-p'-acyloxyazoxybenzenes that were prepared by oxidation of the unsymmetrically substituted azo compounds. One member of this series, (2), exhibited nematic behavior in the range 16-76°C. Nuclear magnetic resonance spectroscopy revealed that these materials were not single compounds but instead consisted of mixtures of two azoxy isomers. The protons of the methoxy group in each isomer showed two different signals.

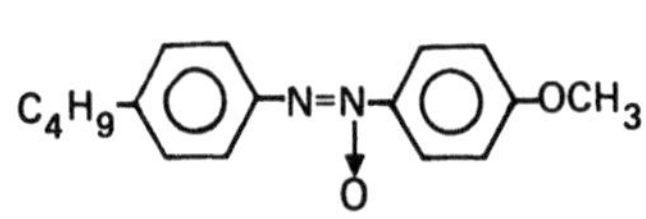


(2)

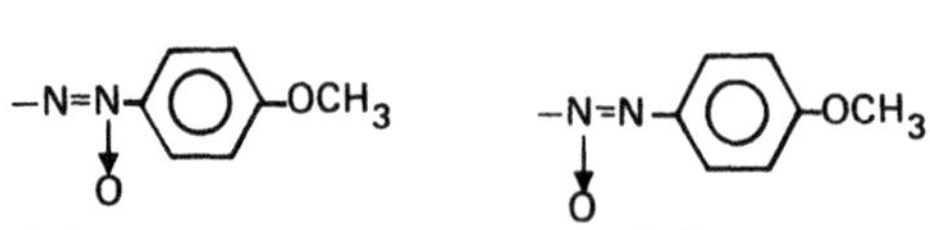


(A) 3.37 ppm (B) 3.30 ppm

The preparation of a series of p-alkoxy-p'-acyloxyazoxybenzenes (3) was also reported recently.[5] The azoxy compounds of this series were generally lower melting than the corresponding azo compounds from which they were derived because the former are mixtures of isomers. The higher nematic isotropic temperature of the azoxy materials compared to the azo compounds and to analogous Schiff base compounds (vide infra) was attributed to broadening of the molecule as a result of the presence of a lateral oxygen atom.

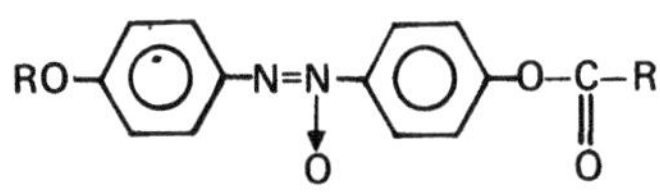


(3)

B. Schiff Bases

Although nematic mesomorphism has been observed in a variety of molecular structures, the large majority of compounds that exhibit the phase are aromatic Schiff base derivatives. These compounds are easily prepared by condensation of ring substituted aldehydes with similarly substituted amines. The discovery that certain of these compounds exhibit the dynamic scattering effect promoted studies directed toward the synthesis of materials with low melting points. Initial studies[6] involved the preparation of benzylideneaniline substituted with a variety of groups in the para positions of both rings. However, only nine of the 21 new compounds exhibited nematic behavior. On the basis of a comparison with three-ring anils, it was concluded that a critical balance of lateral and terminal intermolecular attractive forces must be attained in order for mesomorphism to occur in these two-ring anils. A favorable balance is created by the presence of alkyl, alkoxy and acyloxy groups in the para positions of the rings and a number of compounds containing these groups were prepared. The first Schiff base materials to exhibit nematic behavior at ambient temperatures[7] were mixtures of compounds of this type (4).

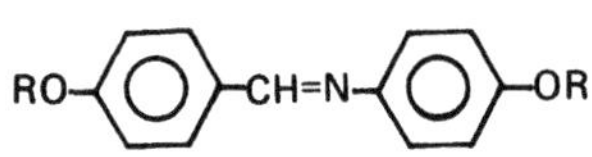


(4)

(R,R = $C_nH_{2n+1}$ or $C_nH_{2n+1}CO$)

For example, an equimolar mixture of compounds (5), (6) and (7) was found to have a nematic range of 22-105°C. Other low-melting binary and ternary mixtures were subsequently reported.

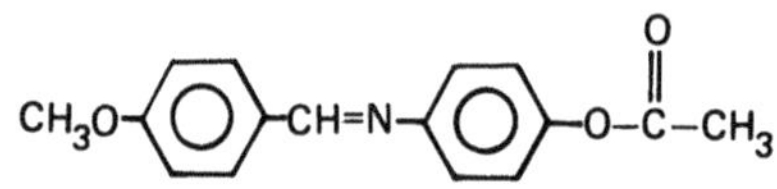

(81° – 110° C)

(5)

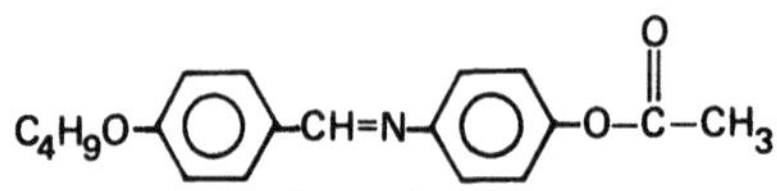

(82° – 113° C)

(6)

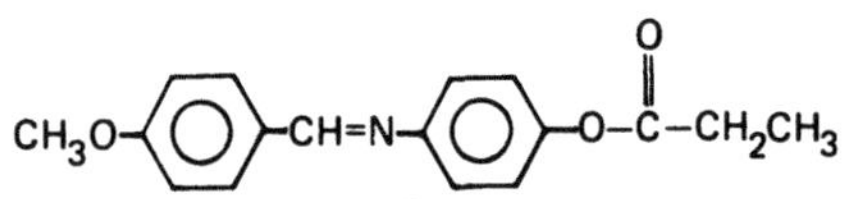

(50° – 113° C)

(7)

The discussion of mixtures is worthy of further comment. It has been known for some time[9] that mixtures of linear, nonmesomorphic molecules with nematic compounds are characterized by a sharp decrease in both the crystal-nematic and nematic-isotropic liquid transitions with increasing concentration of nonmesomorphic component. However, mixtures of two or more nematic compounds which possess subtle differences in molecular structure do not exhibit sharp decreases in the nematic-isotropic liquid transition temperatures with molar composition although eutectic points for the crystal-nematic transitions may be obtained. Thus, the nematic-isotropic liquid transition temperatures form a smooth curve over the entire range of molar composition. Similar smooth curve surfaces are observed in ternary mixtures. This remarkable stability of the nematic mesophase is undoubtedly due to the fact that all the molecules are oriented in the same direction resulting in the formation of a pseudo-lattice type structure. On the other hand, the molecules of a nonmesomorphic guest are randomly oriented in a nematic host and their presence results in the disruption of nematic order.

The first example of a single Schiff base compound to exhibit nematic behavior at ambient temperatures was (8), prepared by Kelker and Scheurle.[10] The electro-optic properties of the new compound, MBBA (p-methoxybenzylidene-p'-butylaniline) were studied shortly thereafter and it was found[11] that the material exhibited dynamic scattering. This compound has since become a model in a number of theoretical studies of the dynamic scattering effect.[12]

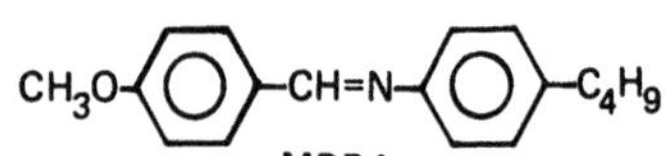

MBBA

(nematic range 22° – 48° C)

(8)

It was found[13,14] that cooperative alignment of certain nematic compounds with strong dipole moments along their long axes could be used for the twisted-nematic field effect. The class of Schiff base compounds of structure (9) was found quite suitable for this effect and a number of derivatives and mixtures were prepared.[15] Several mixtures were found to be nematic from below room temperature to above 90°C.

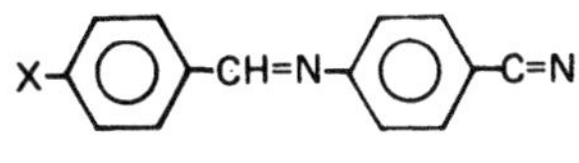

$(X = C_nH_{2n+1}O;\ C_nH_{2n+1};\ C_nH_{2n+1}CO)$ (9)

A number of other Schiff base derivatives have been reported in recent years including the p-alkoxybenzylidene-p'-aminocinnamates,[16] heterocyclic benzylidene-4-amino-4'-methoxybiphenyls,[17] 4-alkoxybenzylidene-2'-alkoxy-5'-aminopyridines[18] and p-alkoxybenzylidene-p'-aminoalkylphenones.[19]

C. Aromatic Esters

In addition to azoxy and Schiff base compounds, certain aromatic esters also exhibit mesomorphic properties. For example, the p-alkylcarbonato-p'-alkoxyphenyl benzoates (10),[20] were prepared in 3 steps. Of the 48 compounds that were synthesized in this series, 41 exhibited nematic mesomorphism. The compound that had $R=C_7H_{15}$ and $R'=C_6H_{13}$ had a nematic range of 36-54°C. In addition, nine mixtures of compounds from the series had crystal→nematic transition temperatures below 30°C.

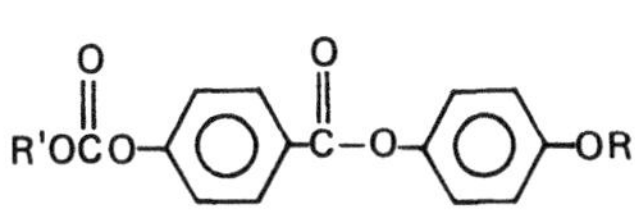

(R,R' = n-Alkyl) (10)

Another group of aromatic esters that show liquid crystalline properties are the p-phenylene esters of hydroquinone (11) and the p-substituted phenyl esters of terephthalic acid (12).[21] Most of the compounds in these series exhibit very broad nematic ranges and high transition temperatures. Lower melting compounds were produced[22] when 2-methylhydroquinone was used in place of hydroquinone as in series (13). This was a result of the decreased lateral interactions brought about by the increased side spacing of the molecules. However, the nematic thermal stability of this series is much lower than that of (12) as a result of the longer alkyl chain.

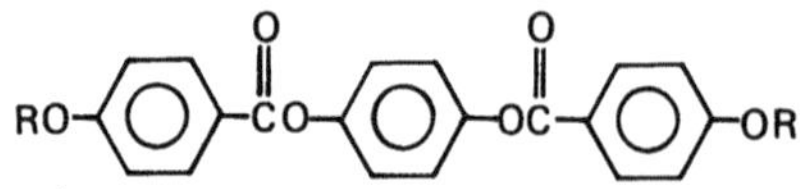

(11)

(e.g., R = $C_6H_{13}$, nematic range 121.5° C)

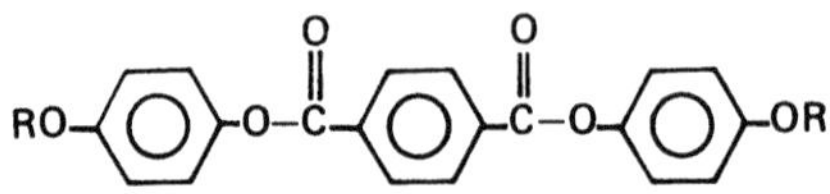

(12)

(e.g., R = $CH_3$, nematic range 205° – 277° C)

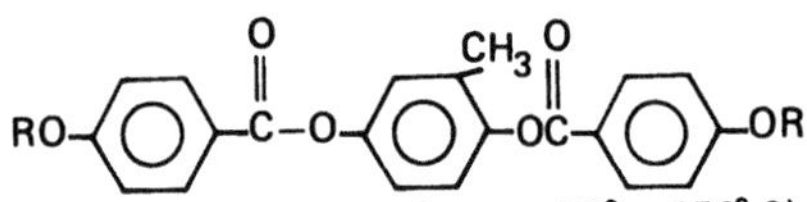

(13)

(e.g., R = $C_8H_{17}$, nematic range 72° – 156° C)

Another series of ester compounds are the 4-(4'-alkylbenzyloxy)-4"-alkylphenylbenzoates (14)[23] and their chloro derivaties (15).[24] Structure (15) has positive dielectric anisotropy and a very low cut-off frequency ($F_c \simeq 2$-3 KHz) compared with other series (i.e., $F_c \simeq 20$-100 KHz).

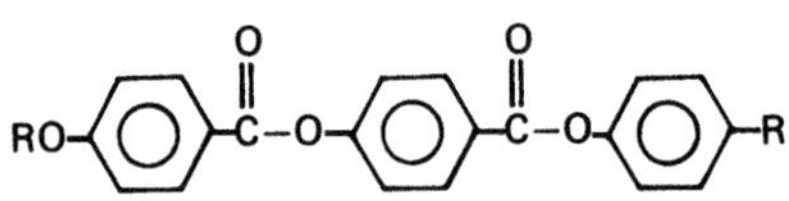

(14)

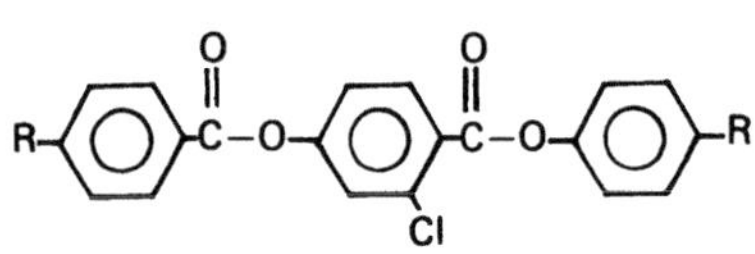

(15)

Cinnamic acid esters of structural formula (16) have strong positive dielectric anisotropy. All members of the series exhibit nematic behavior[25] but the longer chain compounds have the broadest temperature range.

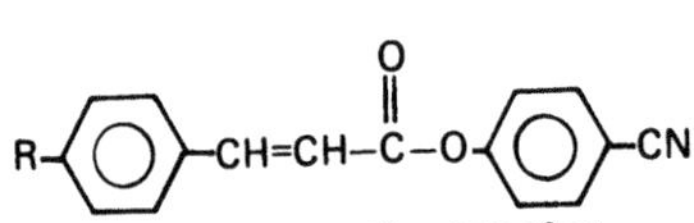

(16)

(R=$C_7H_{15}$; 57.5° – 109.5° C)

Other ester compounds with strong positive dielectric anisotropy are represented by formulas (17)[26] and (18).[27]

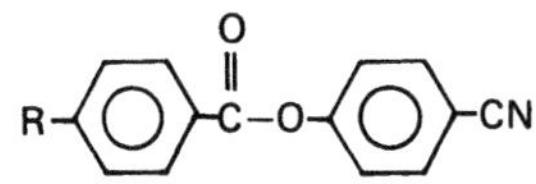

(17)

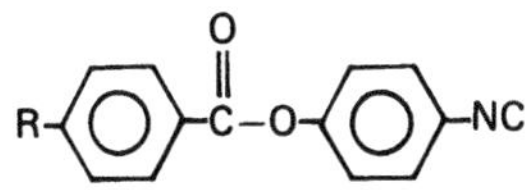

($R=C_7H_{15}$; 38° – 51° C)

(18)

D. Cholesteric Compounds

Compounds which exhibit the cholesteric mesophase are of two types. The first and most common type is the cholesteryl ester with the steroid structure (19). The R group in this structure is generally a long alkyl chain ($C_nH_{2n+1}$) containing from 1 to as many as 18 carbon atoms. These molecules are all optically active since they are derived from the natural product, cholesterol.

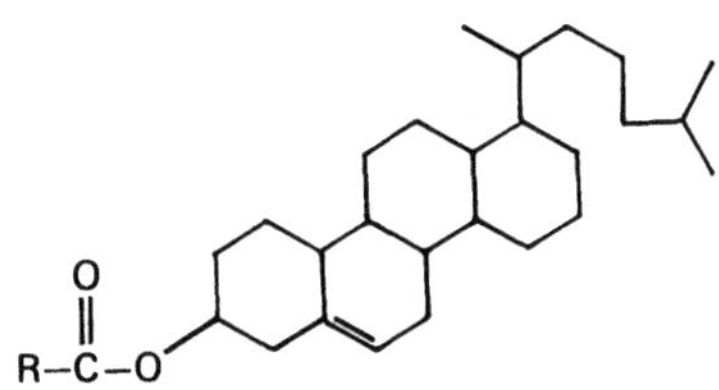

(19)

The second class of compounds which exhibit the "cholesteric" mesophase are not derivatives of cholesterol. These nonsteroidal cholesteric compounds have structures which are nearly identical to those of nematic compounds but they possess an asymmetrically substituted carbon atom. A nonsteroidal cholesteric compound which has been known for some time[9] is represented by structure (20).

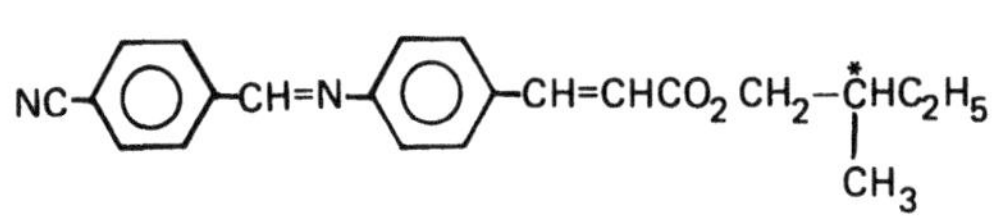

(20)

There are two types of electro-optic effects which occur with cholesteric materials. The materials which possess negative dielectric anisotropy exhibit the so-called reflective optical storage mode while positive materials undergo field-induced cholesteric-nematic phase changes.

A series of optically active aromatic Schiff bases containing p-alkoxy and p-acyloxy groups were prepared and their mesomorphic properties studied.[28] Chirality was achieved by introducing an asymmetric carbon atom into the hydrocarbon chain of either the alkoxy or acyloxy group. Structures (21), (22) and (23) are some examples. Structure (23), with 2 asymmetric centers, has a cholesteric range of nearly 96°!

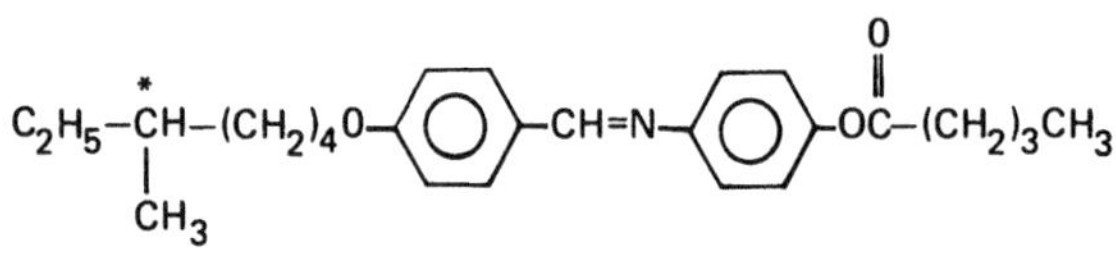

(21)

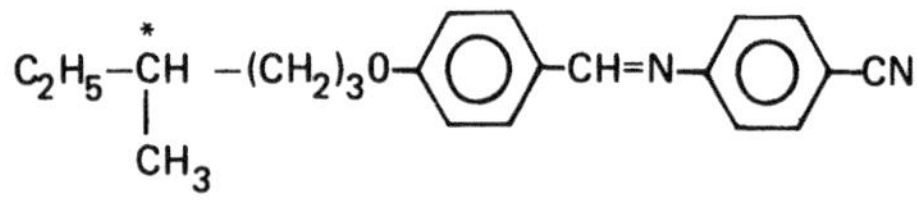

(22)

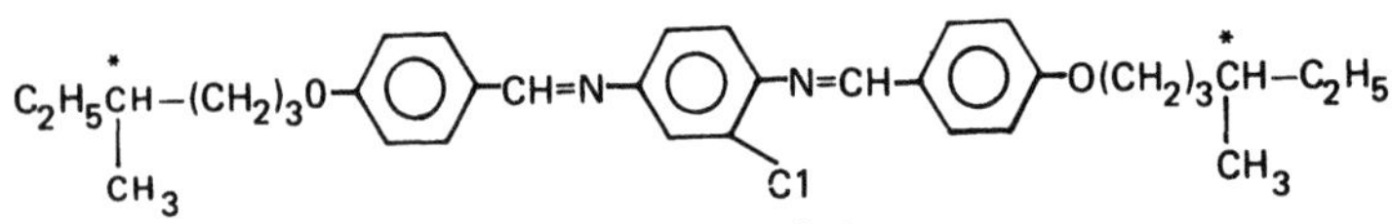

(23)

(Cholesteric, 50 – 146° C)

Most of the compounds exhibited both smectic and cholesteric mesomorphism. Several compounds showed only cholesteric behavior. For example, (24), S(+)-4-methoxybenzylidene-4'aminophenyl P-methylpentanoate was cholesteric between 41 and 78°C. By comparison, the racemic mixture of R and S enantiomers of this compound was nematic between 36 and 78°C.

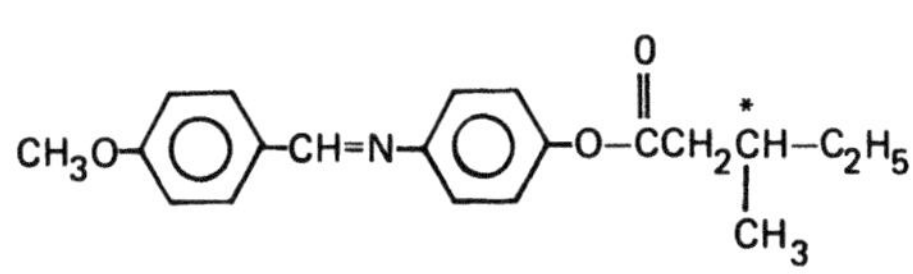

(24)

Finally, examples of biphenyl compounds with optical activity have been reported. Compounds (25) and (26) are widely used as additivies in nematic materials for twisted-nematic displays.

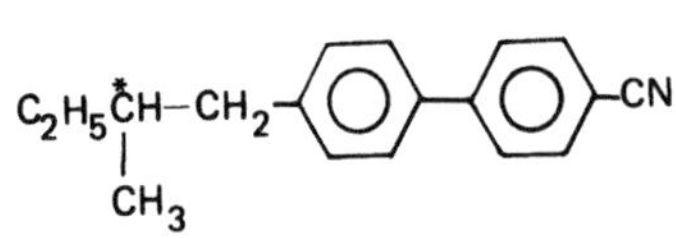

(25)

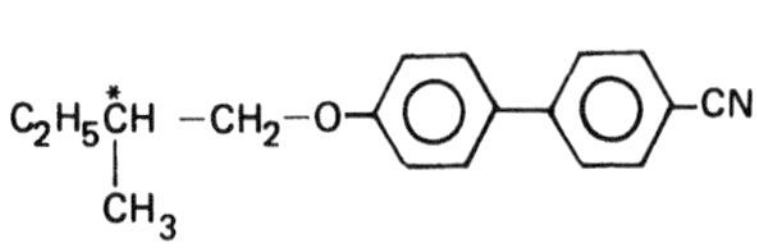

(26)

PART II

The materials that have been discussed so far have answered the basic requirements for room temperature nematics and cholesterics with positive or negative dielectic anisotropy. More recent developments have produced other desirable properties such as good chemical and photochemical stability and have enabled a tailoring of physical properties to meet the ever increasing demands of display devices based on the different electro-optical effects.

The materials discussed in part I can be characterized by a general structure where two para substituted benzene rings are linked by a central group (27) e.g., Schiff's bases. While a central linkage enhances liquid crystalline behavior, it is the cause of the stability problems and color in many materials.

(a)

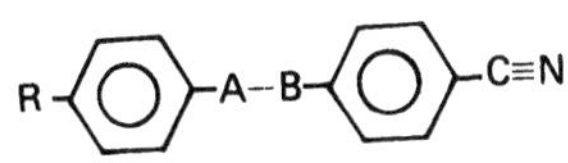


| R | A–B | C-N(C°) | N-1(°C) | Δε | Δn | η |
|---|---|---|---|---|---|---|
| $C_5H_{11}$ | CH=N | 46 | 75 | +18 | 0.28 | 80 |
| $C_5H_{11}$ | CO.O | 60 | (56) | +18.5 | | 46 |

(b)

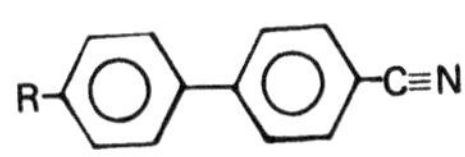


| R | | C-N/$S_A$(°C) | $S_A$-N(°C) | N-1(°C) | Δε | Δn | η |
|---|---|---|---|---|---|---|---|
| $C_5H_{11}$ | | 24 | | 35 | +11 | 0.18 | 26 |
| $C_7H_{15}$ | (K21) | 30 | | 43 | +10 | 0.16 | 35 |
| $C_8H_{17}$ | | 21.5 | 33.5 | 40.5 | +8.5 | | |
| $C_3H_7O$ | | 74.5 | | (64) | | | |
| $C_8H_{17}O$ | | 54.5 | 67 | | | | |

(c)

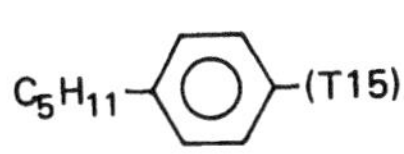


Mixtures

| | C-N/$S_A$(°C) | N-1(°C) | Δε | Δn | η |
|---|---|---|---|---|---|
| Alkyl/Alkoxy (E3) | -2 | 55 | +11 | 0.22 | 52 |
| Alkyl/Alkoxy (E7) + Terphenyl | -10 | 60.5 | +11 | 0.23 | 40 |

Fig. 27.

The cyanobiphenyls[31] have shown that this central linkage may be eliminated successfully (27b). Although the N-I temperatures are reduced when compared with their Schiff's base or ester analogs, (27a) many highly desirable physical properties are observed. Room temperature nematic and smectic A phases are observed in colorless, chemically and photochemically stable materials with fairly low viscosity and positive dielectric anisotropy. The cyanoterphenyl structure[32] provides the required higher N-I temperature materials (27c) through its greater geometric anisotropy and increased polarizability, and enables formulation of wide temperature range mixtures for display devices.[33]

(a)

| | C-N(°C) | N-1(°C) | Δε | Δn | η |
|---|---|---|---|---|---|
| $C_3H_7$ | 42 | 45 | +11.6 | 0.13 | 19 |
| $C_5H_{11}$ | 30 | 55 | +9.9 | 0.12 | 21 |
| $C_7H_{15}$ | 30 | 59 | +8.4 | 0.11 | 27 |
| K21 | 30 | 43 | +10 | 0.16 | 35 |

(b)

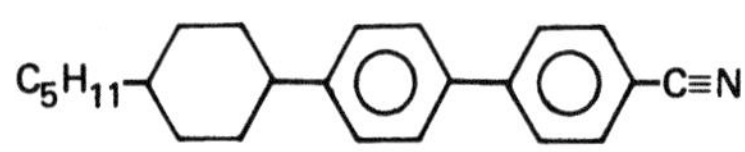

| | C-N(°C) | N-1(°C) | | | |
|---|---|---|---|---|---|
| BCH | 94 | 219 | | | |
| T15 | 131 | 240 | | | |
| Mixtures: | | | | | |
| PCH(ZL11083) | -3 | 51 | +10 | 0.12 | 21 |
| PCH+ BCH(ZL11132) | -6 | 70 | +10 | 0.14 | 20 |

(c)

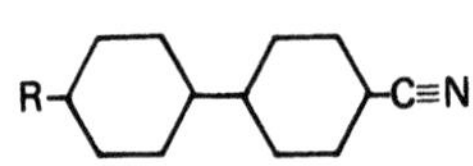

| | C-N(°C) | N-1(°C) | | |
|---|---|---|---|---|
| $C_3H_7$ | 58 | 80 | | |
| $C_5H_{11}$ | 62 | 85 | +3-4 | <0.1 |
| $C_7H_{15}$ | 71 | 83 | | |

Fig. 28.

The fact that the central linkage may be removed, to advantage in many respects, has prompted researchers to seek further structure changes to modify viscosity, birefringence, dielectric anisotropy but still retain the stable, colorless characteristics of the biphenyls. Saturation of one ring[34] to produce the cyanophenylcyclohexanes (PCH) results in some very interesting physical properties (28a). Only the trans compounds exhibit nematic phases and comparison with alkylcyanobiphenyls show an <u>increased</u> N-I, lower viscosity (20-27 cP), and a much lower birefringence (0.11-0.13). By incorporating a second parasubstituted benzene ring (28b), the biphenylcyclohexane compounds (BCH) give a much higher N-I.[35] This family of materials compliments the biphenyls in attractive properties for display devices.[36]

(a)

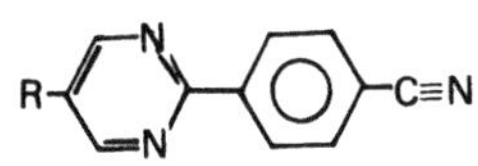

| | C-N(°C) | N-1(°C) | Δε | Δn | η |
|---|---|---|---|---|---|
| $C_5H_{11}$ | 71 | (52) | +21 | | 17(45°C) |
| $C_7H_{15}$ | 45 | 51 | +18 | | 19(45°C) |

(b)

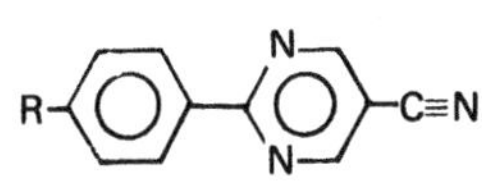

| | C-N(°C) | N-1(°C) | Δε | Δn | η |
|---|---|---|---|---|---|
| $C_3H_7$ | 126 | (106.5) | | | |
| $C_5H_{11}$ | 96 | 109 | +3.4 | | |
| $C_5H_{11}O$ | 98 | 133 | +7.0 | | |

(c)

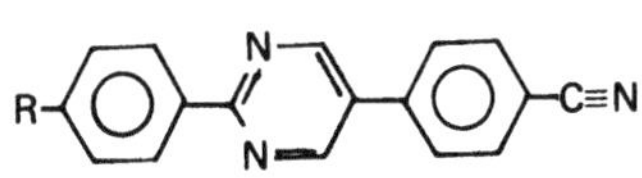

| | C-N(°C) | N-1(°C) | Δε | Δn | η |
|---|---|---|---|---|---|
| $C_5H_{11}$ | 131.5 | 263 | | | |
| $C_7H_{15}$ | 121.5 | 245 | | | |

(d)

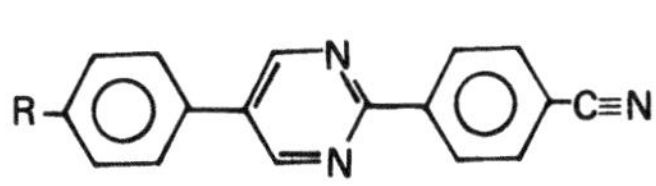

+22(100°C)

| | C-N(°C) | N-1(°C) | Δε | Δn | η |
|---|---|---|---|---|---|
| $C_4H_9$ | 93.6 | 244.7 | | | |

Mixtures:

| | C-N(°C) | N-1(°C) | Δε | Δn | η |
|---|---|---|---|---|---|
| Biphenyl/Pyrimidine (ROTN 430) | | 69.1 | +17.6 | 0.23 | 60 |

Fig. 29.

Although not strictly for display devices, the cyanocyclohexylcyclohexanes (CCH) (28c) show that no aromatic rings are required for nematic behavior.[37] In fact, their higher N-I as compared with biphenyls and PCH's show the importance of structural anisotropy in producing liquid crystalline behavior.

The fact that highly desirable physical properties may be observed in compounds with ring systems other than the benzene ring, focussed attention on heterocyclic systems. Of the enormous number of possibilities, only a few compounds have proved of interest for display devices. The most useful appear to be the cyanophenylprimidines.[38,39] The position of the nitrogens has a dramatic effect on mesophase thermal stability and shown are the more useful materials (29a,b). Compared with the biphenyls and PCH compounds, a higher N-I, dielectric anisotropy, birefringence and viscosity is seen. The dipolar nature of the ring nitrogens obviously has an effect on these properties. As one might expect, the three ring pyrimidines (29c,d) have much higher N-I's than the corresponding terphenyl or BCH compounds.

Unlike the cyanobiphenyls and PCH compounds, the melting points of the pyrimidine compounds are too high for producing useful mixtures by themselves, but they may be incorporated into biphenyls very effectively, e.g., ROTN 430 (Ref. 39).

Liquid crystal materials involving the biphenyl and cyclohexyl ring systems have clearly set guide lines in recent developments for display applications. As may be seen, Fig. 30, numerous classes of compounds have been examined with a view to augmenting the low melting properties of the biphenyls and PCH compounds with wider temperature range materials to enable further tailoring of the different physical properties. Since it is not possible to make large structural changes, e.g., lateral substitution, without destroying liquid crystalline behavior in the simple cyanobiphenyls and PCH compounds, it is necessary to look at terminal substitution and consider possible linking groups, preferably stable ones. The ethyl group,[40] although not rigid or planar, does promote geometric anisotropy and an example is shown of such a system (30a). The N-I temperature is too low to be advantageous in mixtures when compared to the terphenyl system.[32] However, replacement of the aromatic ring of a cyclohexyl ring (30b) increases the nematic thermal stability,[41] and has the further advantage of reducing the viscosity. Similar behavior is seen when the linking group is a methoxy ($CH_2O$) moiety.[41]

(a) 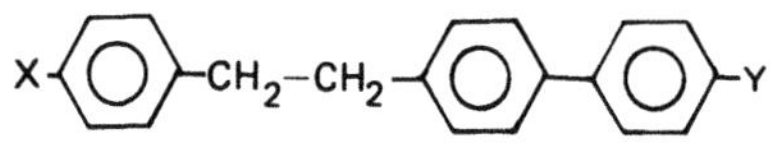

X or Y = Alkyl, Alkoxy, Cyano

C-N 60-120°C N-1 140-170°C

(b) 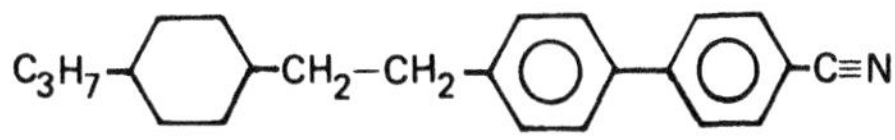

C-N 77°C N-1 194°C

(c) 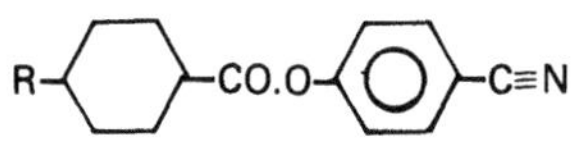

| | C-N(°C) | N-I(°C) | Δε | Δn | η |
|---|---|---|---|---|---|
| $C_5H_{11}$ Phenyl | 60 | (56) | +18.5 | | |
| $C_7H_{15}$ Phenyl | 43.5 | 56 | +20.7 | | 46 |
| $C_5H_{11}$ Cyclohexyl | 48 | 79 | +8 | 0.12 | 30 |
| $C_7H_{15}$ Cyclohexyl | 54 | 82.5 | | | |

(d) 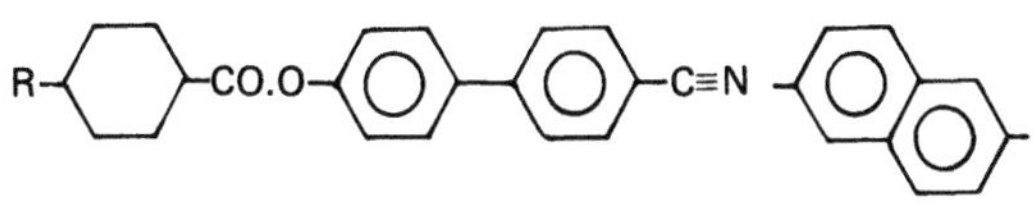

R = Alkyl, Alkoxy, Acyloxy

R = $C_5H_{11}$ C-N 109°C N-1 237.5°C Δε+10

(e) 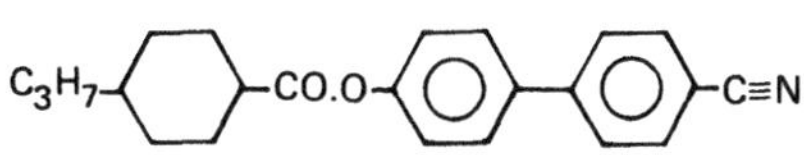

C-N 94°C N-1 248.5°C

Fig. 30.

To produce wider temperature range materials, one is forced to come back to the ester linkage which is perhaps the most versatile and useful group for structural engineering.[42] Benzoate esters are chemically and photochemically quite stable but suffer from the great disadvantage of high viscosity. However, the cyclohexyl group can diminish this problem as a comparison of simple cyano esters shows.[43] In fact, a higher N-I and lower viscosity is observed for the cyclohexyl carboxylate esters (30c). This point is further emphasized by incorporating a biphenyl group. Many compounds[44] have been reported using the three ring cyano ester skeleton (30d) in which R is alkyl, alkoxy and acyloxy. In general melting points are 70-120°C and N-I temperatures are >200°C. While the fully aromatic compounds exhibit high viscosity, the cyclohexyl compounds[41,45] provide very attractive properties as additives to low melting nematogens to raise the clearing point without increasing viscosity too much.[46] Where the biphenyl has been replaced by a 2,6 disubstituted naphthalene ring, a consistent decrease in clearing point is seen.[44]

To date, the emphasis has been placed on strongly positive materials typically produced with a para cyano group. This may be understood in light of the predominance of the twisted nematic effect for display devices. However, it is becoming apparent that weakly positive and negative materials are required, and furthermore, they should exhibit the same high stability, low viscosity, wide temperature range and range of birefringence that can be seen in the cyano compounds.

The terminal cyano group has a strong influence in enhancing nematic stability and giving low melting behavior in many systems. To produce useful compounds without this terminal group, poses quite a challenge as shown in Fig. 31a. Simple disubstituted biphenyls exhibit higher melting points and a strong smectic tendency.[47] In contrast, phenyl pyrimidines, with simple alkyl and alkoxy substitutents,[48] give low melting nematic phases (31b). The ring nitrogens are responsible for the weak positive dielectric anisotropy through a dipole moment parallel to the long molecular axis.

The simple benzoate esters are capable of giving positive or negative materials[49] depending upon the nature of the substituent groups. Their intrinsic high viscosity may be overcome by replacement of one aromatic ring with a cyclohexane ring.[50] In fact, where the saturated ring is incorporated into the acid moiety rather than the phenol moiety, low melting, wide temperature range, low viscosity nematogens result (e.g., 31c).

A range of useful esters (31d) can be obtained in those systems where typically one or more rings are saturated,[51] melting points are typically 60=90°C, N-I temperatures are 150-200°C and viscosities are lower than corresponding fully aromatic esters. These compounds may be designated as PCH esters and are presently used in mixtures with PCH and cyano-biphenyls, (e.g., ZL1 1216, ZL1 1253 and ZL1 1285 (see Ref. 36)). The enhanced liquid crystal behavior of the three ring ester system is shown by the fact that negative nematic materials are possible with a lateral cyano group (Licristal S1014, Ref. 36). An extrapolated $\Delta\varepsilon$ of -4 at room temperature is one of the highest known (31e).

(a) 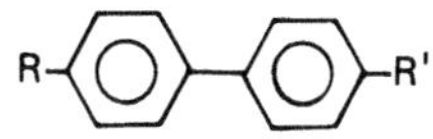

| | | | | | |
|---|---|---|---|---|---|
| $C_2H_5O$ | $C_5H_{11}$ | C-S/N | 72°C | N-1 | 81°C |
| $C_6H_{13}O$ | $C_5H_{11}$ | C-S | 82°C | S-1 | 84°C |

(b) 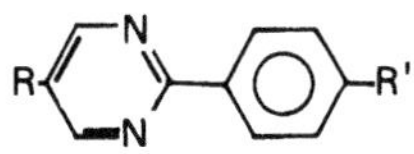

| | | C-N(°C) | N-1(C°) | Δε |
|---|---|---|---|---|
| $C_6H_{13}$ | $OC_6H_{13}$ | 31 | 60.5 | Weak |
| $C_5H_{13}$ | $OC_6H_{13}$ | 36 | 61 | +ve |

(c) 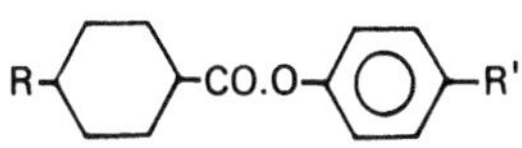

| | | | | |
|---|---|---|---|---|
| $C_3H_7$ | $OC_5H_{11}$ | 38 | 66° | -ve |

(d) 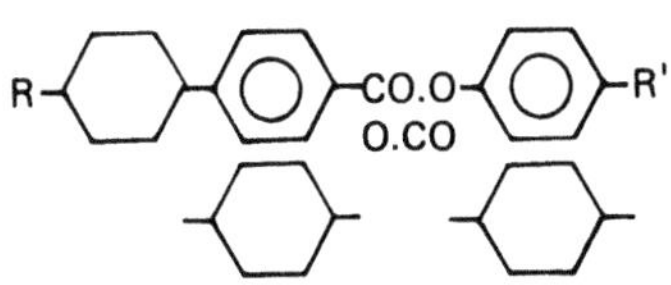

R = Alkyl    Melting points ≥ 60°C
             Clearing points 150-200°C

(e) 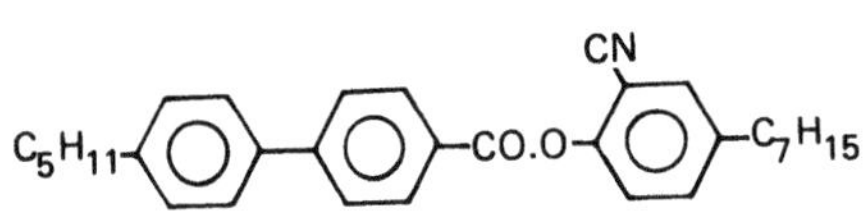

C-N    45°C    N-1    101°C    Δε-4

Fig. 31.

As a result of work to produce stable colorless nematics, the area of nonsteroid cholesteric materials has been opened up with many useful compounds being reported.[52] By introduction of an asymmetric center in the form of a branched alkyl group, specifically the readily available 2-methylbutyl substituent, cholesterics with wide temperature range, different pitch lengths and helical sense may be prepared (e.g., 32a,b,c). An interesting structural influence on sense and pitch is observed. The closer the position of branching to the central core, the tighter the pitch and depending upon the odd or even point of the asymmetric center, a right or left handed sense of the cholesteric results. These materials,[52] in addition, give chiral smectic C mesophases which are of possible display interest.

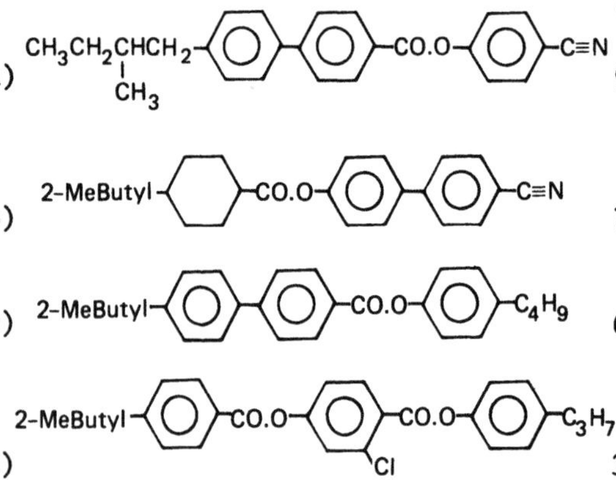


| | C-Ch/S (°C) | S-Ch (°C) | Ch-I (°C) | Pitch μm |
|---|---|---|---|---|
| (a) | 99.5 | | 195.9 | 1.6 |
| (b) | 77.6 | 138 | 190.4 | 4.7 |
| (c) | 66.5 | | 127.5 | 0.2 |
| (d) | 35 | | 92 | 0.32 |

Fig. 32.

As can be seen, the familiar biphenyl, cyclohexane, ester combination provides the building blocks for potentially useful liquid crystals. One structure that provides diverse physical properties in highly mesomorphic compounds is the diester system. For example, Fig. 32d shows a low melting, tight pitch cholesteric[53] and Fig. 33a, shows a very high positive dielectric anisotropy material. In this diester, the ester and chloro dipoles reinforce that of the cyano group. Highly positive materials,[54,55] may be used to lower the threshold in displays using the twisted nematic or cholesteric nematic phase change effects.

Having considered a number of materials' developments, it is interesting to look at their properties in display devices. An important point to be made is that only by considering the materials' properties in conjunction with the display device fabrication can an objective comparison between the different classes of materials be made.

The twisted nematic display is a prime example. Where direct drive for low information content displays is concerned, the primary requirements are stable materials with low thresholds, wide temperature ranges, good contrast over a wide viewing cone and fast response times.

| | | C-N(°C) | N-I(°C) | Δε |
|---|---|---|---|---|
| (a) | 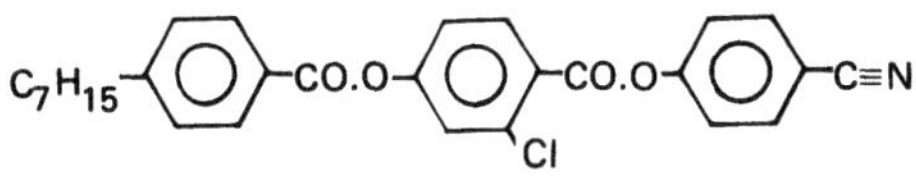 | 69 | 160 | +35(97°C) |
| (b) | 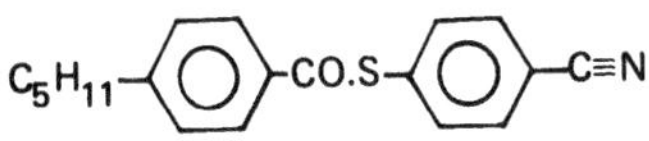 | 76 | 98 | +33 |

Fig. 33.

The commercial mixtures shown in Table 34 exhibit desirable characteristics for directly driven displays. For example, watch displays typically operate at 3V, and these mixtures saturate well below this voltage. The data are for typical 10μ low tilt cells. The PCH mixture has a lower Δn and should give a larger viewing cone.[57] However, the low Δn is a disadvantage where the minimum cell spacing of ~9μ, may offset the low viscosity with respect to response times. The high Δn of the biphenyl/pyrimidine mixture is conducive to thin cells where the higher viscosity is less significant, e.g., cell of 6μ will support the twisted nematic effect. The pure cyanobiphenyl/terphenyl system probably provides the best compromise of physical properties when cell parameters are considered. Lower threshold materials suitable for 1.5V operation are more difficult to achieve with biphenyls or other materials with $\Delta\varepsilon \ngtr +12$. However, by optimizing dielectric behavior, elastic constants and cell fabrication, it is possible to obtain 90% saturation at 1.5V as shown by a biphenyl mixture (e.g., E24LV Ref. 56). In this context the highly positive cyano esters have good characteristics (e.g., ROTN 132 Ref. 39). The question of wider temperature ranges has been answered with mixtures of biphenyls, PCH's or phenylpyrimidines as shown by E37, ZL1 1221 and ROTN 405. The lower end of operation of these materials is principally determined by viscosity and cell spacing and is not lower than -10°C, without an external heater.

As may be seen from Table 34, a choice of stable liquid crystal materials exists for twisted nematic displays in many applications. However, such materials are directed towards low information content, direct drive displays and more difficult material problems exist when the need for multiplexing arises.

In the fast scan r.m.s. addressing mode,[58] the degree of multiplexing is determined by the threshold sharpness, off axis contrast/voltage characteristics and the threshold/temperature variation. All three may be influenced by display construction, but the choice of materials has some bearing.

Table 34. Twisted Nematic - Alignment (High Tilt Low Tilt) Cell Spacing (6μ-12μ)

| Mixture Composition | Cyanobiphenyls + Terphenyl | PCHS + BCH | Pyrimidines + Biphenyls | Schiff Base |
|---|---|---|---|---|
| | E7 | ZL1 1132 | ROTN 430 | ROTN 430 |
| Temp. Range | -10→+60.5°C | -2→+80°C | →69°C | →65°C |
| Threshold 10% | 1.45V | 1.8V | 1.4V | 1.4V |
| Saturation 90% | 2.2V | 2.45V | 2.0V | 1.95V |
| Time On 20°C | 200 ms | 250 ms | 200 ms | 200 ms |
| Time Off 20°C | 120 ms | 80 ms | 150 ms | 250 ms |
| Time On 0°C | ~1s | ~1s | | |
| Time Off 0°C | 500 ms | 500 ms | | |
| Minimum Spacing | 7μ | 9μ | 6μ | 5.5μ |
| Low Threshold | Biphenyls | | | CN-Esters |
| High Tilt Cell | | | | |
| Threshold 10% | 0.8V(12μ) | | | 0.8V(10μ) |
| Saturation 90% | 1.3V | | | 1.29V |
| Wide Temp. Range | E37 | ZL1 1221 | ROTN 405 | |
| | -9→89°C | -11→90°C | →91°C | |
| Threshold 10% | 1.5V | 1.9V | 1.7V | |
| Saturation 90% | 1.9V | | 2.25V | |
| Viscosity (20°C) | 52 | 40 | 84 | |
| | Ref. 56 | Ref. 36 | Ref. 39 | Ref. 39 |

Great attention has been focussed on lower threshold materials and their threshold/temperature characteristics. A comparison of more recent liquid crystal materials for multiplexing has been given by Raynes.[59] Examples of mixtures using biphenyls, PCH's, PCH esters are shown in Fig. 35. These three mixtures may be adequate for up to 1 in 5 duty cycle. The biphenyl mixture is characterized by a sharp threshold while the PCH-ester systems have low temperature-threshold dependence. The exact reason for large differences in threshold/temperature characteristics between the various classes of materials is little understood, e.g., biphenyls are poor and azoxy compounds are excellent. A comparison of threshold sharpness between the materials shows that some progress can be made and the answer appears to be in mixtures of materials of different classes. Such mixtures appear to offer promising results in terms of dielectric behavior and elastic constants, both of which are a key to threshold sharpness.

The off axis viewing can be increased by using lower birefringent materials remembering that the minimum cell spacing must be observed. Although figures of merit[59] give one a good idea of a materials' behavior when multiplexed, a better picture is seen by a polar plot of the viewing cone, i.e., a contour map of contrast ratio as a function of the angle of viewing for a given multiplexed condition.

Table 35. Mixtures for Multiplexing Twisted Nematic off Axis Contrast versus Voltage Threshold/Temperature

| Mixture Type | Biphenyls | PCH/PCH Esters | Biphenyl/ PCH/Esters |
|---|---|---|---|
| Comm. name | E26M | ZL1 1216 | ZL1 1253 |
| Threshold | 1.3V | 1.45V | 1.15V |
| Temp./Threshold (0-40°C) | 10 Mv/°C | 5.3 mV/°C | 6 mV/°C |
| Sharpness | 1.69 | 1.79 | 1.73 |
| Birefringence | 0.22 | 0.13 | 0.13 |
| Min. Cell Spacing | 7μ | 9μ | 9μ |
| | Ref. 56 | Ref. 36 | Ref. 36 |

At the present time, as shown by the pocket calculator, materials are available with adequate electro-optic performance where the duty cycle is 1 in 2, 3 or 4. Where temperature compensation is used, 1 in 7 duty cycle is possible[60] with mixtures of cyanobiphenyls and benzoate esters. For a duty cycle of 1 in 10, temperature compensation is "a must" and the electro-optic performance is determined principally by the threshold sharpness. By way of example, Fig. 36 shows a viewing cone plot for ZL1 1253 (dotted line curves) compared with a sharper threshold material (1.65) (solid line curves) in an 8μ, low tilt cell. By single frequency addressing, a 1 in 10 duty cycle is approaching the maximum for an acceptable performance in a twisted nematic display.

Emphasis has been placed on nematic materials because of the predominance of the twisted nematic effect in display devices. However, its disadvantages have caused considerable interest in the cholesteric-nematic phase change effect and the dye-phase change device in particular.[61,62]

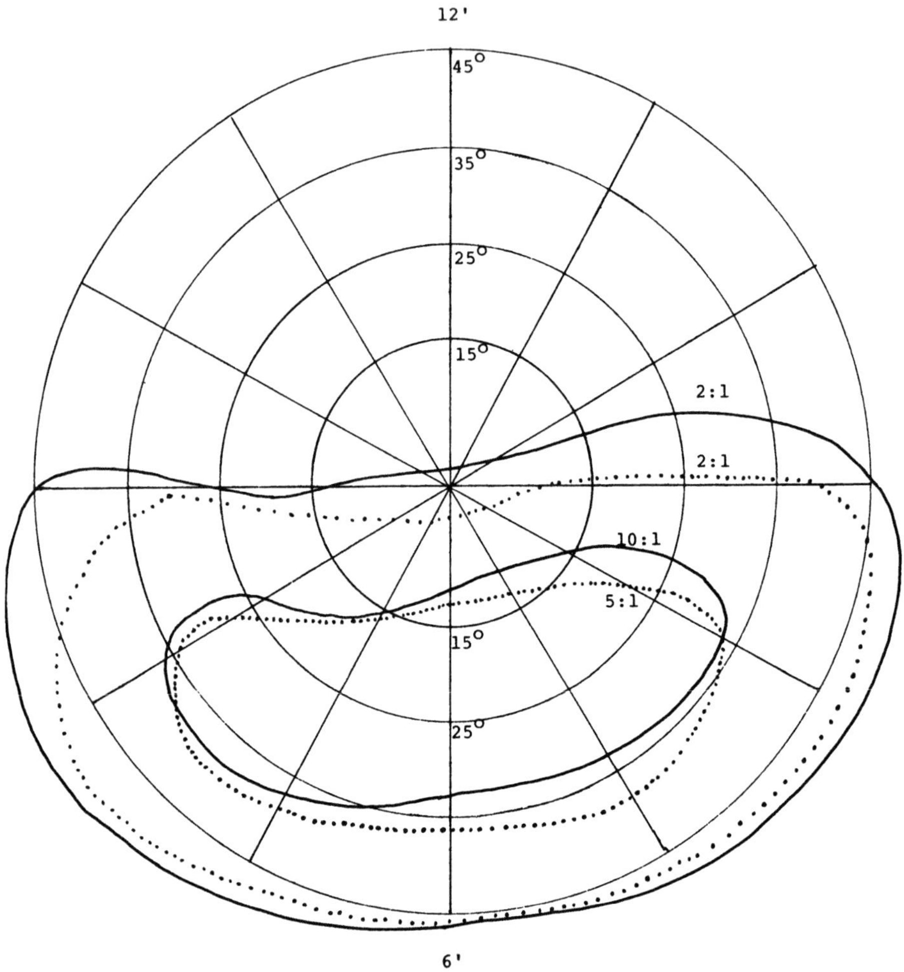


Fig. 36. 10% duty cycle. No crosstalk at 45° off axis.

While the host material presents some problems, the choice of suitable dyes is limited. In addition to being nonionic, soluble, high extinction coefficient and photo-chemically stable, a high order parameter is essential. Two systems (37) have been extensively studied, azo dyes[63] and the anthraquinone dyes.[64] The former has the right structural form and the transition giving rise to the primary absorption peak is along the length of the molecule. However, the only stable dyes are orange or red and the desirable colors are blue, green or black. The anthraquinone dyes answer the color problem and the stability question very well but have a lower order parameter than the azo dyes. The

Dyes

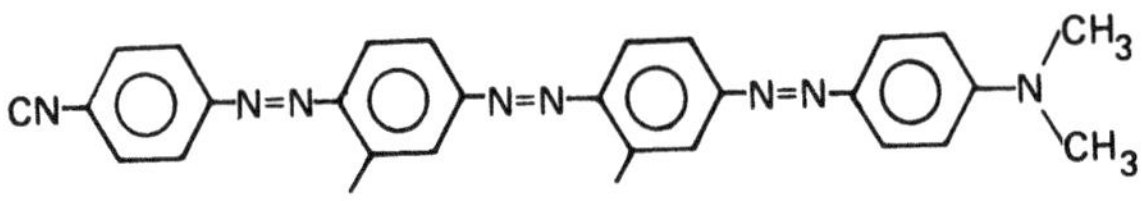


λmax 538 nm, O.P. 0.78

Azo

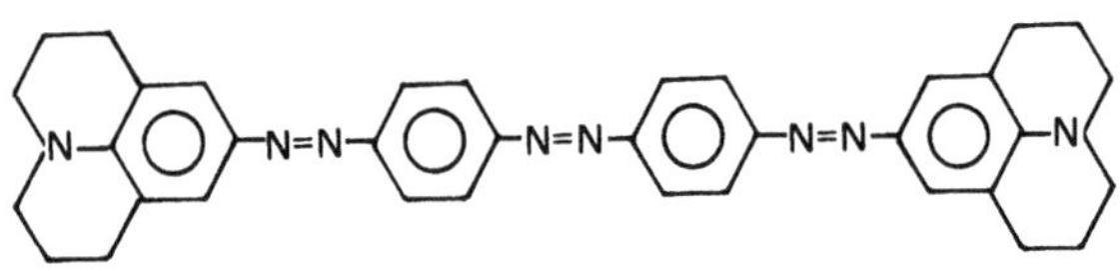


λmax 568 nm, O.P. 0.79

Anthraquinone

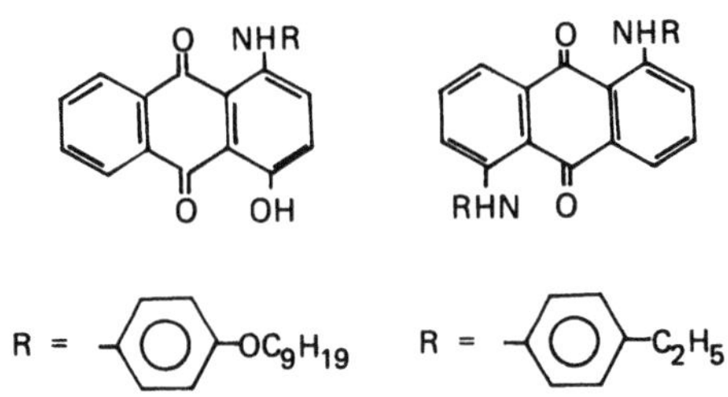


λmax = 596 nm
O.P. = 0.65

λmax = 554 nm
O.P. = 0.67

Host

| | |
|---|---|
| Nematic | High Positive Δε |
| Cholesteric | 1-2 μm Pitch, high positive Δε, low birefringence, high clearing point >80°C |

Fig. 37.

requirements of the host cholesteric may be summarized as follows: High positive dielectric anisotropy, low birefringence, pitch 1-2μ and high clearing point. The PCH compounds would appear to offer suitable host materials, although the threshold may be too high for some applications. It may be difficult to find highly positive, low birefringent materials and some compromise is required.

For display using tuneable birefringence as the electro-optical effect, the physical properties of available materials answer most of the needs although negative materials are less in evidence to do this. The cell fabrication is the primary consideration for such displays.

Smectic phases offer great potential for display devices. For example, the higher order would be good for dyes, and smectic A, C phases may be used for tuneable birefringence.[65] The ferroelectric properties of chiral smectic C are of great interest.

To some extent, development of smectic materials has been slow and usually come as an off shoot of work on nematics. For example, the binary mixture shown in (38) exhibits a room temperature smectic A, a short range nematic, and is of positive dielectric anisotropy.[66] Such properties may be used in thermally addressed displays where a transition from a scattering to a clear state forms the optical effect. This transition may also be effected by an electric field. Obviously, more work on useful smectics is required although their inherent high viscosity is a problem.

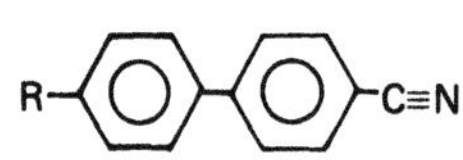


$C_8H_{17}$ 58%

$C_{10}H_{21}$ 42%

C - $S_A$ 5°C $S_A$-N 40°C N-1 43°C

Fig. 38. Smectics

In conclusion, recent materials' developments have tried to answer the needs of current electro-optical effects used in display devices. By concentrating more on desirable physical properties rather than just synthesizing new liquid crystals, considerable progress has been made. However, the materials cannot stand alone and must be linked to device fabrication and addressing to take full advantage of their properties.

REFERENCES

1. R. Williams, J. Chem. Phys. 39:384 (1963).
2. G. Heilmeier, Sci. Am. 222:100 (1970).
3. G. Weygand and R. Gabler, BER. 71B:2399 (1938).
4. R. Steinstrasser and L. Pohl, Tetrahedron Letters 1971:1921; H. Kelker, B. Scheurle, R. Hatz and W. Bartsch, Angew. Chem. 82:987 (1970).
5. M. T. McCaffrey and J. A. Castellano, Mol. Cryst. Liq. Cryst. 18:209 (1972).
6. J. A. Castellano et al., J. Org. Chem. 33:3501 (1968).
7. J. A. Castellano and J. E. Goldmacher, U.S. Patent 3,540,796 (1970).
8. L. Pohl and R. Steinstrasser, III Int'l. Liq. Cryst. Conf., Berlin (1970).
9. G. W. Gray, "Mol. Struct. and Prop. of Liq. Cryst., Academic Press, (1962).
10. H. Kelker and B. Scheurle, Angew. Chem. 81:903 (1969).
11. L. Creagh, D. Jones and S. Lu, Appl. Phys. Lett. 16:61 (1970).
12. Abstracts of Papers presented at III Int. Liq. Cryst. Conf., Berlin, 1970; Il Haller and J. B. Litster, p. 26; Orsay Liq. Cryst. Group, p. 98; W. Haas, J. Adams and J. B. Flannery, p. 109; Koelmans and A. M. van Boxtel, p. 113; R. Borel and F. Tobazeon, p. 117. Full papers in Mol. Cryst. Liquid Cryst. Vol. 13 and following volumes.
13. G. H. Heilmeier and L. A. Zanoni, Appl. Phys. Lett. 13:91 (1968).
14. G. H. Heilmeier, J. A. Castellano and L. A. Zanoni, Mol. Cryst. Liquid Cryst. 8:293 (1969).
15. J. A. Castellano, U.S. Patent No. 3,597,044 (1971); applied for September 1, 1968.
16. G. W. Gray and K. J. Harrison, Mol. Cryst. Liquid Cryst. 13:37 (1971).
17. W. R. Young, I. Haller and L. Williams, Liquid Crystals and Ordered Fluids, Plenum Press, New York, N.Y., 1970, p. 383.
18. C. S. Oh, Abstracts of 162nd Nat. ACS Meeting, Washington, D.C., 1971; Mol. Cryst. Liquid Cryst. in press.
19. J. A. Castellano and M. T. McCaffrey, Liquid Crystals and Ordered Fluids, Plenum Press, New York, N.Y. 1970, p. 293.
20. J. A. Castellano, M. T. McCaffrey and J. E. Goldmacher, Mol. Cryst. Liquid Cryst. 12:345 (1971).
21 M. J. S. Dewar and R. S. Goldberg, J. Org. Chem. 35:2711 (1970).
22. S. L. Arora, J. L. Gergason and T. R. Taylor, J. Org. Chem. 35:4055 (1970).
23. R. Steinstrasser, Agnew. Chem. 84:636 (1972).
24. J. P. VanMeter and B. H. Klanderman, Mol. Cryst. Liq. Cryst. 22:285 (1973).
25. S. Scherrer and A. Boller, U.S. Patent No. 3,927,066 (1975).

26. A. Boller et al., Proc. IEEE, 60:1002 (1972).
27. H. Scherrer and A. Boller, U.S. Patent No. 3,925,444 (1975).
28. J. A. Castellano, C. Oh, M. T. McCaffrey, Mol. Cryst. Liq. Cryst. 27:417 (1973).
29. G. W. Gray et al., Electronic Letters 11:556 (1975).
30. G. W. Gray et al., Electronic Letters 8 (1973).
31. G. W. Gray, K. J. Harrison and J. A. Nash, Electronic Letters, 9:130 (1973).
32. G. W. Gray, K. J. Harrison and J. A. Nash, J.C.S. Chem. Comm. 431 (1974).
33. "Liquid Crystals," BDH Chemicals Ltd. 1-29 (1978).
34. L. Pohl et al., Physics Letters 60A(5):421 (1977).
35. R. Eidenschink, J. Krause and L. Pohl, U.S. Patent No. 4, 4,154,697.
36. Supplied by E. M. Laboratories, Inc., Elmsford, New York.
37. L. Pohl et al., Physics Letters 65A(2):169 (1978).
38. A. Boller et al., Mol. Cryst. 42:215 (1977).
39. Supplied by Roche Chemical Division, Hoffmann-LaRoch Inc., Nutley, New Jersey.
40. D. Coates and G. W. Gray, U.S. Patent No. 4,035,056.
41. G. W. Gray and D. G. McDonnell, Seventh International Liquid Crystal Conference, Bordeaux, France (1978) paper A02.
42. D. Demus, H. Demus and H. Zaschke, "Flussige Krsitalle in Tabellen," VEB Deutscher Verlag fur Grundstoff Industrie, Leipig (1974).
43. H-J. Deutscher et al., Z. Chem. 17(2):64 (1977).
44. For example, see (1) D. Coates and G. W. Gray, Mol. Cryst. Liq. Cryst. 37:249 (1976). (2) D. M. Gavrilovic, U.S. Patent No. 3,951,846.
45. D. Coates, G. W. Gray and D. G. McDonnel, U.S. Patent No. 4,113,647.
46. G. W. Gray, "Advances in Liquid Crystal Materials for Applications," BDH Publication (1978).
47. Licristal S1009 and S1010, see Ref. 36.
48. H. Zaschke, J. Prakt. Chem. 317:617 (1975).
49. D. Demus, in: "Nonemissive Electrooptic Displays," (edited by A. R. Kmetz and F. K. Van Willisen, Plenum Press, New York, 1976), pp. 83-119.
50. H-J. Deutscher et al., J. Prakt. Chem. 320(2):191 (1978).
51. R. Eidenschnik et al., Seventh International Liquid Crystal Conference, Bordeaux, France, 1978, paper A04.
52. G. W. Gray and D. G. McDonnell, Mol. Cryst. Liq. Cryst. 37:189 (1976).
53. B. H. Klandeman and T. R. Criswell, J.A.C.S. 97(6):586 (1975).
54. J. P. Van Meter, R. T. Klingbeil and D. J. Genova, Solid State Comm. 16:315 (1975).
55. Licristal S1016, see Ref. 36.
56. Supplied by BDH Chemicals Ltd. Poole, Dorset, England.
57. D. J. Meyerhofer, Applied Phys. 48(3):1179 (1977).

58. A. R.Kmetz, in: "Nonemissive Electrooptic Displays," (edited by A. R. Kmetz and F. K. Van Willisen, Plenum Press, New York, 1976), pp. 261-289.
59. E. P. Raynes, IEEE/SID Biennial Display Research Conference 8-11 (1976).
60. K. Odawaru et al., SID Digest p. 13:6 (1979).
61. D. L. White and G. N. Taylor, J. Appl. Phys. 45:4718 (1974),
62. M. G. Clark, Displays 1:17 (1979).
63. J. Constant et al., J. Phys. D: Appl. Phys. 11:479 (1978).
64. J. Constant et al., Seventh International Liquid Crystal Conference, Bordeaux, France, 1978, p. DP33.
65. D. Coates et al., Mol. Cryst. Liq. Cryst. 41 (letters):151 (1978).
66. G. W. Gray and A. Mosley, J.C.S. Chem. Comm. 147 (1976).

# SYNTHESIS OF TRANS-4-ALKYLCYCLOHEXANE CARBOXYLIC ACIDS AND THEIR NEMATOGENIC ESTERS

Curtis Sheley

Hewlett Packard Company
1000 NE Circle
Corvallis, Oregon 97330

Sun Lu and Leo Carlino

Exxon/Kylex
420 Bernardo Avenue
Mountain View, California

## INTRODUCTION

The rapidly expanding technology of liquid crystal devices requires increasingly sophisticated liquid crystalline materials tailored to the specific requirements of a research and development effort. Many excellent materials are commercially available, but in order to obtain other potentially useful ones, electronics firms must frequently rely upon what internal synthetic capabilities they may muster. To this end, we present the following practical information concerning the syntheses of several trans-4-alkylcyclohexane carboxylic acids and their esters.

## RESULTS AND DISCUSSION

Brief accounts of the synthesis of trans-4-alkylcyclohexane carboxylic acids[1a,c] and their para-cyanophenyl esters[1b,d] have already appeared. This paper details a convenient synthesis of the parent nematogenic acids and reports on some of their para-alkylphenyl esters as well as those of Demus et al.[1c]

4-Alkylbenzoic acids were synthesized as shown in Fig. 1a. This procedure, though time-consuming, usually gave quite clean product in good yield. Hydrogenation of the aromatic acids (Fig. 1b) was performed on a 0.1 molar scale. The results of several experiments suggested that the desired trans acid product is favored by high temperature and hydrogen pressure. Propyl benzoic acid reduced at 250°C and 1400 psig (initial) pressure gave a product ratio of 80/20 (c/t). Reduction of the acid at 280°C (1700 psig) resulted in a ratio of 85/15 (c/t). A reduction of 4-pentyl benzoic acid at 1400 psig and 160°C returned a mixture of 50/50 cis/trans.

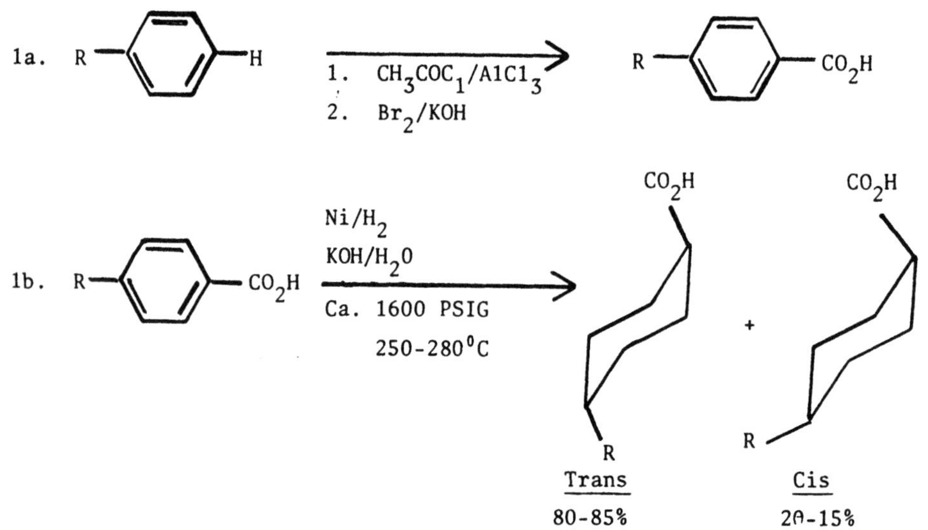


Fig. 1.

Previous papers on the synthesis of these compounds report the use of n-pentanol/sodium and $PtO_2/H_2$ for the hydrogenation step (Fig. 2).

Sodium-pentanol reductions require a time consuming addition of sodium pellets to the refluxing alcohol, frequently in ratios of 5 to 10 gm atom Na per mole of acid. Also, during work-up of the reaction mixture, the pentanol is most effectively removed by steam distillation: not generally a practical procedure in the electronics industry.

Adams catalyst ($PtO_2/H+$) reductions of the substituted benzoic acids are generally rapid but usually give product ratios characteristic of kinetic controlled reactions. Reductions in strong base with Raney nickel catalyst appear to give the equilibrated (thermodynamic) product.

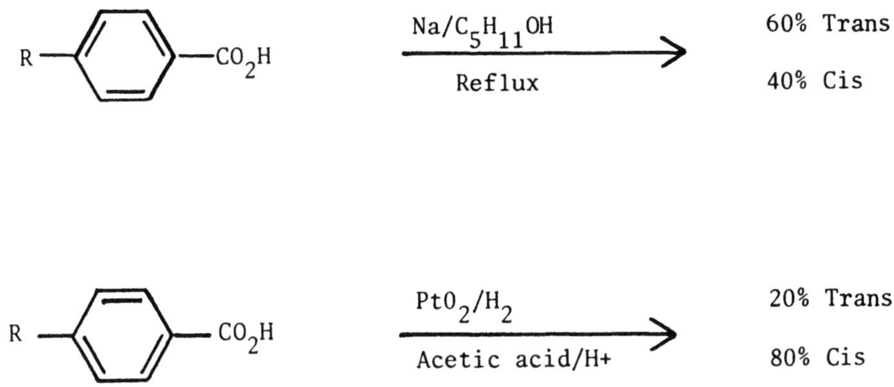


Fig. 2.

Schubert et al.[1a] reported that the mixture of saturated acids derived from hydrogenation over Adams' catalyst may be equilibrated as per Fig. 3. Two attempts by us to confirm this failed to give this result, which, if true, would certainly eliminate the necessity for high pressure apparatus.

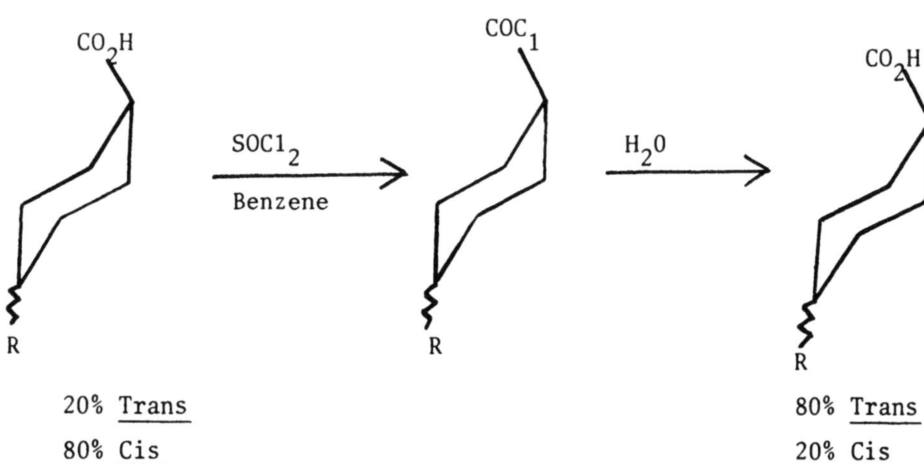


Fig. 3.

All the nematogenic esters of trans-4-alkylcyclohexane carboxylic acids listed in Table 1 were synthesized in our laboratory. The para-cyanophenyl esters have also been reported by Demus et al.;[1a] the para-alkyl and para-alkyoxy phenyl esters are to the best of our knowledge, being reported in print for the first time.

Melting point data of Table 1, comparing the esters of unsaturated[2] vs. saturated acids, are quite illustrative of the reasons for our interest in this latter series of compounds: they possess attractively low C→N transition temperatures and fairly wide ranges for simple mono esters.

## EXPERIMENTAL

The following two examples serve to illustrate the general synthetic approach utilized.

### 4-Alkyl Benzoic Acids

A mixture of 400 mls dry methylene chloride and 77g (0.58m) of fresh aluminum chloride was chilled to 0°C under dry nitrogen and a drying tube. Addition of acetyl chloride (30g, 0.38m) produced a clear, tan solution. To this was added with stirring over 1 1/2 hours time 50g (0.36m) freshly distilled butyl benzene. The stirred mixture warmed to room temperature overnight and was refluxed 15-20 minutes to complete reaction. Distillation of the crude product at 100°C ($10^{-4}$ Torr) gave 57.6g (89%) of >99% pure 4-butyl acetophenone.

Sodium hydroxide (168g, 4.2m) in 700 ml water was chilled to 2°C. Cold bromine (172g, 1.07m) was added over 30-40 minutes, never allowing the temperature of the mixture to exceed 5°C. 4-Butyl acetophenone (57.6g, 0.33 m) in 200 ml dioxane was added in one portion and the solution stirred at 5-10°C for four (4) hours before reaction was evident from disappearance of the organic phase. After being stirred overnight at room temperature, the mixture was acidified to pH 1 with concentrated $H_2SO_4$, excess hypobromite discharged with sodium meta bisulfite and chilled in an ice bath to obtain crystalline product. Two recrystallizations of the crude product from hexane at 0°C yielded 43g (80%) of 4-butyl benzoic acid.

### Trans-4-Alkyl Cyclohexane Carboxylic Acids

Hydrogenation of the aromatic acids was accomplished in a 600 ml capacity Parr Company mini-reactor (type 4563). Catalyst action was provided by a commercial (Alfa Inorganics) Raney nickel

Table 1. Nematogenic Esters of Trans-4-Alkyl Cyclohexane Carboxylic Acids

| R | R' | C→N | N→I | ΔT | Unsat'd Analog Nematic Range | ΔT |
|---|---|---|---|---|---|---|
| $C_3H_7$ | CN | 55 | 69 | 14 | Monotropic | |
| $C_4H_9$ | CN | 56 | 68 | 12 | Monotropic | |
| $C_5H_{11}$ | CN | 47 | 77 | 30 | Monotropic | |
| $C_6H_{13}$ | CN | 49 | 70 | 21 | 46-48 | 2 |
| $C_7H_{15}$ | CN | 53 | 81 | 28 | 45-57.6 | 12.6 |
| $C_3H_7$ | $OC_5H_{11}$ | 37 | 65 | 28 | 45-50 | 5 |
| $C_5H_{11}$ | $OC_6H_{13}$ | 32 | 79 | 47 | 35-60 | 25 |
| $C_7H_{15}$ | $OC_6H_{13}$ | 45 | 74 | 29 | No data | |
| $C_5H_{11}$ | $C_6H_{13}$ | 40.5 | 43.8 | 3.3 | Monotropic | |
| $C_5H_{11}$ | $C_5H_{11}$ | 35.8 | 38.2 | 2.5 | Monotropic | |
| $C_3H_7$ | $C_5H_{11}$ | | No data | | 14-19 | 5 |
| See Fig. 5 | | 88 | 178 | 90 | No data | |

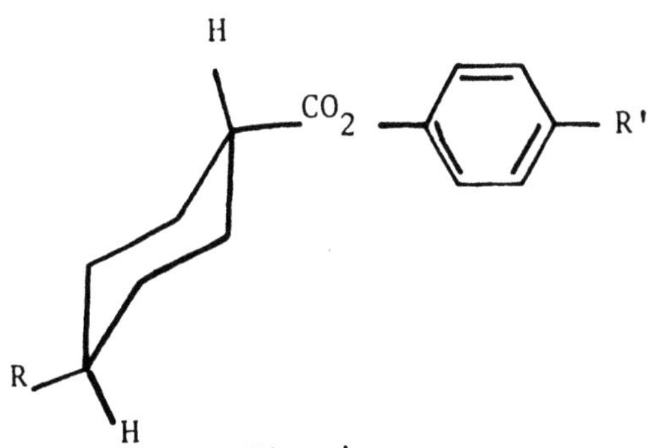


Fig. 4.

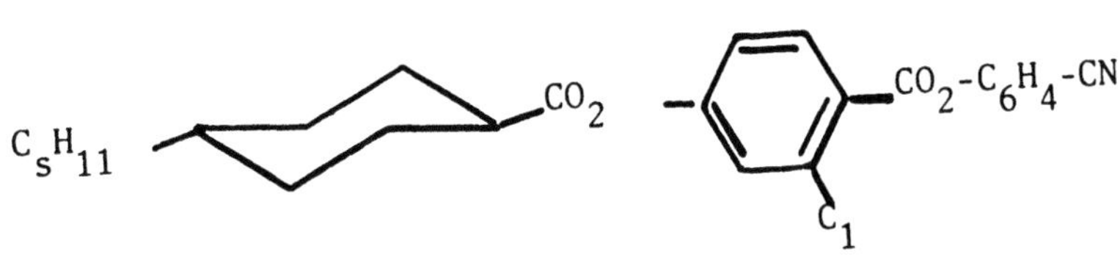


Fig. 5.

preparation. A solution of 0.10 to 0.15m of the acid with 1.5 equivalents of KOH in ca. 300 mls $H_2O$ together with 15-20g catalyst in the sealed reactor was hydrogen purged and heated to 250°-280°C. Hydrogen was admitted to a total pressure of 1700 psig; hydrogen uptake was generally fairly rapid. As pressure falls to 800-900 psig, the reactor was repressurized to 1700 psig and the reaction continued for a total of four (4) hours. It was sometimes necessary to effect the addition of another 10-15g of catalyst if 0.15-0.20m of aromatic acid were being reduced. The cooled, filtered reaction mixture was acidified to pH-1 with concentrated HCl, chilled to 0°C and the crystalline product isolated. Solid, dimeric trans-4-alkyl cyclohexane carboxylic acid was isolated by two or three recrystallizations of the product from pentane at -20°C. Monomeric cis acid was isolated as an oil from the mother liquors. Yields of trans product were generally on the order of 40-50%. Product analysis was accomplished by glc of the trimethyl silyl derivatives of the acids. Using a 50% phenyl silicon column, elution order was observed to be cis→trans→aromatic (acids).

Using the above procedure, 25g (0.152m) of 4-propyl benzoic acid returned a completely reduced mixture of 84.4% trans and 15.6% cis acids. Two recrystallizations from pentane gave 10.5g (48%) of 99.5% pure trans-4-propyl cyclohexane carboxylic acid.

REFERENCES

1. (a) H. Schubert et al., Z. Chem. 12(b):219 (1972).
   (b) D. Demus et al., ibid., 17(2):64 (1977).
   (c) J. Billard et L. Mamlok, Mol. Cryst. and Liq. Cryst. Let. 41(8):217 (1978).
   (d) G. W. Gray, "Advances in Liquid Crystal Materials for Applications," British Drig House Chemicals, Ltd. publication, March 1978.
2. J. P. Van Meter and B. Klanderman, Mol. Cryst. and Liq. Cryst. 17:285 (1973) and references cited therein.

# PREPARATION AND CHARACTERIZATION OF A SERIES OF LIQUID CRYSTALLINE DIESTERS

Robert J. Cox

IBM Research Laboratory
San Jose, California 95193

## INTRODUCTION

Continuing our interest in the relationship between molecular structure and liquid crystalline properties of compounds, we have prepared three series of aromatic diesters and examined their thermodynamic properties. Two of these are symmetric diesters of trans 1,4-cyclohexanedicarboxylic acid; the other is a series of diesters derived from terephthalic acid. The preparation of the cyclohexane esters is carried out starting from trans 1,4-cyclohexanedicarboxylic acid, converting to the acid chloride and reacting with the appropriately substituted phenol in pyridine, or more conveniently by direct esterification of the acid using dicyclohexycarbodimide as condensing agent and 4-dimethylaminopyridine as catalyst.[1] The esters of terephthalic acid must be prepared through the acid chloride as the acid is too insoluble to react directly. These reactions are summarized in Fig. 1. The compounds were all purified by recrystallization at least two times, followed by column chromatography. The purity was then checked by thin layer chromatography using two different elution solvents. The structure was also verified by NMR spectrometry. Because slight impurities can be of so much importance, particularly when measuring transition enthalpies of the magnitude found in mesomorphic transitions, the purity was also determined using the differential scanning calorimetric technique. The merits of this method have recently been discussed by Barrall and Johnson.[2] A further check was through combustion analysis. Measurement of the transition temperatures was carried out using hot stage polarized light microscopy. This technique was also used to identify the mesophase type. Differential scanning calorimetry was used to check the transition temperatures

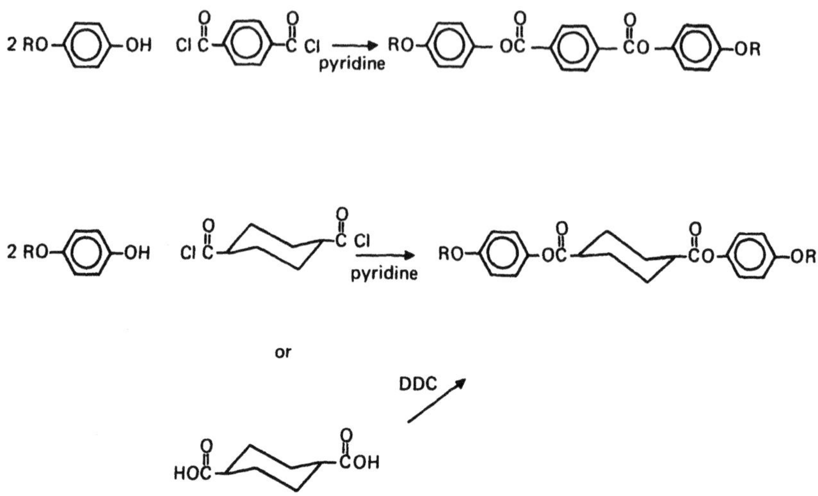


Fig. 1. Preparation of diesters.

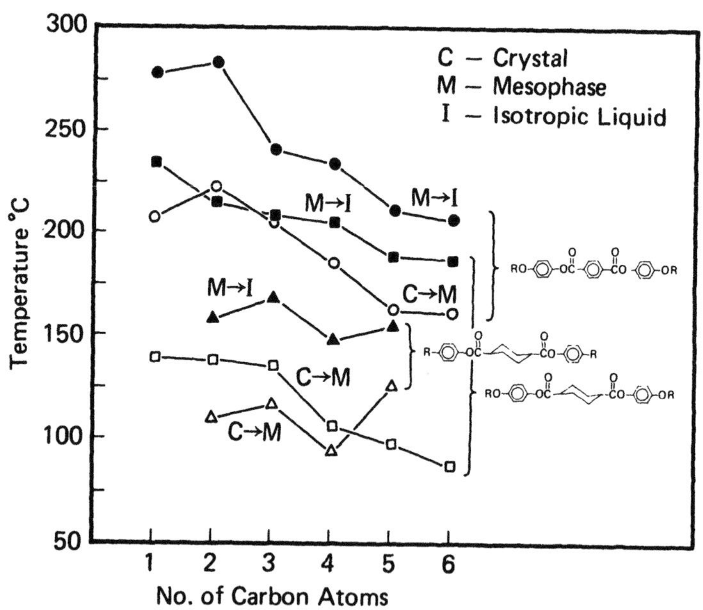


Fig. 2. Transition temperatures versus number of carbon atoms in the alkyl or alkoxy groups of diesters.

and to measure the heats and purities. The instrumentation used for these latter determinations was a Perkin-Elmer DSC-2 which had been computer interfaced. This system, which has been described in detail,[3] allows maximum accuracy in determination of transition heats and temperatures and in the purity calculations. A Dupont DSC-990 was also used in several experiments.

Aromatic diesters reverse their dielectric anisotropy as the frequency of an applied field is changed.[4] This is potentially useful when the liquid crystal is to be used in a display mode. Esters of terephthalic acid, however, have transition temperatures which are much too high for practical display purposes. It has been shown that replacement of an aromatic ring with its less rigid saturated analogue lowers the transition temperature, and in fact this is the case for the first member of this particular series.[5] We therefore decided to complete the series and study carefully the thermodynamic transitions in order to gain a more complete insight into the effect of replacement of an aromatic with a saturated ring in liquid crystals. The availability of individual members of a series coupled with the knowledge of transition heats and temperatures also makes possible the formulation of low-melting eutectic mixtures which would greatly facilitate the measurement of dielectric anisotropies. The effect on the dielectric anisotropy of changing an aromatic to a saturated ring has recently been discussed.[13]

In this paper the transition temperatures, heats and entropy changes are reported for individual members of the three series, as well as the thermodynamic relationship between the series.

In the p-alkoxyphenyl esters of cyclohexanedicarboxylic acid, the series was prepared and thermally examined in which the alkoxy group was varied from methoxy through hexyloxy.[14] The thermal properties and analytical results for this series are listed in Table 1. The first two members of the series have only nematic mesophases, while the propoxy through the hexyloxy have both nematic and smectic phases. As the length of the alkyl tail increases, so does the complexity of the transitions. There are four distinct solid polymorphs in the pentyloxy and three in the hexyloxy compound. The hexyloxy also has three smectic phases as well as the nematic. These three transitions were identified as smectic by microscopy, however no attempt was made to assign them to a particular class. The heats of two of these transitions were too small to measure with any degree of accuracy and only the temperatures were recorded. Solid polymorphism, such as encountered in these compounds where the transitions occur at very close temperatures, prevents the application of the differential scanning calorimetric method for determination of purity.

# Table 1

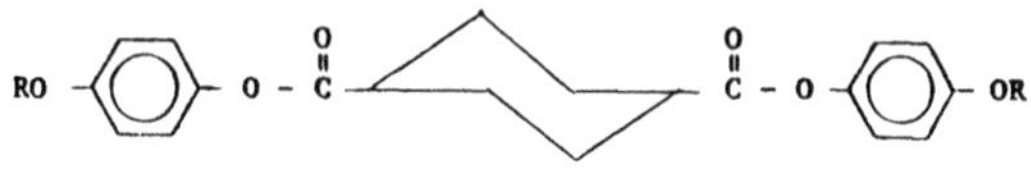

| R | TRANSITION | T°C | ΔH cal/mole | ΔS cal/mole/°K | Mole % Purity | Analysis Found C | Analysis Found H | Analysis Calculated C | Analysis Calculated H |
|---|---|---|---|---|---|---|---|---|---|
| $CH_3$ | $C \rightarrow N$ | 139.0 | 7560 | 18.4 | 99.8 | 68.60 | 6.11 | 68.8 | 6.25 |
| | $N \rightarrow I$ | 232.0 | 231 | 0.46 | | | | | |
| $C_2H_5$ | $C \rightarrow N$ | 138.6 | 8630 | 20.9 | 99.6 | 69.70 | 6.71 | 69.9 | 6.79 |
| | $N \rightarrow I$ | 215.6 | 211.6 | 0.43 | | | | | |
| $C_3H_7$ | $C \rightarrow N$ | 136.0 | 6190 | 15.2 | | | | | |
| | $S \rightarrow N$ | 106.7 | | | 99.7 | 70.92 | 7.24 | 70.9 | 7.27 |
| | $N \rightarrow I$ | 209.9 | 83 | 0.17 | | | | | |
| $C_4H_9$ | $C \rightarrow S$ | 106.0 | 6180 | 16.4 | 99.5 | 71.66 | 7.50 | 71.8 | 7.69 |
| | $S \rightarrow N$ | 154.0 | 74 | 0.17 | | | | | |
| | $N \rightarrow I$ | 206.7 | 227 | 0.47 | | | | | |
| $C_5H_{11}$ | $C_1 \rightarrow C_2$ | 80.0 | 3020 | 8.57 | | | | | |
| | $C_2 \rightarrow C_3$ | 85.8 | 131 | 0.36 | | | | | |
| | $C_3 \rightarrow C_4$ | 95.7 | 1390 | 3.7 | | | | | |
| | $C_4 \rightarrow S$ | 99.1 | 283 | 0.76 | | | | | |
| | $S \rightarrow N$ | 169.4 | 255 | 0.57 | | | | | |
| | $N \rightarrow I$ | 189.0 | 250 | 0.54 | | | | | |
| $C_6H_{13}$ | $C_1 \rightarrow C_2$ | 74.0 | 246 | 0.71 | | | | | |
| | $C_2 \rightarrow C_3$ | 79.0 | 3010 | 8.55 | | 72.50 | 8.14 | 73.3 | 8.40 |
| | $C_3 \rightarrow S_1$ | 89.5 | 2100 | 5.79 | | | | | |
| | $S_1 \rightarrow S_2$ | 173.0 | 307 | 0.70 | | | | | |
| | $S_2 \rightarrow S_3$ | 176.0 | 200 | 0.44 | | | | | |
| | $S_3 \rightarrow N$ | 185.0 | | | | | | | |
| | $N \rightarrow I$ | 188.0 | | | | | | | |

# Table 2

| R | TRANSITION | T°C | ΔH cal/mole | ΔS cal/mole/°K | Mole % Purity | Analysis Found C | Analysis Found H | Analysis Calculated C | Analysis Calculated H |
|---|---|---|---|---|---|---|---|---|---|
| $C_2H_5$ | $C \rightarrow N$ | 109.6 | 7264 | 18.9 | 99.6 | 75.66 | 7.44 | 75.8 | 7.37 |
| | $N \rightarrow I$ | 158.9 | 150 | 0.34 | | | | | |
| $C_3H_7$ | $C \rightarrow N$ | 115.1 | 7300 | 18.8 | 99.3 | 76.60 | 8.04 | 76.5 | 7.84 |
| | $N \rightarrow I$ | 168.8 | 134 | 0.30 | | | | | |
| $C_4H_9$ | $C \rightarrow S$ | 92.1 | 5620 | 15.4 | | 77.00 | 8.30 | 77.1 | 8.26 |
| | $S \rightarrow N$ | 106.3 | 156 | 0.41 | | | | | |
| | $N \rightarrow I$ | 149.4 | 175 | 0.41 | | | | | |
| $C_5H_{11}$ | $C \rightarrow S_1$ | 93.3 | 8145 | 22.2 | | 77.44 | 8.62 | 77.6 | 8.62 |
| | $S_1 \rightarrow S_2$ | 92.9 | | | | | | | |
| | $S_2 \rightarrow S_3$ | 91.6 | | | | | | | |
| | $S_1 \rightarrow N$ | 128.5 | 182 | 0.45 | | | | | |
| | $N \rightarrow I$ | 153.0 | 252 | 0.59 | | | | | |

C - crystal
N - nematic
S - smectic
I - isotropic liquid

Table 2 gives the thermal and analytical properties of the series of p-alkylphenyl esters of 1,4-cyclohexanedicarboxylic acid. Four members of this series were prepared in which the alkyl group was varied from ethyl through n-pentyl. The first two members, the ethyl and propyl substituted compounds, have straightforward thermal transitions, both showing only enantiotropic nematic mesophases. A smectic phase first appears in the n-butyl member and the butyl and pentyl compounds are both smectic and nematic. The solids in this series are less complex than in the alkoxy series, and there is no indication of solid polymorphism. The n-pentyl does become more complex in its mesophase and three smectics are observed, two of which are monotropic. The transition temperatures of these are too close to allow resolution of the peaks on the DSC, and no determination of the separate heats was made.

The thermodynamic data for the third series, the p-alkoxyphenyl esters of terephthalic acid, are shown in Table 3.

Table 3

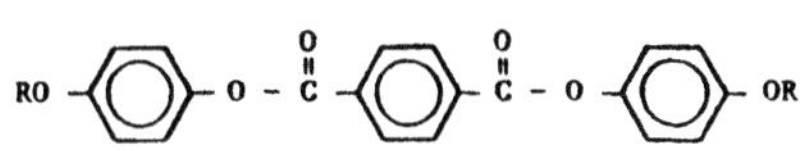

| R | TRANSITION | T°C | ΔH cal/mole | ΔS cal/mole/°K | Mole % Purity | Analysis Found C | Analysis Found H | Analysis Calculated C | Analysis Calculated H |
|---|---|---|---|---|---|---|---|---|---|
| $CH_3$ | C ⟶ N<br>N ⟶ I | 209.4<br>279.5 | 11,160<br>235 | 23.3<br>0.43 | 99.9 | 69.80 | 4.73 | 69.8 | 4.76 |
| $C_2H_5$ | C ⟶ N<br>N ⟶ I | 220.3<br>282.0 | 13,100<br>358 | 26.7<br>0.69 | 99.9 | 70.79 | 5.50 | 70.9 | 5.42 |
| $C_3H_7$ | C ⟶ N<br>N ⟶ I | 205<br>240 | —<br>— | —<br>— | | | | | |
| $C_4H_9$ | C ⟶ N<br>N ⟶ I | 186.6<br>233.9 | 9751<br>288 | 21.2<br>0.57 | 99.9 | 72.51 | 6.49 | 72.7 | 6.49 |
| $C_5H_{11}$ | $C_1$ ⟶ $C_2$<br>$C_2$ ⟶ $S_1$ | 137.3<br>162.5 | 1586<br>8860 | 3.86<br>20.3 | 99.9 | 73.21 | 6.96 | 73.5 | 6.94 |
| | $S_1$ ⟶ $S_2$<br>$S_2$ ⟶ I | 175.1<br>211.2 | —<br>227 | —<br>0.47 | | | | | |
| $C_6H_{13}$ | C ⟶ $S_1$<br>$S_1$ ⟶ $S_2$<br>$S_2$ ⟶ I | 161.2<br>181.0<br>208.0 | 8640<br>—<br>310 | 20.1<br>—<br>0.65 | 99.9 | 73.73 | 7.32 | 74.1 | 7.3 |

C - crystal
N - nematic
S - smectic
I - isotropic liquid

The alkoxy group in this series was varied from methoxy through hexyloxy. The propoxy, however, was not prepared and the values given are from the literature.[6] In this series, which has been examined by several authors,[7,8,9] the first four members from methoxy through butoxy are straightforward, and the only mesophase observed was enantiotropic nematic with no solid polymorphism. The last two members, the pentyloxy and the hexyloxy, each show two smectic phases and in the case of the pentyloxy there are two crystal forms. In both these cases the entropic change going from one smectic to another is very small, indicating a great similarity of the order in the two phases.

Figure 2 plots the number of carbon atoms in the alkyl or alkoxy chain as a function of the transition temperatures for the three series. It is instructive to compare the crystal to mesophase temperatures of the two series in which the only change was substitution of a rigid aromatic ring for the more flexible saturated one. The compounds with the aromatic rings have a 50-60 degree higher temperature for this transition. The trend in both series is for the transition temperature to decrease as the alkoxy chain length increases, both series decreasing with about the same slope. The mesomorphic to isotropic transition temperatures for these two series also decrease, but appear to be converging as the alkoxy tail gets longer. A comparison of the two series of 1,4-cyclohexanedicarboxylic acid esters, in which an alkyl group is substituted for the alkoxy, shows the crystal to mesophase temperatures to be close, although in general they are somewhat lower for the alkyl compounds. The mesophases in this latter group, however, are less stable by 30-40 degrees. Each of the series shows a trend toward the familiar odd-even effect, although no more so for the mesomorphic transitions than for the crystal transitions which usually conform to this odd-even pattern to a less extent than the mesomorphic transitions.

An interesting feature when comparing these series is the early onset of the smectic phase in the cyclohexane series. The alkoxy substituted cyclohexanes have a monotropic smectic appearing in the propoxy member. The corresponding unsaturated esters do not show smectic mesophases until the pentyloxy. This is not what one would necessarily expect, and in fact comparison of a terphenyl series to the corresponding series in which the central benzene ring is replaced by a cyclohexyl ring shows smectic appearing in the terphenyl at $C_4$[10] and not in the saturated derivative until $C_8$.[11]

Figure 3 plots the total entropy change in going from the solid to the mesophase as a function of the length of the alkyl chains for the three series. Comparison of the two alkoxy series indicates a much higher degree of order present in the completely aromatic crystal than in the saturated analogue. There apparently is better crystal packing for these rigid molecules. The two cyclohexyl series have, except for one anomalous point, about the same order in the crystal phase. The $C_5$ derivative has a larger entropy change than the other members of either of the cyclohexyl series. The ether linkage does not play an important role in the overall packing of the crystal. Barrall and Johnson have suggested that the relationship between transition entropy and alkyl chain length should be regular in a series if all the solid forms are accounted for,[12] and indeed a regular odd-even alternation is evident in these series.

Figure 4 compares the total entropy change versus number of carbon atoms for the mesophase to isotropic transition for the three series. The difference between the aromatic and the cyclohexyl series is due partially to the early onset of the smectic phase in the saturated case, and partially to the smaller than normal entropy changes found in the aromatic series even when the smectic does appear. It is possible that the smectic formed in this latter case is of a different, less ordered class than found in the cyclohexyl series. Positive classification would be necessary to establish this.

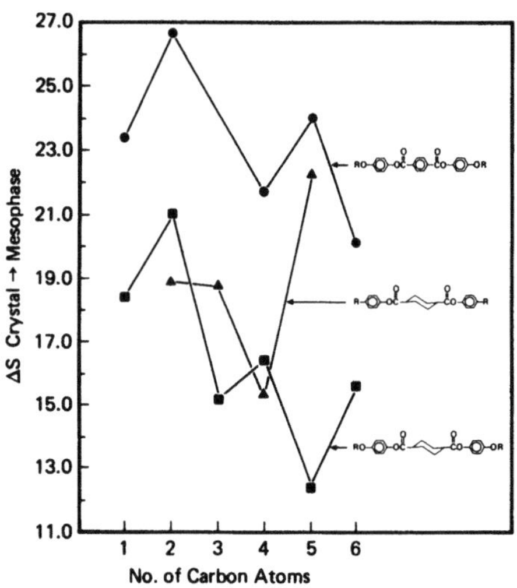


Fig. 3. Entropy change for the crystal mesophase transition versus number of carbon atoms in the alkyl or alkoxy groups of diesters.

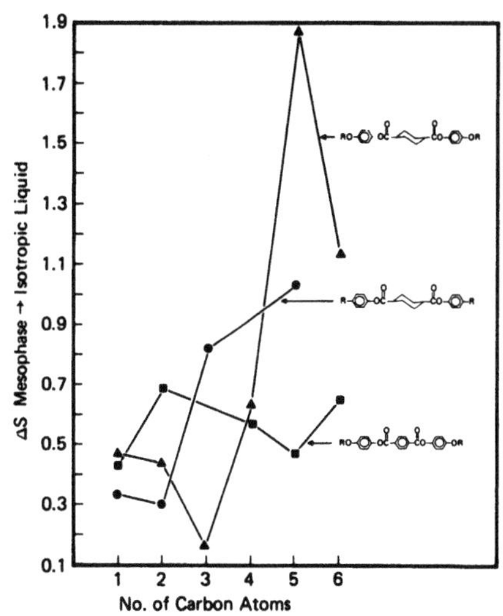


Fig. 4. Entropy change for the mesophase isotropic liquid transition versus number of carbon atoms in the alkyl groups of diesters.

Examination of these series has allowed several generalizations:

1. Replacement of the rigid aromatic ring with a flexible saturated analogue lowers the transition temperatures in both the mesophase and the crystal.

2. Substitution of an alkyl for an alkoxy end group lowers the mesomorphic transition temperature markedly, but the crystal transition only slightly.

3. The entropy change for the solid to mesomorphic mesophase is greater for the aromatic compounds that for the cyclohexyl, i.e., the aromatic crystals are more ordered.

4. The place in the series at which the smectic phase appears does not seem to bear any simple relationship to the presence or absence of rigid ring structures in these three ring systems.

## REFERENCES

1. A. Hessner and V. Alexanian, Tetrahedron Lett. 46:4475 (1978).
2. E. M. Barrall and J. F. Johnson in: "Liquid Crystals and Plastic Crystals," G. W. Gray and P. A. Winsor, eds., Vol. II, E. Horwood, Chichester, England (1974), p. 303.
3. A. Doelman, A. R. Gregges and E. M. Barrall in: "Analytical Calorimetry," R. S. Porter and J. F. Johnson, eds., Vol. 4, Plenum Press, New York, N.Y. (1976), p. 1.
4. H. K. Bücher, R. T. Klingbiel and J. P. Van Meter, Appl. Phys. Letters 25:186 (1974).
5. M. J. S. Dewar and R. M. Riddle, J. Am. Chem. Soc. 97:6658 (1975).
6. D. Demus, H. Demus and H. Zaschke, Flüssige Kristalle in Tabellen, VEB Deutscher Verlag für Grundstoffindustrie, Leipzig (1974), p. 87.
7. M. J. S. Dewar and R. S. Goldberg, J. Org. Chem. 35:2711 (1970).
8. L. Verbit and R. L. Tuggey, Mol. Cryst. Liq. Cryst. 17:49 (1972).
9. H. Kelker and B. Scheurle, J. Physique 30:C4,104 (1969).
10. Ref. 6, p. 225.
11. Ref. 6, p. 35.
12. Ref. 2, p. 260.
13. L. Pohl, R. Eidenschink, J. Krause and G. Weber, Abstracts, Seventh International Congress of Liquid Crystals, Bordeaux, France, DP35, (1968).

14. The preparation and study of the thermodynamic transitions for this series was reported by M. E. Neubert and J. P. Ferrato, Abstracts, Seventh International Congress of Liquid Crystals, Bordeaux, France, AP16, (1968).

# EFFECT OF MOLECULAR STRUCTURE ON MESOMORPHISM. 10[1]. DESIGN OF MATERIALS HAVING SMECTIC-ISOTROPIC MAXIMA IN BINARY MIXTURES.[2]

Anselm C. Griffin, Neal W. Buckley and Stephen J. Havens

Department of Chemistry, Box 8448 Southern Station
University of Southern Mississippi
Hattiesburg, Mississippi 39401

Julian F. Johnson, Gail J. Bertolini and Linda Thomas

Department of Chemistry and Institute of Materials Science
University of Connecticut
Storrs, Connecticut 06268

## INTRODUCTION

Use of liquid crystalline materials for display devices demands a wide mesomorphic temperature existence range, above and below ambient temperatures.[3,4] There are numerous specifications of physical properties such as birefringence, viscosity, dielectric anisotropy, etc. which must be met, but the temperature range of nematic or smectic A materials is among the most crucial. Most binary liquid crystalline (mesomorphic) mixtures exhibit eutectic melting behavior and mesophase-isotropic transition temperatures which are linearly dependent on composition. Eutectic melting lowers the temperature of entry into the mesomorphic phase. If, however, the mesophase-isotropic temperature could be enhanced (increased) in mixed systems this would permit operation of the device at higher temperatures. Numerous recent papers[5] have been devoted to experimental and/or theoretical investigation of enhanced mesophases and several models dealing with intermolecular interactions at the molecular level have been proposed to account for this non-ideal mixture behavior. In this paper we will report data on the effect of alkoxy chain length on mesophase enhancement. There have been previous reports[5a,s,t] of similar studies with a rather limited scope but none as extensive as that described here.

We chose to study the three homologous series of liquid crystals shown below.

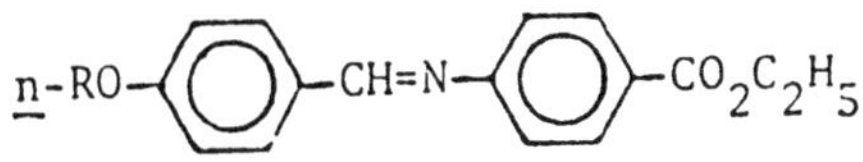


1

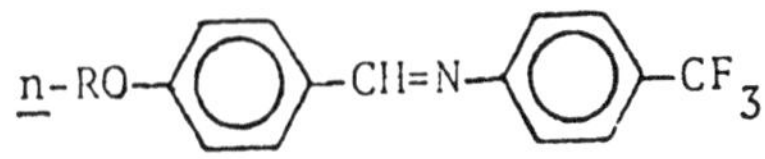


2

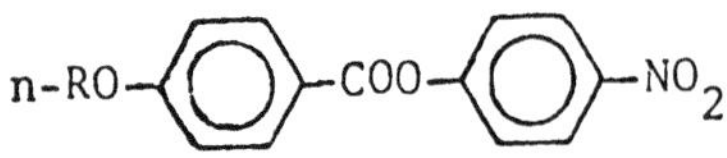


3

Compounds in series 1 were known[5s,t] to form enhanced smectic A mesophases when mixed with compounds in series 2 (1:2 mixtures) or series 3 (1:3 mixtures). With sufficient members of the homologous series of 1, 2 and 3 on hand it seemed to us an obvious and important extension would be to form binary mixtures of each available 1:2 and 1:3 combination. By noting the maximum in the mesophase-isotropic transition temperature curve for each combination it should be possible to evaluate the effect of alkoxy chain length on enhancement of these temperatures. An accurate assessment of chain length factors would obviously aid in design of liquid crystalline materials for device use, particularly for the smectic storage displays.[6] We also wanted to study one binary system in detail by use of DSC (differential scanning calorimetry), DLI (depolarized light intensity) and optical microscopy to allow construction of an accurate phase diagram which should be representative of those combinations which we studied.

## EXPERIMENTAL

Compounds in the three series have been reported previously, series 1,[7,5s] Series 2[8] and series 3.[5s,t] DSC curves were obtained on a duPont 990 Thermal Analysis System equipped with a DSC cell and with a two stage mechanical cooler for low temperature work. Calibration was based on sapphire standards and enthalpic reference standards. Thermograms were run from -50°C to about 15°C above the mesophase-isotropic temperature. Prior to obtaining the thermogram the samples were held isothermally at -50°C for one hour in the DSC cell. Depolarized Light Intensity (DLI) curves were obtained on an instrument specially designed and constructed at the University of Connecticut.[9] The method gives results in which the measured birefringence can be quantitated if calibrated with known standards. We, however, used DLI to obtain qualitative birefringence data as a function of sample temperature.[10] The temperature axis of the DLI apparatus was calibrated by fusion of materials with accurately known melting points.

Determination of maxima in the mesophase-isotropic curves for each binary combination was made either by microscopical observation of a Kofler contact preparation of the pure components using a Thomas micro hot stage or by a technique described previously[5t] using a Thomas-Hoover melting point apparatus.

## RESULTS AND DISCUSSION

Data in Tables 1 and 2 show maximum mesophase-isotropic transition temperatures for 1:2 and 1:3 binary combinations, respectively. Temperatures in these Tables are given only if the mixed mesophase is more thermally stable than either of the pure components. It is seen that along a column or along a row there is a general trend for a maximum value of smectic-isotropic transition temperature to occur when the alkoxy chain lengths of both components are equal or nearly so.

There is a wealth of information in Tables 1 and 2 and there are numerous ways in which the data could be presented. We have chosen to use four plots for this purpose (Fig. 1). In each of these plots we have considered one homologous series as the basis set (primary series) and then plotted, as a function of alkoxy chain length in this basis set, maximum mesophase-isotropic transition temperatures of binary mixtures of this primary series with selected members of the other (secondary) series - a short-, a medium- and a long-chained member of this second series. Plots 1a and 1b are plots of mixtures of the same two compounds; the primary and secondary sets have been exchanged. Plots 1c and 1d are related to each other in a similar fashion. It is obvious from these plots that for a given member of the secondary series the mesophase-isotropic transition temperatures vary in a regular, often pronounced, fashion; but this variation is often considerably different from the plots of the other two members of the secondary series. For example in 1a with 2 as the primary series, the n=2, 5 and 8 members of the secondary series shown distinctly different types of curves. That for n=2 rises quickly, then rapidly descends; n=5 rises quickly, then levels off; and that for n=8 rises slowly. The various alkoxy chains of the primary series homologues are interacting with the three secondary series members (one short chain, one medium chain and one long chain) in very different ways. If the disparity of chain lengths of the components is too great there is no enhanced mesophase seen. Transition temperatures in these plots show a clear tendency for maximization at or near equal chain lengths of the components in the binary mixture. All smectic-isotropic

Table 1. Maxima in Smectic A→Isotropic Transition Temperature (°C) obtained by the contact method for each mixture shown. A number indicates a $S_A$→I temperature higher than the mesophase→Isotropic temperature of either pure component.

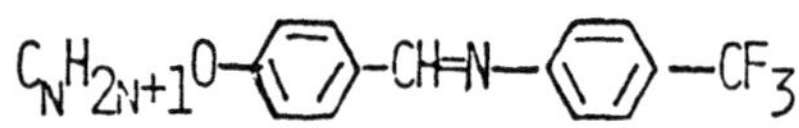

| N | 1 | 2 | 3 | 4 | 5 | 6 | 7 | 8 | 10 | 12 |
|---|---|---|---|---|---|---|---|---|---|---|
| 1 | -- | -- | 92 | 94 | 88.5 | 87.5 | 84 | 83.7 | 79.5 | -- |
| 2 | -- | -- | 103.9 | 108 | 105.5 | 104.5 | 101.5 | 99 | 94.5 | -- |
| 3 | -- | -- | 102 | 106.7 | 102.8 | 101 | 98.5 | 97 | 92.5 | 88 |
| 4 | -- | -- | 102.7 | 108 | 106 | 106 | 105 | 103.5 | 100.7 | -- |
| 5 | -- | -- | 99.5 | 105 | 103 | 103 | 102.7 | 102 | 99 | -- |
| 6 | -- | -- | 99 | 103 | 101 | 103 | 102.7 | 103 | 101.1 | 99 |
| 7 | -- | -- | 100 | 101 | 100 | 101.5 | 101.8 | 101 | 101 | 100 |
| 8 | -- | -- | -- | 100 | 99 | 100 | 101 | 102.9 | 102 | 101.5 |
| 10 | -- | -- | -- | -- | -- | -- | -- | 100 | 101 | 101.5 |
| 12 | -- | -- | -- | -- | -- | -- | -- | -- | -- | 100 |

Table 2. Maxima in Mesophase→Isotropic Transition Temperature (°C) obtained by Contact Method for each mixture shown. A number indicates a Mesophase→Isotropic temperature higher than the Mesophase→Isotropic temperature of either pure component. All values refer to Smectic A→Isotropic Transitions except those in parentheses which refer to Nematic→Isotropic Transitions.

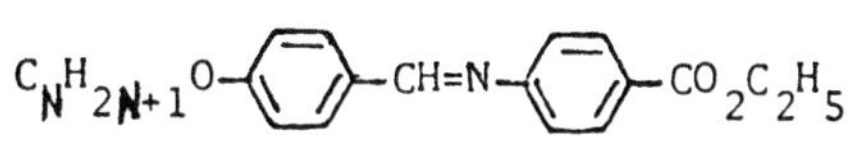

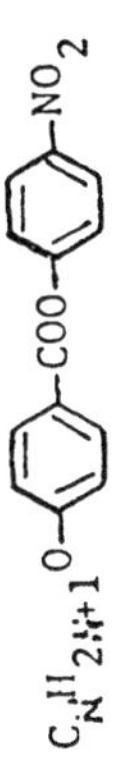

| N \ N | 1 | 2 | 3 | 4 | 5 | 6 | 7 | 8 | 10 | 12 |
|---|---|---|---|---|---|---|---|---|---|---|
| 1 | — | — | — | — | — | — | — | — | — | — |
| 2 | — | — | — | — | — | — | — | — | — | — |
| 3 | 97 | (92) | 94 | 102 | 101 | 100.5 | 100.5 | 100.5 | 100 | 98.5 |
| 4 | 87 | (92.5) | 95 | 104 | 104 | 104 | 103.5 | 103.5 | 102.5 | 100.5 |
| 5 | — | (91) | 94 | 103.5 | 103.5 | 104 | 104.5 | 103.5 | 103 | 101 |
| 6 | — | (91) | 90 | 101.5 | 102.5 | 104 | 106 | 105.5 | 105 | 103.5 |
| 7 | — | (92.5) | 91 | 103 | 103.5 | 105 | 105.5 | 105 | 105 | 103 |
| 8 | — | (92) | (84.5) | 100 | 101.5 | 104 | 104.5 | 106 | 106 | 104.5 |
| 10 | — | (90.5) | (86.8) | (97.2) | 97.7 | 102.5 | 104.4 | 104.8 | 106.9 | 105.8 |
| 12 | — | (89) | — | — | — | 98.5 | 101 | 103.5 | 105.5 | 105 |
| 14 | — | — | — | — | — | — | 97.5 | 101 | 103.5 | 103.5 |
| 16 | — | — | — | — | — | — | — | 98.5 | 102 | 102 |

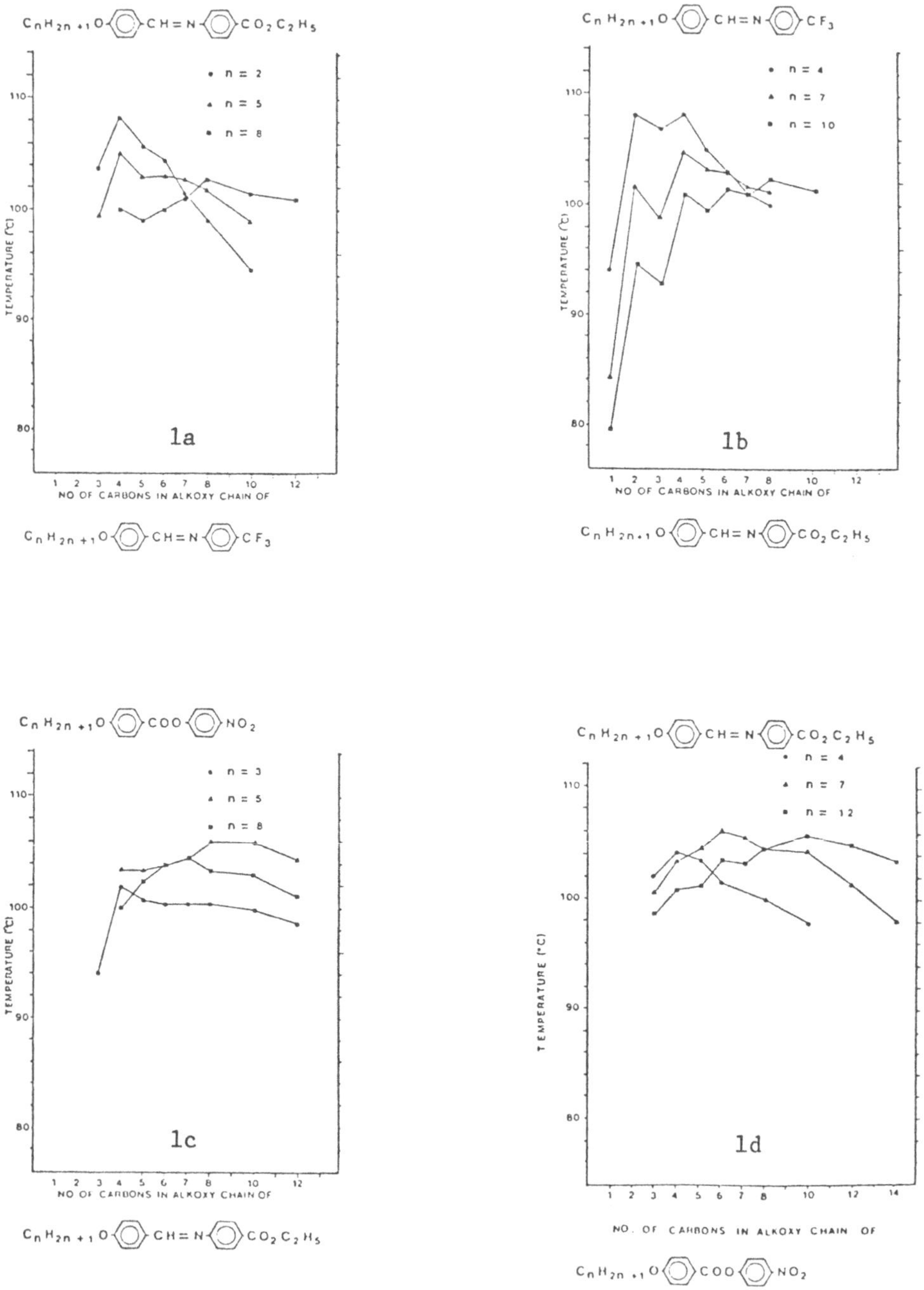


Fig. 1. Plots of maximum smectic A$\rightleftharpoons$isotropic temperatures of selected 1:2 and 1:3 binary mixtures.

transitions occur from a smectic A ($S_A$) mesophase. The tendency to form enhanced $S_A$ phases is pronounced for these mixtures since compounds of series 2 show only smectic B phases and the short chained members of series 3 are only nematic. Although the data is not graphed in plot 1c, there are several enhanced nematic phases in these mixtures. This observation should prove useful for applications such as smectic storage devices in which a short range nematic above a smectic A is needed.

The binary, isobaric phase diagram for a mixture of ethyl 4-dodecyloxybenzylidene-4'-aminobenzoate(A) and 4-octyloxybenzylidene-4'-trifluoromethylaniline(B) is presented in Fig. 2. This mixture has a mildly enhanced smectic A phase.

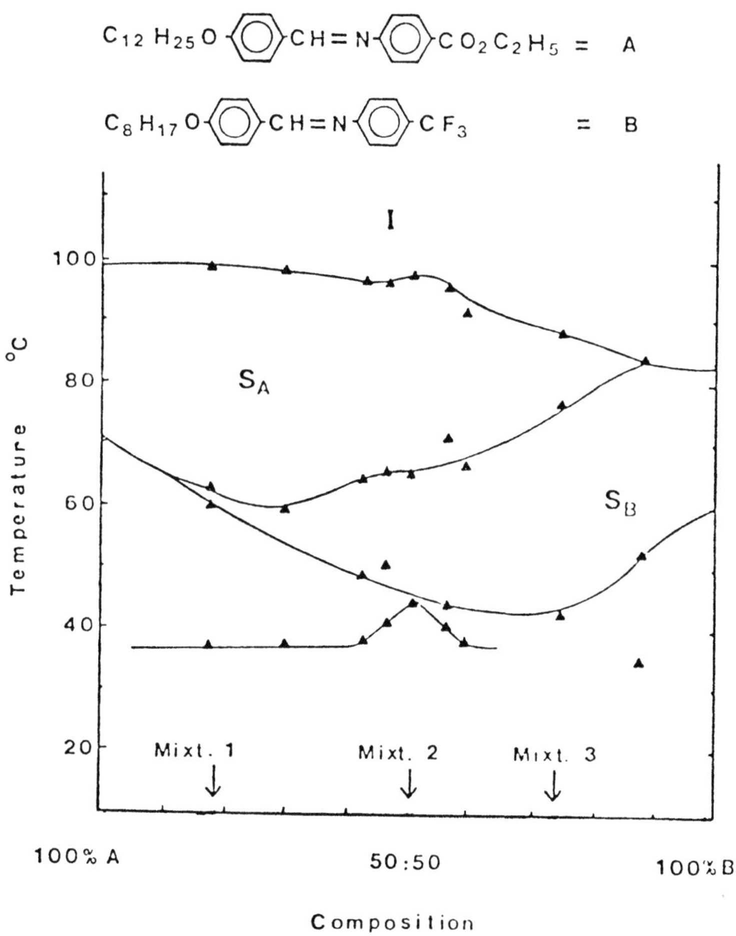


Fig. 2. Binary, isobaric phase diagram for a mixture of ethyl 4-dodecyloxybenzylidene-4'-aminobenzoate(A) and 4-octyloxybenzylidene-4'-trifluoromethylaniline(B).

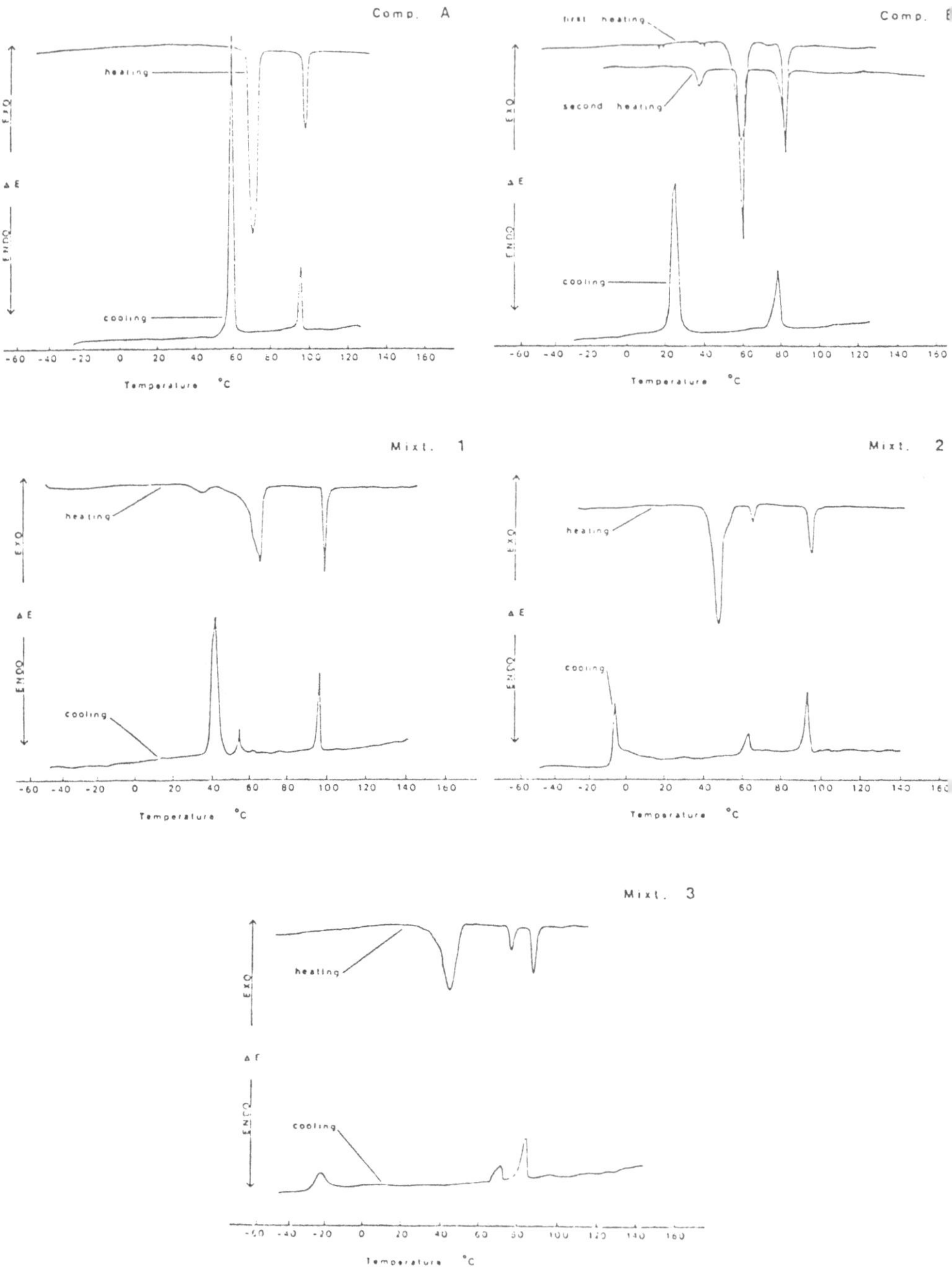


Fig. 3. Differential scanning calorimetry curves for end members and selected compositions of the phase diagram in Fig. 2.

Construction of the phase diagram followed from DSC, DLI and optical microscopy of a number of individual compositions of the two end members. DSC curves for the end members and for three selected mixed compositions are presented in Fig. 3. The mixed compositions were chosen for presentation since they represent points at low, medium and high concentrations of one component in the other. Heating curves only were used in construction of the phase diagram. Sample handling prior to calorimetry, as described in the Experimental Section, precluded our observation of metastable states.

DLI curves for the end members and for the same three mixed compositions are presented in Fig. 4. Data from DLI observations correlated well with that from DSC and optical microscopy. This technique permits enhanced sensitivity when one wants to observe transitions that may involve a small enthalpy change but a discernible change in birefringence. In particular DLI is advantageous when solid-phase polymorphism is present in a pure compound or a mixture.

The phase diagram itself reveals an enhanced smectic A phase with a smectic B phase which extends far into regions of high concentrations of (A). Of particular interest is the indication of complex formation between (A) and (B) at 50-50 mole percent composition. This is in general agreement with a recent report by Engelen[5c] in which he suggests double eutectics should be expected for mixtures of terminal polar and terminal non-polar liquid crystals. Although the chemical nature of the complexation is not completely clear to us the formation of a molecular complex is quite clear from the diagram. In Fig. 5 are presented enthalpic data for this binary mixture. Eutectic heats and melting heats of excess components are not plotted separately in Fig. 5, but are included in calculation of total heats. This plotting omission is due to difficulty in obtaining necessary resolution to evaluate separate enthalpies for the two transitions. The combined heats for these two transitions are, however, easily quantitated and are included in our calculations. Of particular interest in the enthalpic diagram is the large increase in total heat of the composition at near 50-50 mole percent. The major contributor to this dramatic deviation in the total heat curve is the heat of melting (Solid→Smectic B; 18.4 cal/gram) for this composition. We feel this enthalpy data along with the temperature maximum in the melting curve at this same composition implicates the formation of a molecular complex with a 1:1 stoichiometry. Optical microscopy was not sufficient to accurately determine phase behavior around the solidus points. DLI and DSC were most helpful at these low temperatures.

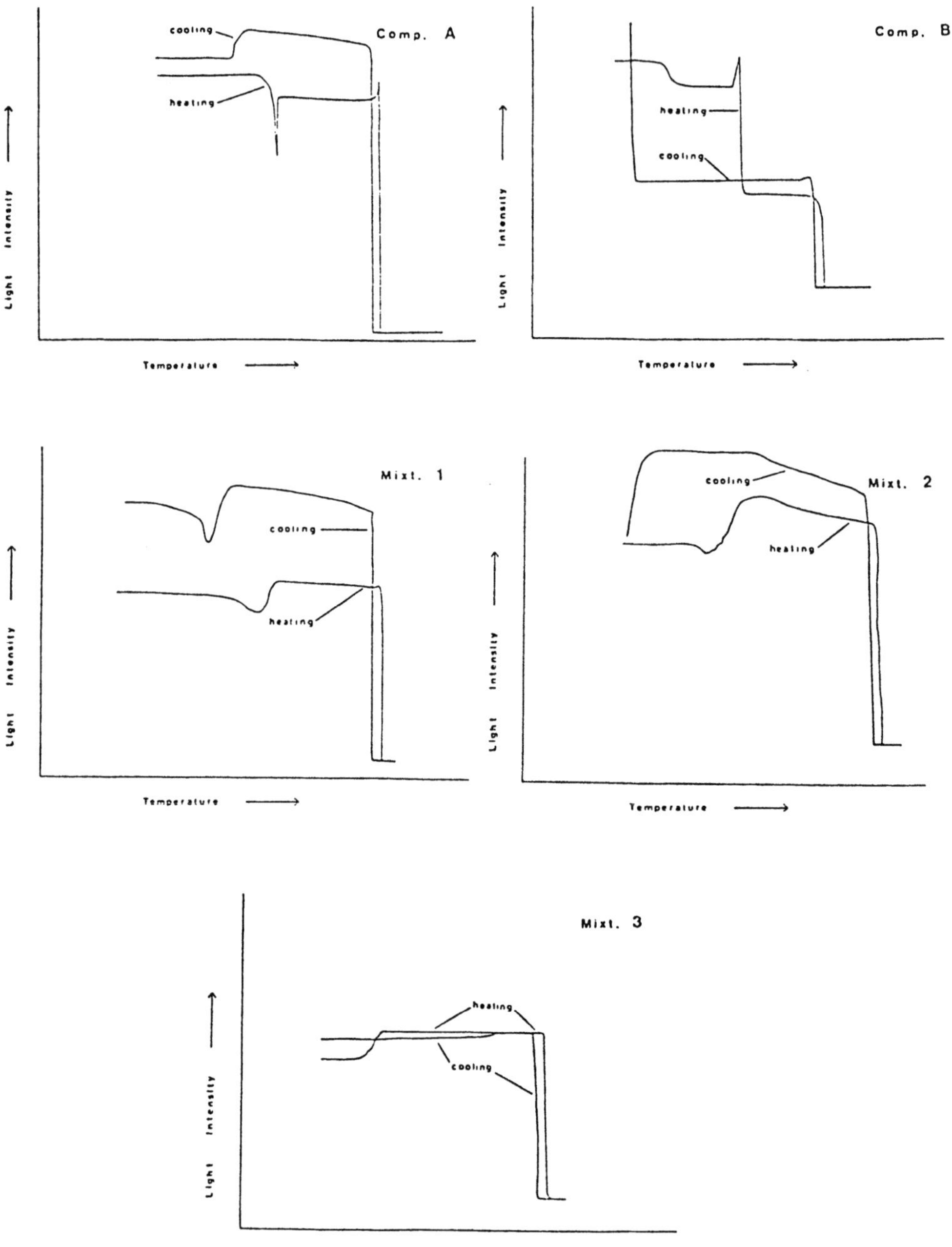


Fig. 4. Depolarized light intensity curves for end members and selected compositions of the phase diagram in Fig. 2.

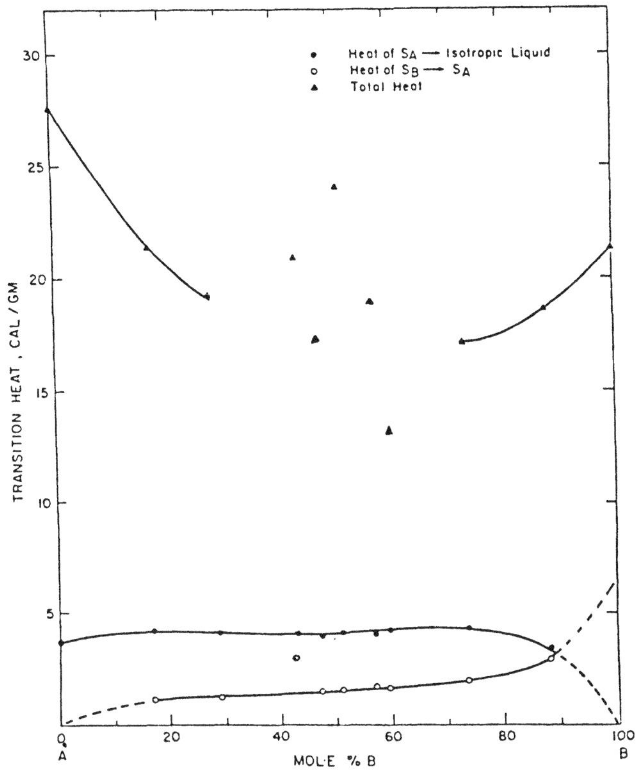


Fig. 5. Enthapy versus composition for the binary mixture: ethyl 4-dodecyloxybenzylidene-4'-aminobenzoate(A) and 4-octyloxybenzylidene-4'-trifluoromethylaniline(B).

## CONCLUSION

It has been shown that alkoxy chain length of the components of a binary mixture can alter the maximum mesophase-isotropic temperature in a regular, predictable manner. The effect of chain length is clearly a secondary one, but nevertheless, a contributing factor to non-ideal behavior in mixtures. Maximum enhancement of a mesophase occurs when the chain lengths of the components are similar. By careful matching of end members, it appears possible to design mixtures which possess the requisite temperatures and sequence of mesophases for use in display devices. A detailed study by DSC, DLI and optical microscopy of a binary mixture allowed construction of a phase diagram in which an enhanced $S_A$ phase is present. There is a maximum in the melting curve for this mixture. We feel this represents non-linear behavior of the solid phases in this mixture. Studies of similar binary phase diagrams are in progress.

ACKNOWLEDGMENT

This work was supported by a grant from the National Science Foundation (DMR 7805284).

REFERENCES

1. For Part 9 in this series see A. C. Griffin, M. L. Steele, J. F. Johnson and G. J. Bertolini, "Nouveau Journal de Chimie," submitted.
2. Presented at the Symposium on the Physics and Chemistry of Liquid Crystal Devices, San Jose, California, February, 1979.
3. L. Creagh, Proc. IEEE, 61:814 (1973).
4. R. Steinstrasser and L. Pohl, Angew. Chem. 85:706 (1973).
5. (a) G. Heppke and E.-J. Richter, Z. Naturforsch. 33a:185 (1978); (b) B. Engelen, G. Heppke, R. Hopf and F. Schneider, Ann. Physique 3:403 (1978); (c) B. Engelen and F. Schneider, Z. Naturforsch. 33a:1077 (1978); (d) M. Bock, G. Heppke, E.-J. Richter and F. Schneider, Mol. Cryst. Liq. Cryst. 45:221 (1978); (e) B. Engelen, G. Heppke, R. Hopf and F. Schneider, Mol. Cryst. Liq. Cryst. (Letters), in press; (f) J. W. Park, C. S. Bak and M. M. Labes, J. Am. Chem. Soc. 97:4398 (1975); (g) J. W. Park and M. M. Labes, Mol. Cryst. Liq. Cryst. Letters 34:147 (1977); (h) C. S. Oh, Mol. Cryst. Liq. Cryst. 42:1 (1977); (i) H. Sackmann and D. Demus, Z. Phys. Chem. (Leipzig), 224:177 (1963); (j) ibid. 230:285 (1965). (k) J. S. Dave, P. R. Patel and K. L. Vasanth, Ind. J. Chem. 4:505 (1966); (l) J. S. Dave, P. R. Patel and K. L. Vasanth, Mol. Cryst. Liq. Cryst. 8:93 (1969); (m) J. S. Dave and P. R. Patel, J. Ind. Chem. Soc. 47:815 (1970) (n) J. P. Schroeder and D. C. Schroeder, J. Org. Chem. 33:591 (1968); (o) H. Kozawaguchi and M. Wada, Mol. Cryst. Liq. Cryst. 44:97 (1978); (p) M. Domon and J. Billard, J. de Phys. in press; (q) F. Hardovin, G. Sigand, M.-F. Archard and H. Gasparoux, Ann. Phys. 3:381 (1978); (r) A. C. Griffin and J. F. Johnson, J. Am. Chem. Soc. 99:4859 (1977); (s) A. C. Griffin, T. R. Britt, N. W. Buckley, R. Fischer, S. J. Havens and D. W. Goodman, "Liquid Crystals and Ordered Fluids, Vol. 3", R. S. Porter and J. F. Johnson, eds., Plenum Press, New York, New York (1978); p. 61. (t) A. C. Griffin, R. F. Fischer and S. J. Havens, J. Am. Chem. Soc. 100:6329 (1978); (u) P. E. Cladis, R. K. Bogardus and D. Aadsen, Phys. Rev. A18:2292 (1978); (v) G. R. van Hecke, J. Phys. Chem. in press.
6. (a) F. J. Kahn, Appl. Phys. Letters 22:111 (1973); (b) M. Hareng, S. LeBerre and L. Thirant, Appl. Phys. Lett. 25:683 (1974); (c) ibid. 27:575 (1975); (d) D. Coates, W. A. Crossland, J. H. Morrissy and B. Needham, J. Phys. D: Appl. Phys. 11:2025 (1978); (d) D. Coates, W. A. Crossland, J. H. Morrissy and B. Needham, J. Phys. D: Appl. Phys. 11:2025 (1978).

7. D. L. Fishel and P. R. Patel, Mol. Cryst. Liq. Cryst. 17:139 (1972).
8. A. C. Griffin and Neal W. Buckley, Mol. Cryst. Liq. Cryst. Letters 41:141 (1978).
9. J. A. Reffner, Ph.D. Thesis, University of Conneticut, 1975.
10. For a description of the technique, see E. M. Barrall, II and J. F. Johnson, Thermochem. Acta. 5:41 (1972) and references therein.

# THERMODYNAMICS OF THE LIQUID TWO-PHASE REGIONS IN BINARY MIXTURES OF MESOGENIC MATERIALS

Gerald R. Van Hecke

Department of Chemistry
Harvey Mudd College
Claremont, California 91711

## INTRODUCTION

The phase behavior of liquid crystalline mixtures has been studied for a variety of reasons, but for the purposes of the discussions here, two are predominant: (1) use of binary mixtures for miscibility studies to identify mesophases, (2) use of eutectic mixtures to enhance the temperature range of mesomorphic stability, typically the nematic phase. While many phase diagrams have been reported, only very few have been discussed in a quantitative manner, that is, calculated using thermodynamic relationships.[1-4] Further, very few calculations have been reported for phase diagrams with maxima or minima, azeotrope-like behavior in the liquid phase regions of the diagram.[5-6] The occurrence of such maxima or minima cannot be described thermodynamically in terms of ideal solution behavior. However, regular solution theory can and does allow convenient calculation of such azeotrope-like behavior increasingly exhibited in reported phase diagrams of liquid crystalline mixtures. We will introduce here the method of calculation and demonstrate its usefulness by summarizing the results obtained for the calculation of the phase diagrams for a homologous series of symmetric azoxybenzenes and for three mixtures of cholesterol esters with a common reference Schiff base. Also, we will comment on some procedures for estimating the parameters required by the regular solution theory.

## IDEAL SOLUTIONS

The thermodynamic formulism required to describe ideal two-phase regions begins with the chemical potential of component i in phase $\emptyset$ at constant T and P given as

$$\mu_{i\emptyset} = \mu^{\circ}_{i\emptyset} + RT \ln X_{i\emptyset} \tag{1}$$

where $\mu^{\circ}_{i\emptyset}$ is the chemical potential of pure i at T and P. Noting that for a binary system of components 1 and 2 in equilibrium between a phase $\alpha$ (e.g., nematic, smectic), and a $\beta$ phase (e.g., isotropic, nematic), standard thermodynamic treatments yield two equations that describe the equilibrium phase compositions as

$$\ln(x_{1\beta}/x_{1\alpha}) = \Delta H^{\circ}_{1\alpha\beta}(T-T_1)/RTT_1 \tag{2a}$$

$$\ln(x_{2\beta}/x_{2\alpha}) = \Delta H^{\circ}_{2\alpha\beta}(T-T_2)/RTT_2 \tag{2b}$$

where $\Delta H^{\circ}_{i\alpha\beta}$ is the $\alpha \rightleftharpoons \beta$ transition enthalphy at $T_i$, assumed independent of temperature. Equations (2a) and (b) are just the Schroeder-van Laar equations. Noting $x_{1\emptyset}+x_{2\emptyset}=1$, Eqs. (2a) and (b) can be solved explicitly for the compositions in $\alpha$ and $\beta$ phases yielding

$$x_{1\alpha} = \left(1-e^{H_2}\right)/\left(e^{H_1}-e^{H_2}\right) \tag{3a}$$

$$x_{1\beta} = x_{1\alpha}\, e^{H_1} \tag{3b}$$

where

$$H_i = \Delta H^{\circ}_{i\alpha\beta}(T-T_i)/RTT_i \;. \tag{4}$$

Equations (2a) and (b) allow calculation of ideal solution phase diagrams such as those illustrated in Figs. 1b through g. The various types of ideal two phase spindles that can exist depend on the magnitudes of the enthalpies and temperatures for the pure materials and were noted some time ago by van Laar and more recently quite thoroughly discussed by Reisman, Domon and Billard.[8,1] However, the azeotrope-like behavior of liquid crystal systems cannot be treated ideally as a brief calculation will demonstrate. Suppose $H_1$ and $H_2$ were small enough to expand the exponential as $\exp(H_i)\simeq 1+H_i$ then

$$x_{1\alpha} \simeq \left[1 - \frac{\Delta H^{\circ}_{1\alpha\beta}\; T_2(T-T_1)}{\Delta H^{\circ}_{2\alpha\beta}\; T_1(T-T_2)}\right]^{-1} \;. \tag{5}$$

Since $0\leq x_{1\alpha}\leq 1$, the permissible values of T are such that for $T_2>T_1$, T must be less than $T_2$ or greater than $T_1$, otherwise $x_{1\alpha}>1$ which is not possible. An ideal two phase region must have one of the spindle shapes described originally by van Laar and be bounded by the pure component transition temperatures $T_1$ and $T_2$.

Thus, the phase diagrams in Figs. la and h illustrate that maximum or minimum azeotrope behavior violates the T criteria above and must therefore be discussed by the introduction of non-ideality. Because a diagram has one of the spindle shapes of van Laar it is, however, no guarantee of its ideality. In Fig. 2 is presented the calculated and observed phase diagram for a binary mixture of homologous azobenzenes. Clearly the observed diagram is not ideal even though it has one of the most classically ideal spindle shapes, and the diagram would have to be calculated introducing non-ideality.

## NON-IDEAL SOLUTIONS

Non-ideality as it has been found in vapor-liquid and alloy systems, is usually treated in terms of excess functions. Thus, an excess chemical potential term is added to (1) giving

$$\mu_{1\alpha} = \mu^{\circ}_{1\alpha} + RT\ln x_{1\alpha} + \mu^{E}_{1\alpha} \tag{6}$$

where $\mu^{E}_{1\alpha}$ is the excess chemical potential of $\alpha$ phase for component 1. Proceeding through a treatment similar to that mentioned earlier, the following equations are obtained:

$$\ln(x_{1\beta}/x_{1\alpha}) + \left(\mu^{E}_{1\beta}-\mu^{E}_{1\alpha}\right)/RT = H_1 \tag{7}$$

$$\ln(x_{2\beta}/x_{2\alpha}) + \left(\mu^{E}_{2\beta}-\mu^{E}_{2\alpha}\right)/RT = H_2\ . \tag{7b}$$

Regular solution theory substitutes for the excess chemical potential of component 1 in $\alpha$ phase

$$\mu^{E}_{1\alpha} = A_{\alpha}\, x^{2}_{2\alpha} \tag{8}$$

where $A_{\alpha}$ is a positive parameter with units of free energy, independent of temperature and composition, and dependent only on the characteristics of the phase.[9,10] With liquid crystalline azeotropes, it is convenient to use a parameter for each phase. Substituting such regular solution parameters for the excess chemical potentials in Eqs. (7) gives

$$\ln(x_{1\beta}/x_{1\alpha}) + \left(A_{\beta}x^{2}_{2\beta}-A_{\alpha}x^{2}_{2\alpha}\right)/RT = H_1 \tag{9a}$$

$$\ln(x_{2\beta}/x_{2\alpha}) + \left(A_{\beta}x^{2}_{1\beta}-A_{\alpha}x^{2}_{1\alpha}\right)/RT = H_2\ . \tag{9b}$$

While Eqs. (3a) and (b) can be solved explicitly for the composition as a function of temperature, Eqs. (9a) and (b) must be solved by some iterative procedure. Again noting that $x_{1\alpha}+x_{2\alpha}=1$ and $x_{1\beta}+x_{2\beta}=1$, Eqs. (9a) and (b) can be simplified to

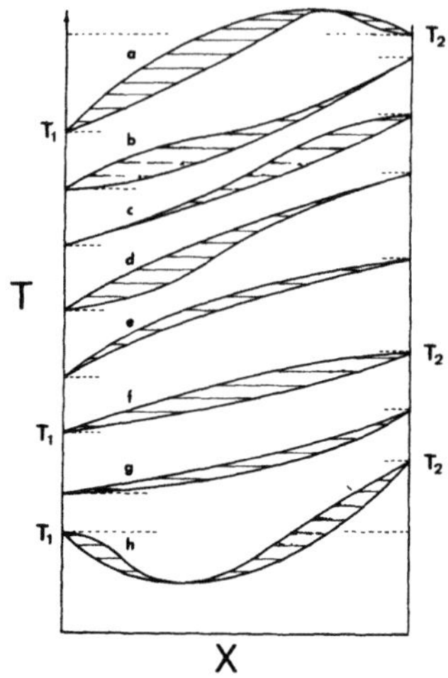


Fig. 1. Possible spindle shapes for two phase regions in binary diagrams where neither phase is a pure compound. (After Domon and Billard, Ref. 1).

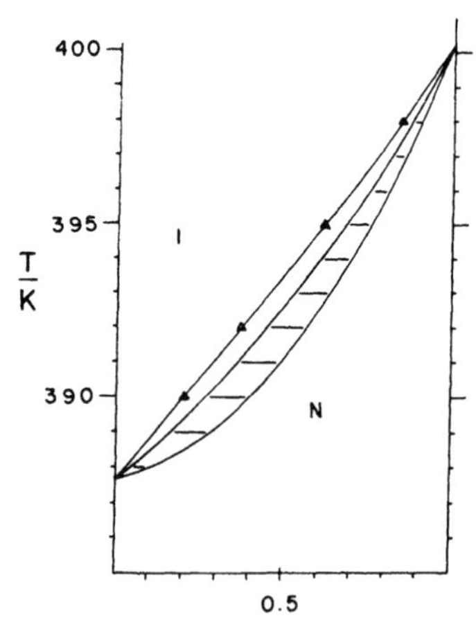


Fig. 2. Isobaric phase diagram for the binary pair of azobenzenes: 4-ethoxy-4'-n-pentanoyloxyazobenzene and 4-ethoxy-4'-n-nonanoyloxyazobenzene, Ref. 7.

$$\left(\frac{1-x_{1\alpha}\exp\left[H_1+\frac{A_\alpha(1-x_{1\alpha})^2-A_\beta(1-x_{1\beta})^2}{RT}\right]}{1-x_{1\alpha}}\right)\left(\exp\left[\frac{A_\beta x_{1\beta}^2-A_\alpha x_{1\alpha}^2}{RT}\right]\right)=\exp H_2 \tag{10}$$

where $H_i$ is as defined earlier. Iterative solution of (10) is possible for $x_{1\alpha}$ and $x_{1\beta}$ once given T, the properties of the pure components, and some estimates for $A_\alpha$ and $A_\beta$.

In the event that one phase can be treated as ideal, for example, a liquid-solid equilibrium where the solid can be treated as ideal, Eq. (10) can be much simplified, since by taking $A_\alpha=0$ ($\alpha$ phase ideal) $x_{1\alpha}=1$ and $x_{2\alpha}=0$, we obtain

$$\ln x_{1\beta}+A_\beta(1-x_{1\beta})^2/RT=H_1 \; . \tag{11}$$

This equation still requires some kind of iterative solution for $x_{1\beta}$. In most cases, the changes brought about by use of Eq. (11) as opposed to assuming ideality are small. Use of Eq. (11) does have significant favorable effect on the values calculated for eutectic temperatures and compositions and for that reason it was justified (vide supra).

With the appearance of a minimum or maximum point in the phase diagram, Eqs. (9a) and (b) simplify a great deal, since at the azeotrope point $T=T_m$, $x_{1\alpha}=x_{1\beta}=x_{1m}$, and $x_{2\alpha}=x_{2\beta}=x_{2m}$ which results in

$$(A_\beta-A_\alpha)x_{2m}^2=RT_mH_1 \tag{12a}$$

$$(A_\beta-A_\alpha)x_{1m}^2=RT_mH_2 \; . \tag{12b}$$

Moreover a relationship for $x_{1m}$ depending only on $T_m$, $H_1$ and $H_2$ can be found by taking the ratio of (12a) to (12b) and again noting that $x_{1m}+x_{2m}=1$

$$x_{1m}=\left[1+(H_1/H_2)^{1/2}\right]^{-1} \; . \tag{13}$$

Further, the difference in the regular solution parameters can also be found from knowledge of only $T_m$ by

$$(A_\beta-A_\alpha)=RT_mH_2\left[1+(H_1/H_2)^{1/2}\right]^2 \; . \tag{14}$$

Thus, to calculate the composition of the azeotrope point and the difference in the regular solution parameters, only $T_m$ and pure component transition temperatures and enthalpies are required. That so much can be accomplished with only knowledge of $T_m$ is very convenient, since the commonly used contact method for determining binary phase diagrams of mesogenic mixtures will only provide temperature information.

Once knowing $T_m$, $(A_\beta - A_\alpha)$ can be estimated as mentioned above, but conversely, if an estimate for $(A_\beta - A_\alpha)$ can be made, then $T_m$ and $x_{1m}$ can be calculated. Thus, we can derive a condition for predicting whether or not a minimum (or maximum) in the phase diagram would be expected based on knowledge of pure component transition temperatures, enthalpies and the $(A_\beta - A_\alpha)$ estimate. In the limit, the minimum occurs at pure 1, $x_{1m}=1$, $T_m=T_1$, and Eq. (14b) becomes

$$(A_\beta - A_\alpha)_{x_{1m}=1} = \Delta H_{2\alpha\beta}(T_1 - T_2)/T_2 \ . \tag{15}$$

For a diagram with the minimum very close to pure 1, $x_{1m} \simeq 1$ and $T_m < T_1$ whence Eq. (12b) becomes

$$(A_\beta - A_\alpha)_{x_{1m}\approx 1} = \Delta H_{2\alpha\beta}(T_m(T_m - T_2)/T_1 T_2 \ . \tag{16}$$

The ratio gives

$$\frac{(A_\beta - A_\alpha)_{x_{1m}=1}}{(A_\beta - A_\alpha)_{x_{1m}\simeq 1}} = \frac{T_1(T_1 - T_2)}{T_m(T_m - T_2)} \ . \tag{17}$$

The right-hand-side of Eq. (17) can be shown to be greater than 1 (for reasonable temperatures at least) which means

$$(A_\beta - A_\alpha)_{x_{1m}\simeq 1} < (A_\beta - A_\alpha)_{x_{1m}=1} = H_{2\alpha\beta}(T_1 - T_2)/T_2 \ . \tag{18}$$

Thus if the value of the estimated parameter difference is less than the right-hand-side of Eq. (18), a minimum phase diagram is suggested.

As mentioned above, the experimentally observed $T_m$ will only allow the difference in the regular solution parameters to be calculated but to calculate the phase diagram the individual parameters have to be known. For simple minimum or maximum diagrams where two non-ideal liquid phases appear, for example, a nematic-isotropic equilibrium, two regular solution parameters are necessary. To obtain the individual values once knowing their difference, values are choosen equally spaced around the difference that is,

$$A_N = -1.5(A_I - A_N) \tag{19}$$

$$A_I = -0.5(A_I - A_N)$$

where for a minimum $(A_I - A_N) < 0$ but all A parameters must be greater than zero. This is an arbitrary procedure chosen for convenience. For constant parameter differences, different choices for the individual parameters, that is, not using the above criterion, merely changes the convexity of the minimum curve around the minimum point. In the case of complex maxima or minima, and here complex is taken to mean more than two liquid phases, still one parameter per phase is required and again criteria for choosing parameters are required. The complex maxima systems discussed below exhibit three liquid phases, isotropic, cholesteric and smectic A. In each case, two maxima are observed allowing $(A_I - A_{Ch})$ and $(A_{Ch} - A_{S_A})$ differences to be calculated. Individual parameters are chosen by

$$A_{S_A} = 0.5(A_{Ch} - A_{S_A}) \tag{20}$$

$$A_{Ch} = 1.5(A_{Ch} - A_{S_A}) \tag{20b}$$

$$A_I = A_{Ch} + (A_I - A_{Ch}) \tag{20c}$$

where note, since these refer to maxima, the parameter differences are greater than zero and still $A_I > A_{Ch} > A_{S_A}$. Again this is an arbitary procedure keyed, however, to maintaining the parameter differences fixed at the values calculated from $T_m$ values.

## APPLICATIONS

### Simple Minima

The binary phase diagrams of homologous 4,4'-*di*-alkoxyazoxybenzenes have been reported by Hsu and Johnson,[11,12] and Demus *et al*. present phase diagrams for all the fifteen possible binary combinations of the homologues in the series from 4,4'-*di*-hexyloxyazoxybenzene (mixtures will be designated 1-6, e.g., meaning a methoxy-1 and hexyl-6 mixture). Of the fifteen possibilities, six show minimum azeotrope behavior. Figure 3 presents the experimental points (estimated from published figures), the regular solution calculated phase coexistence lines, and points calculated assuming ideal solution behavior for the reported minimum forming systems. The agreement between calculated and observed is excellent. The width of the two phase region deserves brief comment, for, even though it is small for these diagrams ≃0.003 (mole fraction scale), it does exist. This is in keeping with the calculations of Cox and Johnson for ideal systems and accounts for the frequently encountered difficulty of experimentally observing (and therefore reporting) the two phase region.[4]

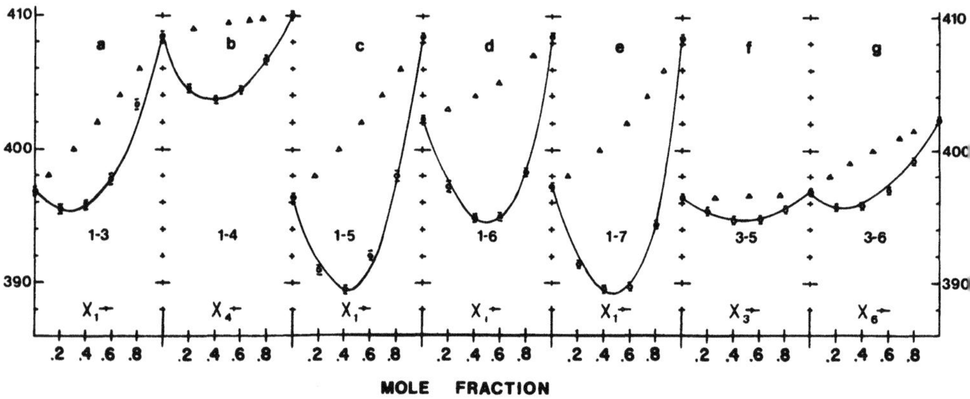


Fig. 3. Calculated and observed nematic-isotropic phase diagrams for binary mixtures of homologous 4,4'-*di*-*n*-alkyloxyazoxybenzenes. Solid line is calculated by regular solution approximation. Δ denotes point calculated assuming ideal solution. Φ denotes experimentally observed point. 3-6 means 4,4'-*di*-*n*-propyloxy in mixture with 4,4'-*di*-*n*-hexyloxyazoxybenzene. $X_i$ is mole fraction of ith component.

Curiously a good correlation between the observed regular solution parameter differences and the ratio of the molar volumes of the binary components appears to exist as shown in Fig. 4. Use of this correlation allows prediction of the required regular solution parameters for the remaining binary phase diagrams of this homologous series. The complete details of the results of such calculations have been described elsewhere, but to illustrate the technique the example of the 1-7 binary diagram will be mentioned.[5] The regular solution parameters have been estimated for this pair based on a molar volume ratio of 0.52 which by Fig. 4 estimates the parameter difference to be -11.40K which in turn, predicts a minimum formation by use of Eq. (18). The phase diagram is given as Fig. 3e. Hsu and Johnson reported the phase diagram for this mixture and the agreement with the calculated results as seen in Fig. 3e is excellent.[12] Thus for binary mixtures of homologs, the correlation between the regular solution parameters and the molar volume ratio appears to be quite useful.

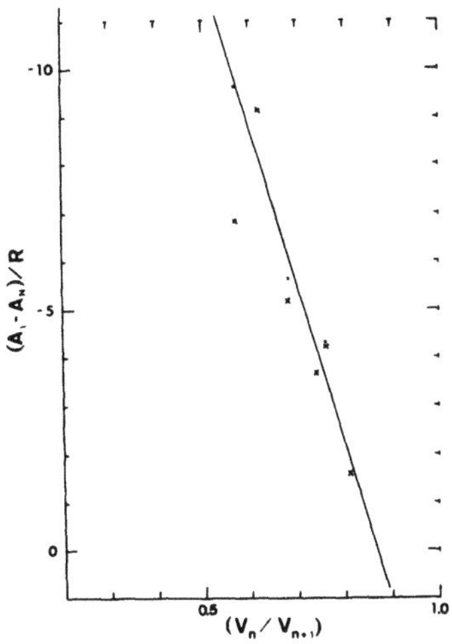


Fig. 4. Plot of molar volume ratio ($V_n/V_{n+1}$) versus difference in the regular solution parameters for the isotropic and nematic phases ($A_I - A_N$) in multiples of the gas constant R for the six minimum forming systems in binary mixtures of homologous 4,4'-n-alkloxyazoxybenzenes. O Data of Hsu and Johnson, Ref. 11. X Data of Demus et al. Ref. 13.

## Complex Maxima

The phase diagrams reported for three binary mixtures of p-nitrobenzalaminocinnamic acid 2-methyl hexyl ester with cholesterol nonanoate, tetradecanoate (myristate) and octadecanoate (stearate), have been calculated and are presented in Figs. 5a, b and c, respectively. These diagrams were obtained experimentally using the contact method and no detailed composition data is available. The observed temperatures are indicated on the figures by the dashed lines and in all cases the agreement between observed and calculated is very good. Qualitatively the diagrams match essentially exactly with experiment. The greatest disagreements come in calculating the eutectic temperature. It should be mentioned that calculating the eutectic temperatures using ideal equation gave worse agreement than using the regular solution equation. In any case, these temperatures are quite sensitive to uncertainties in the transition temperatures and enthalpies of the pure compounds, particularly methyl hexyl ester. Further, it should be pointed out that several of the phase transitions involved for the pure materials are, following the nomenclature of Domon and Billard, virtual transitions for which, in most cases, the transition temperatures and enthalpies had to be calculated via thermodynamic cycles.[1] The details have been described elsewhere.[6]

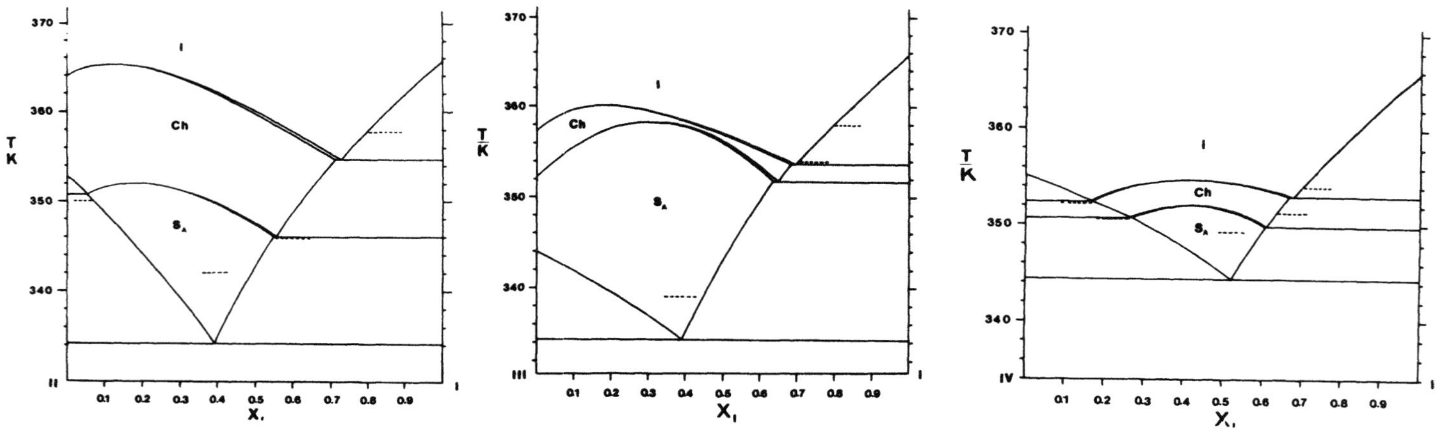


Fig. 5. Calculated phase diagrams for binary mixtures of p-nitrobenzalaminocinnamic acid 2-methyl hexyl ester(I) with cholesterol nonanoate(II) (a); cholesterol tetradecanoate(III) (b); cholesterol octadecanoate(IV) (c). Dashed lines indicate observed temperatures.

Since each of these phase diagrams involves three liquid phases taken to be non-ideal, three parameters are required for the calculations. The parameters used, as determined from the observed $T_m$ and using the criteria mentioned above are given in Table 1. While there appears to be a general increase in any given parameter, such as $A_I$ with molecular weight of the second component in going from nonanoate to octadecanoate, no simple linear relationship is apparent. Thus, while the diagrams can be calculated, unfortunately, no simple correlation for the parameters with some physical property of the pure materials has yet been discovered.

Table 1.

| Phase Diagram | $A_I/R$ K | $A_{Ch}/R$ K | $A_{S_A}/R$ K |
|---|---|---|---|
| I-II | 46.359 | 22.652 | 7.551 |
| I-III | 60.941 | 41.172 | 13.724 |
| I-IV | 67.506 | 44.840 | 14.947 |

CONCLUSIONS

Regular solution theory has been successfully applied to the calculation of binary phase diagrams of mesomorphic materials. Correlation of the parameters required by the theory with simple physical properties of the mixture's components, may offer an empirical way to predict the phase diagrams for mixtures involving members of homologous series.

REFERENCES

1. M. Domon and J. Billard, "Pramama," Suppl. No. 1, 131 (1975).
2. H. T. Peterson and D. E. Martire, Mol. Cryst. Liq. Cryst. 25:89 (1974).
3. G. W. Smith, Mol. Cryst. Liq. Cryst. 30:101 (1975).
4. R. J. Cox and J. F. Johnson, IBM J. Res. Develop. 22:51 (1978).
5. G. R. Van Hecke, J. Phys. Chem. 83:2344 (1979).

6. G. R. Van Hecke, T. S. Cantu, M. Domon and J. Billard, presented at the VIIth International Liquid Crystal Conference, Bordeaux, France, July 1978 in press.
7. R. D. Loft and G. R. Van Hecke, unpublished experiments.
8. A. Reisman, "Phase Equilibria," Academic Press, New York, New York 1970, Ch. 28.
9. J. H. Hildebrand and R. L. Scott, "Solubility and Non-Electrolytes," 3rd ed., Reinhold, New York, New York 1950, p. 46.
10. E. A. Guggenheim, "Thermodynamics," 5th Ed., North-Holland, Amsterdam 1967, p. 197.
11. E. C-H Hsu and J. F. Johnson, Mol. Cryst. Liq. Cryst. 20:117 (1973).
12. E. C-H Hsu and J. F. Johnson, Mol. Cryst. Liq. Cryst. 25:145 (1974).
13. D. Demus, C. H. Fiethau, R. Schubert and H. Kehlen, Mol. Cryst. Liq. Cryst. 25:215 (1974).

# THE ANALYSIS OF LIQUID CRYSTAL MIXTURES[8]

Robert L. Hubbard

Tektronix Laboratory
Tektronix, Inc.
Beaverton, Oregon 97077

## INTRODUCTION

Liquid crystals have become a widely used material for information display. There have been many references to the use of liquid crystals as media for analytical investigations including use as a stationary phase in Gas Chromatography (GC)[1,2,3,4,5,6] but few have mentioned the chromatography of liquid crystals themselves.[7]

Since the early liquid crystals were of the relatively unstable azoxy and Schiff base types the analysis procedures had to include care to prevent degradation from light, heat, oxygen, or moisture. The use of GC was therefore somewhat risky. The more obvious choice of a separation technique would be HPLC since no heat is necessary and samples can be treated very gently. There is also no practical limit to the molecular features present in a sample. Very polar as well as very non-polar moieties are easily eluted with appropriate conditions. The most complex samples require solvent gradients at the worst. The liquid crystals developed more recently and that have become the most widely used are the more stable biphenyl (BP), phenylcyclohexane (PCH) and ester types. These compounds are much more stable under harsh conditions and would be expected to survive analysis by GC, for example. With the advent of preparative liquid chromatography, many manufacturers have found in it an ideal method for the purification on a kilogram scale of stable liquid crystals.[7]

Since liquid crystals are used in mixtures to achieve the required device specifications, the analysis of such complex mixtures would begin with HPLC to assure that all components be eluted. The advantage of GC is the common interface with mass spectrometers for component identification. Other identification methods, such as Nuclear Magnetic Resonance (NMR), Infrared Spectroscopy (IR) and Ultraviolet Spectroscopy (UV) are typically used on individual components.

The commercial liquid crystals of the 4-cyano-biphenyl type from BDH have a systematic shortened name system. The K series are the 4-n-alkylbiphenyls; the M series are 4-n-alkoxybiphenyls; and the T series are the 4-n-alkylterphenyls. The numbering represents three times the actual number of carbons in the alkyl or alkoxy chains. Components from E. Merck Labs of Darmstadt have the notations of either S or NP with numbers that are orderly, but not systematic. The Hoffmann-LaRoche materials include esters (EX), alkylcyanobiphenyls (BX), alkoxycyanobiphenyls (BOX), alkylcyanoterphenyls (TX), alkylcyanopyrimidines ($P_3X$) and terpyrimidines ($T_3PX$). The X represents the actual number of carbons in the side chain.

When a component is identified from a commercial display as being "K15", for example, this does not imply that the component was necessarily purchased from BDH, especially since Hoffmann-LaRoche is also selling that same component as "RO-CB-5115."

## EXPERIMENTAL

### Apparatus

The liquid chromatograph used was a Perkin Elmer Series 3 with an LC-55S scanning UV detector. Samples were injected on a Rheodyne 7105 sample-loop high pressure injection port. The columns used were the Dupont Instruments Zorbax ODS and Zorbax C8.

The gas chromatograph/mass spectrometer was a Hewlett Packard 5985A with a one meter OV-101 column.

The infrared spectrophotometer was a Perkin Elmer 467 grating instrument. The FT-NMR was JEOL FX-60Q with a $^{13}C/^{1}H$ switchable microprobe (1.7 mm diameter tubes). The proton spectra were obtained on a JEOL C60HL continuous wave instrument.

### Reagents

Solvents were HPLC grade and were filtered through 0.4μ millipore teflon filters before use. Samples for HPLC were also filtered through these filters before use. Chromatography grade gases were used for the GC/MS.

Liquid crystals were obtained from EM Laboratories, Gallard Schlesinger, and Hoffmann LaRoche. Unknown mixtures were obtained from commercially available displays, broken open under nitrogen and the contents extracted.

## HPLC of Commercial Mixtures

The liquid chromatograms and UV spectra of the components are presented by the structural groups biphenyls, pch's, esters, and others. The UV spectra were obtained by stopping the flow of the chromatograph as the component passed through the UV cell and scanning from 200 nm to 380 nm. The solvent absorption was subtracted automatically at each point.

The biphenyls E7 (BDH) and 563 (HLR) and the pch's S1132 (EM) and 1218 (EM) are represented by Fig. 1, respectively. The esters S1081 (EM) and 331E (HLR) and the mixtures NP935 (EM) and TN132 (HLR) are represented by Fig. 2, respectively.

All of these mixtures could be resolved using methanol/water solvents in various concentrations or in continuous gradients in the reverse phase mode.

## Correlations of UV Spectra

The homologues within each structural type should have very similar UV absorbance maxima and this is found to be the case. It can be seen that the biphenyls absorb as a group at around 280-300 nm. It should be noted, however, that most liquid crystal mixtures include at least one three-ring structure (e.g., T15) to increase the clearing point of the mixture as a whole. This compound will typically absorb 20-50 nm higher than the rest of the group. This is particularly evident in the spectra from pch mixture S1132. As would be expected, the cyclohexane ring does not contribute much as a chromophore to the cyanophenyl nucleus. The maxima of pch's are therefore typically lower and are in the 230-240 nm range. The exception at 280 nm in S1132 is the three-ring 4'n-pentylcyclohexyl-4-cyanobiphenyl which is commonly named biphenylylcyclohexane (BCH). Most of the ester maxima center around 240-260 nm with the exceptions of three-ring additives. The two azoxies in NP5 and NP935 absorb at 350-360 nm. The thioesters in NP935 give two maxima each at 260 nm and 280 nm.

It is not very difficult therefore to get a rough idea of the structural types present in an unknown mixture very quickly.

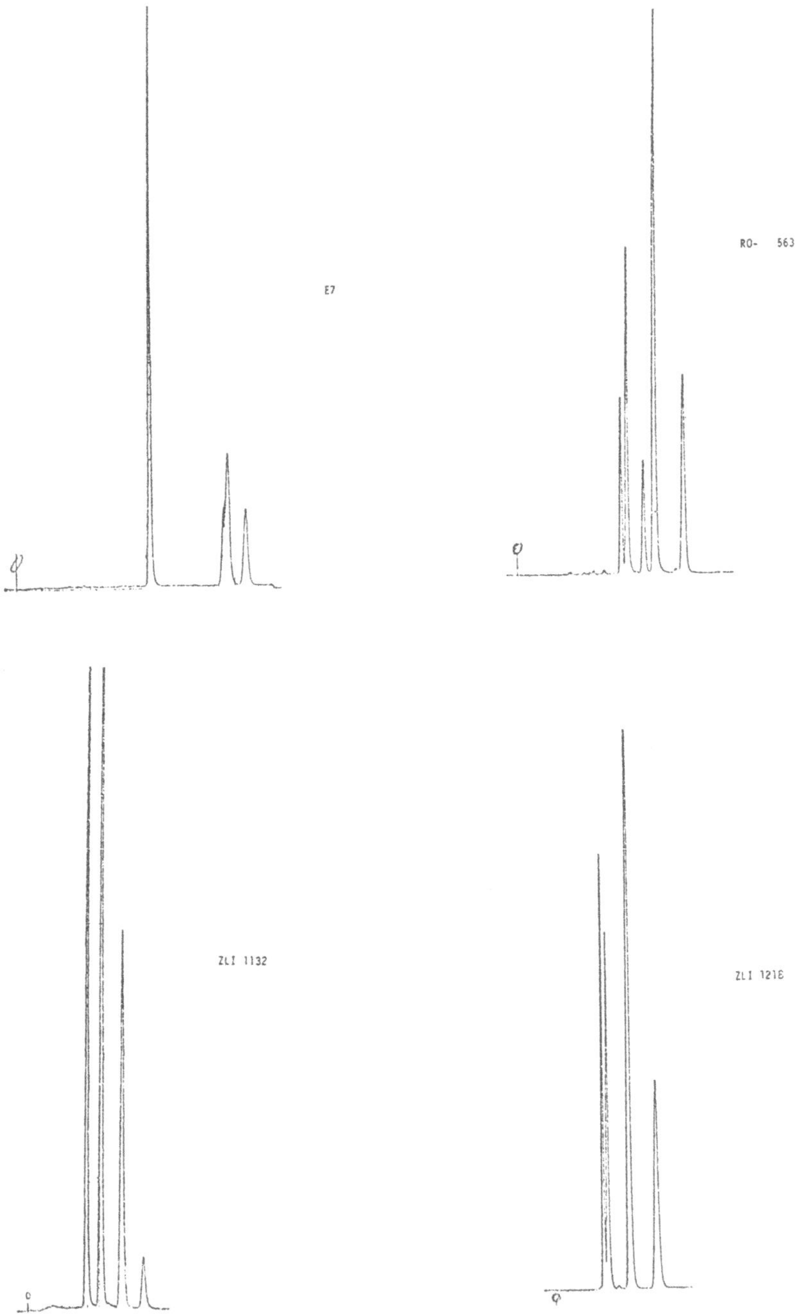


Fig. 1.

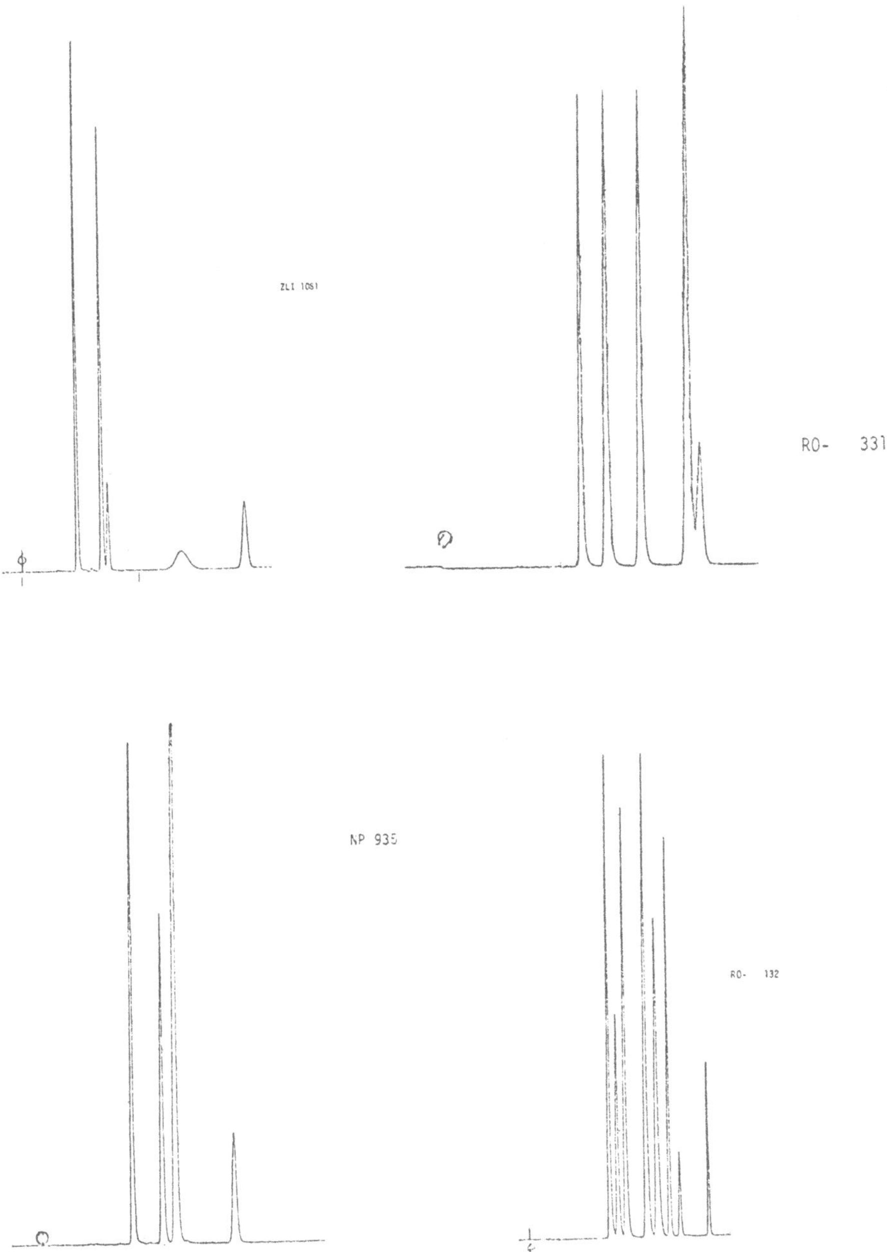


Fig. 2.

The 230-240 nm range generally indicates pch's; the 240-260 nm range generally indicates esters; the 280-300 nm range generally indicates biphenyls; and 350 nm and up indicates the highly conjugated azoxies. The thioesters can be identified by their characteristic double maxima. Some care should be exercised with this method due to the recently announced components which include mixed ester-biphenyls, ester-phenylcyclohexyls and biphenyl-cyclohexyls.

## GC/MS of Commercial Mixtures

The use of combined GC and MS allows the analysis of very small samples of complex mixtures and provides structural information simultaneously. Any GC restrictions (heat lability, for example) also apply to GC/MS, however. By programming the column temperature from 100-250°C at various rates most of the commercial mixtures were analyzed. Some difficulty was obtained with very later eluting esters in the form of band spreading. One has to be careful to not keep the temperature high for a prolonged period, as many late eluting compounds will give double peaks and/or tailing. This is most pronounced with the Schiff bases like TN 200 (not reported on here).

The alkylbiphenyls (Fig. 3) were found to consistently give a fragment at m/e=192. This is due to the tropylium rearrangement mechanism:

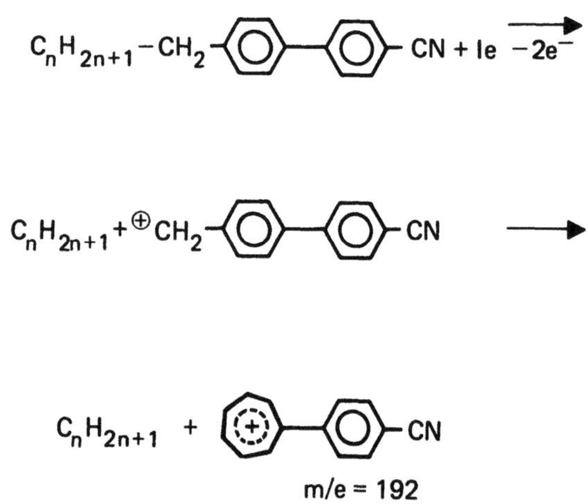

Since the only strong peaks in the spectrum are the tropylium product and the molecular ion, then the difference between those masses easily gives the weight of the R group and thereby the total structural identification. For an alkylterphenyl component, the predominant fragment should have an additional weight of 76 units. This gives a tropylium-like fragment of m/e=268. Any alkoxybiphenyl would be expected to give a strong fragment at m/e=195 from the phenol-like rearrangement. Again, the complete structure can be deduced from the size of the alkene fragment.

Using this method and the assumption that no branched alkyls or alkoxys are encountered provides the analyses in Table 1 for biphenyls. As in all of the following analyses, the actual compositions were verified with the manufacturer and found to be correct. All of the components are listed in the order of elution.

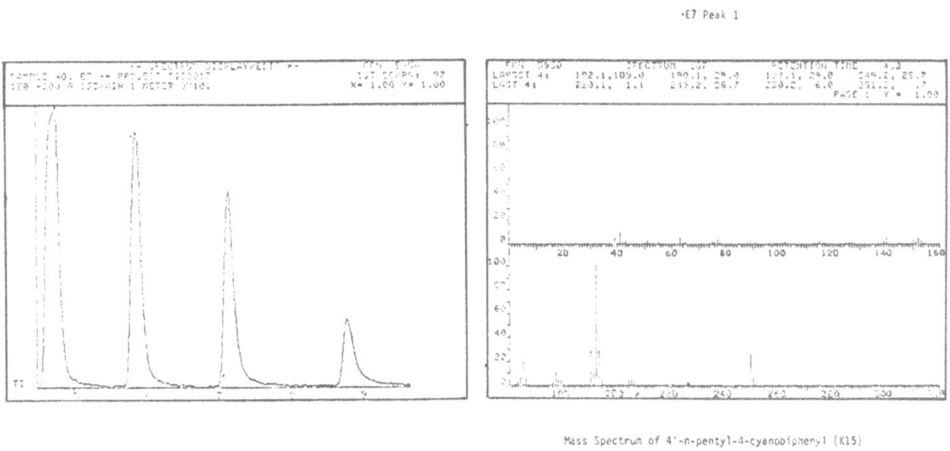


Fig. 3.

Table 1.

| | | | |
|---|---|---|---|
| E4: | M24, M21, K21, M15 | MEX1: | M9, K18, M15, K24, T15 |
| E7: | K15, K21, M24, T15 | MEX4: | M9, K21, M21, K30, T15 |
| E8: | M9, K15, M15, M24, T15 | E20: | K15, K21, M24, T15 |
| | | E25M: | M9, K18, M15, K24, T15 |

ZLI 1132 Peak 1

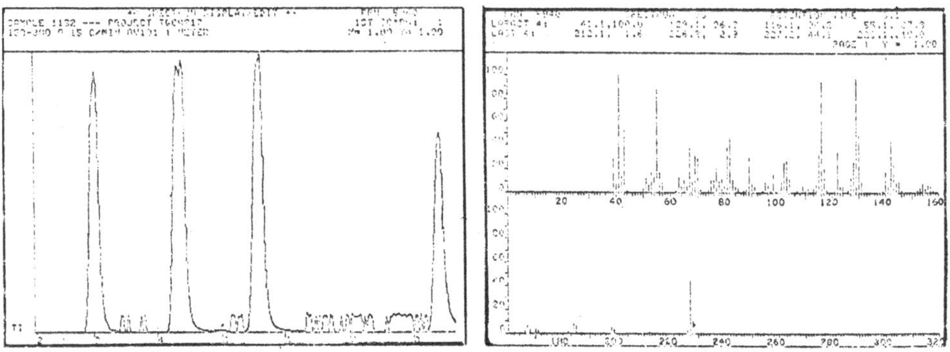

Mass Spectrum of trans-4-propyl-(4-cyanophenyl)-cyclohexane (PCH3)

Fig. 4.

Table 2.

PCH 3:

PCH 5:

PCH 7:

1222:

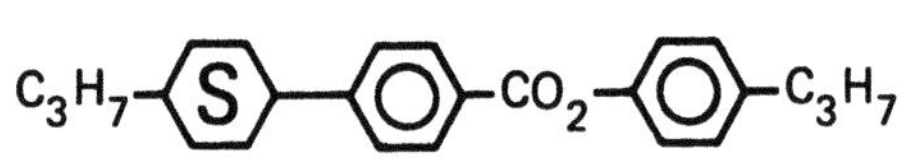


1223:

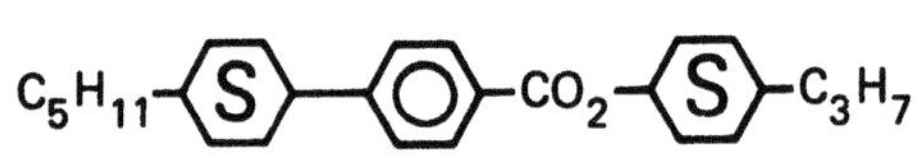


1224:

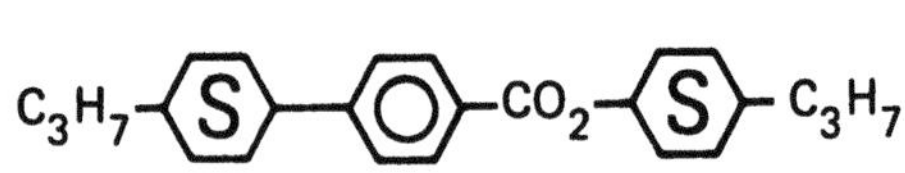

Table 3.

| | |
|---|---|
| ZLI 1132: | PCH 3 + PCH 5 + PCH 7 + BCH 5 |
| ZLI 1237: | PCH 2 + PCH 3 + PCH 5 + BCH 5 |
| ZLI 1216: | PCH 3 + PCH 5 + 1222 + 1223 + 1224 |
| | 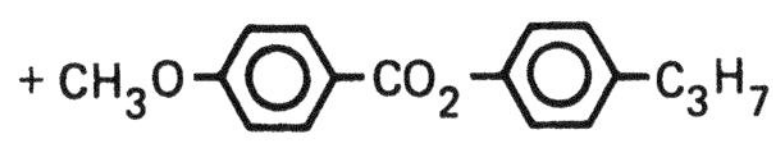  |
| ZLI 1218: | PCH 1 + PCH 3 + 1222 + 1224 |
| ZLI 1221: | PCH 3 + PCH 5 + PCH 7 + BCH 5 + 1223 + 1224 |

As would be expected, the phenylcyclohexanes (Fig. 4) were found to give very elaborate fragmentation due to the many possible fragment patterns of the cyclohexyl ring. Fortunately the molecular ion is quite strong on the pch's. The use of "PCH 3" as an abbreviation of the propyl derivative should explain the common naming system of the manufacturer (EM). New phenylcyclohexane esters have been produced by E.M. Laboratories and have the structures listed in Table 2. The fragmentation mechanism expected of the S1222-4 compounds should be similar to any other ester. The carbonyl cation is so stable that the other fragment is usually not seen. Even the molecular ion is often absent. To obtain the identity of the alcohol moiety it is often necessary to use chemical ionization (CI) mass spectroscopy which uses a reagent gas to donate stability to the molecular ion. By running the normal electron impact (EI) spectrum to obtain the carbonyl cation fragment and running the CI spectrum to obtain the molecular weight from the M+1 peak, the total structure can be deduced. A feature that sets apart the S1222-4 esters, however, is the abundance of small cyclohexyl ring fragments at m/e=40-150. A curious feature of these spectra is the very large peak 18 mass units above the carbonyl cation peaks in S1222-4. The explanation is not obvious. Table 3 gives the breakdown of the pch and pch-esters deduced from the reasoning above.

Other ester mixtures were analyzed with the same techniques of EI/CI described above, but with further delineation. The carbonyl cation fragment usually includes an aromatic ring so that the remainder of the fragment can be deduced by subtracting either 104 or 120 mass units representing the ring plus carbonyl or oxygen (from alkoxy side group) plus ring plus carbonyl. The remainder should be a common hydrocarbon fragment. The alcohol portion is deduced by subtracting 92 units for the other ring and alcohol oxygen. Occasionally an additional 76 units indicate an additional ring on either side of the ester linkage. Care should be taken to not mistake a $C_7H_{15}O$-grouping for a $C_5H_{11}CO_2$-grouping as is found in an ester in S1081. The difference is clearly noted by the even stronger m/e=121, which represents the p-hydroxy phenylcarbonyl cation. Ester mixtures are listed in Table 4.

As it turns out, azoxy materials exhibit many fragments in the 40-120 unit region as do the pch's. The thioesters have the same carbonyl cations as esters, of course, but have alcohol fragments of 16 mass units higher. The analysis of pyrimidines from Hoffmann-Laroche is assisted by the strong odd molecular ion peak which indicates an odd number of nitrogens. The expected appearance of a tropylium-type ion is noted at m/e=194. The same calculations can be made for the pyrimidines as were made for the alkylbiphenyls and alkylterphenyls. The data on these other types of mixtures is found in Table 5.

## NMR and IR of Mixtures

The components of mixtures can be easily collected from the effluent of a liquid chromatograph and analyzed by NMR and IR spectroscopies. As an example, a sample of S1132 was injected repetitively and the four components collected separately. The small amounts obtained are quite suitable for infrared samples. New Fourier Transform NMR spectrometers equipped with "micro" sized probes are now available for small samples. The $^1H$ and $^{13}C$ spectra were obtained in this manner. Combined with UV and MS data these compounds are completely characterized.

## Display Analyses

By the use of the techniques listed above it ought to be possible to determine the identity of an unknown liquid crystal from a display.

The first step is to carefully break open the display in a glove box and extract the 1-2 µl fluid by capillary pipet. The washing of the fluid out with solvent is not recommended since soluble adhesive components may be included. A small amount of solvent (usually methanol) is added for HPLC or GC/MS injection convenience. If GC/MS is used first, it is recommended that less than 50 µl of solvent be used to avoid solvent peak interferences.

Table 4.

ZLI 1052:

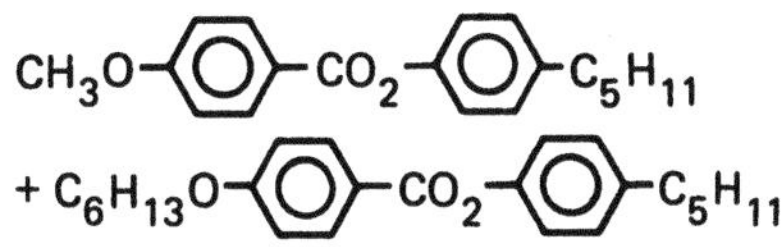

ZLI 1081:

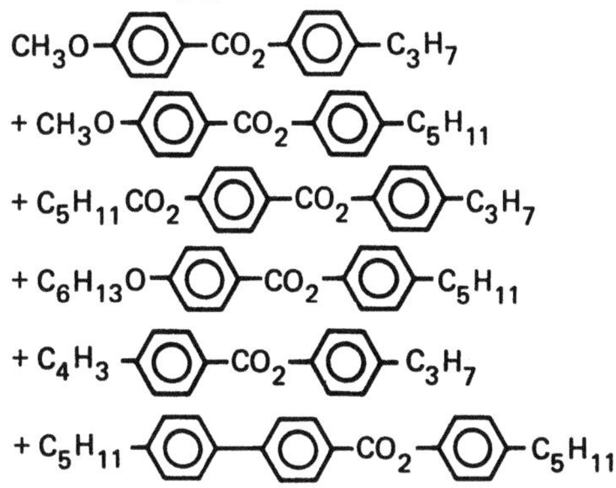

ZLI 1089: PCH 3 +

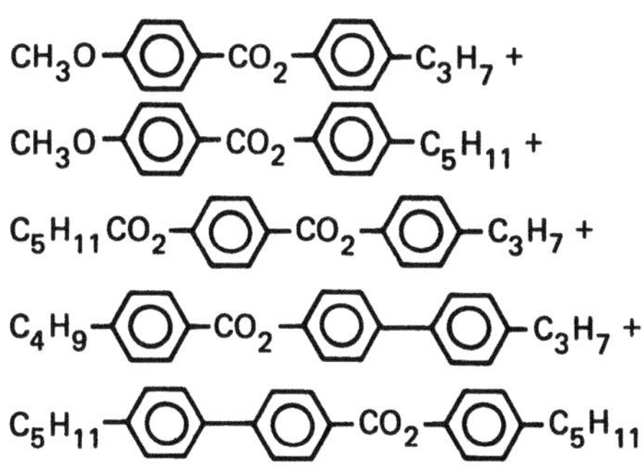

TN - 100:

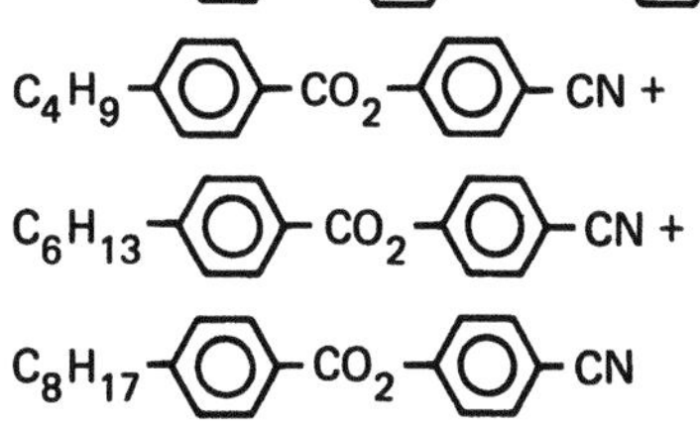

Table 5.

NP 5:

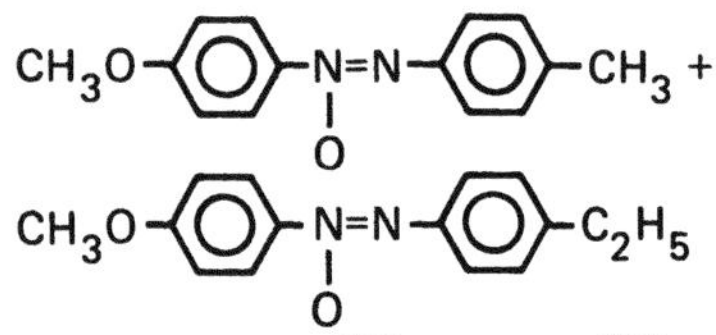

NP 935: NP 5 +

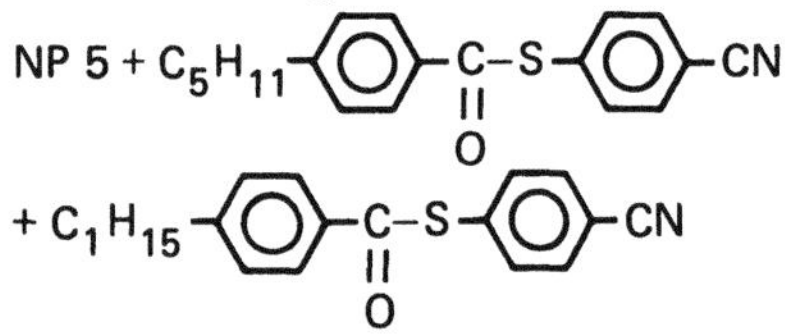

TN 404: $P_35 + M15 + P_37 + M24 + TP_34 + T15$

TN 405: $P_35 + M15 + K21 + P_37 + TP_34 + T15$

TN 132: $P_35 + E4 + E5 + P_37 + E6 + E7 + TP_34$

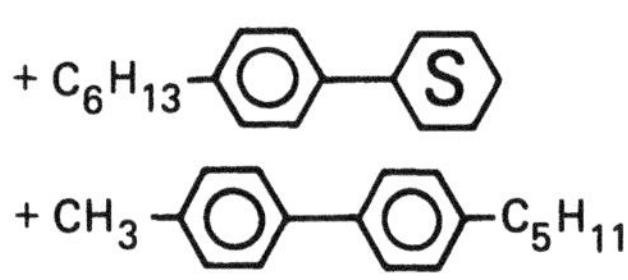

Table 6.

Display A: K15 + RO - CE 1500 + K21 + M24 + T15

Display B: K15 +

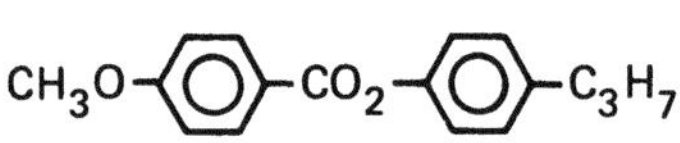

+ K18 + K21 + ZLI245 + M39 + SZLI 1004 + T15

Display C: K15 + K21 + RO - Ce 1500 +

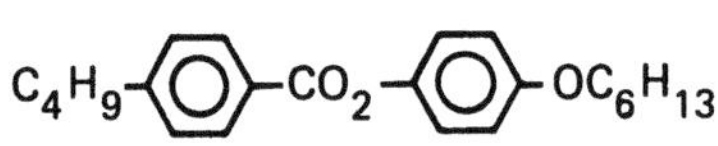

Display D: M9 +

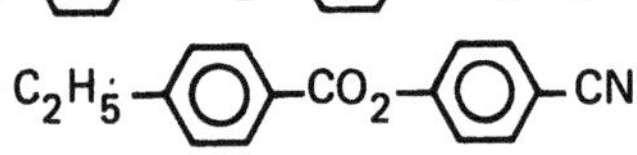

+ K15 + RO - CE 1500 + T15

Commercially obtained displays from different manufacturers were analyzed. Liquid chromatograms with UV spectra (obtained simultaneously), as well as their infrared and $^1$H-NMR spectra were obtained on many of the collected peaks from these displays, but sample size limitations prevented analyses of the smaller components. For this reason the conditions necessary for GC separations were devised and the samples subjected to GC/MS analysis.

The results obtained are summarized in Table 6. Even though GC/MS is a very high sensitivity technique, the high temperatures necessary for these analyses did not allow for the signal to noise ratio necessary to the analysis of components in quantities of less than 1-2%. With this minor consideration accepted, it can be seen that GC/MS provided substantial information.

## DISCUSSION

The use of HPLC on commercial liquid crystal mixtures provided high resolution of all components. These components could be collected for further identification by FT-NMR and IR if sufficient quantities were available. Since only about 6 to 8 milligrams of fluid is found in a standard display, the number of displays necessary would exceed a dozen. Since this is not usually practical, the GC/MS technique has been shown to give acceptable results from one display. The limitations inherent in the GC/MS technique revolve around the limited volatility and lability of many of the liquid crystal components. One should also be wary of using only the MS of a substance as positive identification of structure. When the newer technique of HPLC/MS is refined, the advantages of both techniques would suit the requirements of display analysis more closely.

## ACKNOWLEDGMENTS

I would like to thank Sarah Campbell, Paul Fields, Steven Slaughter and Larry Venable for their assistance; Tom Jackson at SunMark for the mass spectra; Teri Daniel, Mary Nute and Carolyn King for typing.

## REFERENCES

1. G. M. Janini, G. M. Muschik, J. A. Schroer, W. L. Zielinski, Jr. Anal. Chem. 48:1879 (1976).
2. H. Kelker and E. von Schivizhofen, "Advances in Chromatography," Dekker, Vol. 6, (1968).

3. M. J. S. Dewar and J. P. Schroeder, J. Amer. Chem. Soc. 86:5235 (1964).
4. H. Kelker, Z. Anal. Chem. 198:254 (1963).
5. W. L. Zielinski, et al., Anal. Chem. 48:907 (1976).
6. M. Hall and D. N. B. Mallen, J. Chromatogr. 118:268 (1976).
7. Waters Associates Bulletin H 81: "Ultra Purity in Gram Quantities Liquid Crystal Formulations."
8. This work was completed at the Central Research Laboratories, Texas Instruments, Inc., Dallas, Texas.

AUTHOR INDEX

## SUBJECT INDEX

GPSR Compliance

*The European Union's (EU) General Product Safety Regulation (GPSR) is a set of rules that requires consumer products to be safe and our obligations to ensure this.*

*If you have any concerns about our products, you can contact us on ProductSafety@springernature.com*

In case Publisher is established outside the EU, the EU authorized representative is:

Springer Nature Customer Service Center GmbH
Europaplatz 3
69115 Heidelberg, Germany

Batch number: 10620143

Printed by Printforce, the Netherlands